Wu 1984 Feb.

WESTERN GEOPHYSICAL

HANDBOOK OF GEOPHYSICAL EXPLORATION

SECTION I. SEISMIC EXPLORATION

VOLUME 14A
VERTICAL SEISMIC PROFILING

PART A: PRINCIPLES

HANDBOOK OF GEOPHYSICAL EXPLORATION

I. SEISMIC
II. ELECTRICAL
III. GRAVITY
IV. MAGNETIC
V. WELL-LOGGING
VI. RADIOMETRIC
VII. REMOTE SENSING
VIII. GEOTHERMAL

SECTION I. SEISMIC EXPLORATION

Editors: Klaus Helbig and Sven Treitel

SEISMIC EXPLORATION

Volume 14A

VERTICAL SEISMIC PROFILING
PART A: PRINCIPLES

by

BOB A. HARDAGE

Chief Geophysicist
Seismic Stratigraphy
Phillips Petroleum Company
Bartlesville, Oklahoma, U.S.A.

GEOPHYSICAL PRESS
London — Amsterdam   1983

GEOPHYSICAL PRESS LIMITED
Westcombe House
56-58 Whitcomb Street
London WC2H 7DR
U.K.

Worldwide Distributors:

EXPRO SCIENCE PUBLICATIONS
Brouwersgracht 236
1013 HE Amsterdam
The Netherlands

ISBN 0-946631-00-X (Vol. 14A)
ISBN 0-946631-25-5 (Series)

Printed by St. Catherine Press, Bruges, Belgium

Of making many books there is no end; and much study is a weariness of the flesh.

Solomon
(Ecclesiastes 12:12)

PREFACE

Vertical seismic profiling (commonly abbreviated to the shorter name VSP) is one of the rapidly developing areas of geophysical technology in the Western hemisphere. The measurement basically involves recording the total upgoing and downgoing seismic wavefields propagating through a stratigraphic section by means of geophones clamped to the wall of a drilled well. In most seismic measurements, both the energy source and the receivers are positioned on the earth's surface. What happens to the seismic wavelet as it propagates from the source to a subsurface reflector and back to the receivers is mostly a matter of inference based on the characteristics of the source and on the properties of the wavefield measured at the surface. Vertical seismic profiling replaces much of this inference with several closely spaced direct physical measurements of the seismic wavefield in the real earth conditions that exist between the earth's surface and the subsurface reflector. These measurements are proving to be invaluable in structural, stratigraphic, and lithological interpretations of the subsurface and are particularly valuable when combined with surface-recorded seismic data covering a prospective area around a VSP well.

Vertical seismic profiling technology has been widely used in the USSR for at least the past 25 years, and Soviet geophysicists have advanced its concepts beyond the research stage and into active and valuable seismic exploration applications. Many non-Soviet geophysicists are just becoming aware of the value of VSP measurements, and most American and European oil companies are now engaged in a competitive race to incorporate vertical seismic profiling into their hydrocarbon and mineral exploration and development activities. It is therefore expedient to share with all explorationists and production people some of the facts presently known about vertical seismic profiling.

Much of my interest in vertical seismic profiling must be credited to Dr. E. I. Gal'perin of the Institute of the Physics of the Earth, Moscow. Dr. Gal'perin has been a leading researcher in vertical seismic profiling for more than 25 years, and one of his books on the subject has been translated into English by the Society of Exploration Geophysicists (Gal'perin, 1974). Phillips Petroleum Company hosted Dr. Gal'perin on a visit to the United States for several weeks in October and November, 1979. During that time, I was able to have many pleasant and informative discussions with him on the technical aspects and exploration potentials of vertical seismic profiling.

This book is intended to be a practical description of the theory and applications of vertical seismic profiling and is written so that it can be easily read by field explorationists, who are the people that will likely benefit the most from understanding how to use VSP technology. I have found that field explorationists are usually so pressured in their job assignments of evaluating acreage before bidding deadlines, developing prospects at a rapid pace so that drill rigs will not be idle, deciding which leases should be relinquished or farmed-out, and attending endless partnership meetings, that they have precious little time left to read technical books. If a book has a considerable amount of difficult mathematics or is not well illustrated, it is usually put aside by these people, regardless of whether or not it is technicially valuable, simply because there is not sufficient time to review it. Consequently, mathematics is kept simple and to a minimum in this text, and all concepts and applications are supported by illustrations even when word explanations would suffice. Hopefully, the material is easy to read and understand as a result. Hopefully too, there is enough insight provided into most aspects of vertical seismic profiling so that geophysical researchers who desire more rigorous and theoretical investigations than given here can use this material as a foundation from which to build their own VSP studies.

Many people need to be publicly acknowledged for their contribution to this book. I particularly want to thank the management of Phillips Petroleum Company for permitting me to use some of the drafting, photographic, and word processing capabilities of Phillips in preparing this text. If the material has any value to the worldwide geophysical community, then explorationists both inside and outside of Phillips Petroleum Company owe a debt of gratitude to W. W. Dunn, Vice President of Exploration, S. E. Elliott, Manager of Worldwide Geophysics, W. E. Ryker, Manager of Phillips Exploration and Production Technology Branch, and R. M. Allyn, Director of Seismic Stratigraphy, who allowed me to work on this project in addition to my regular duties at Phillips. Also K. D. Wyatt and J. G. Gallagher of the Geophysics Branch, Phillips Research and Development, provided several illustrations and critiques which have been directly incorporated into the text. H. L. Clark of Phillips Field Measurements Section was a valuable counselor on matters related to VSP equipment and field procedures.

Discussions with representatives of VSP service companies were invaluable in assessing the status and future trends in VSP equipment, field procedures, and data processing. The time spent with Peter Kennett, R. L. Ireson, and A. A. Fitch of Seismograph Service (England) Limited is gratefully acknowledged, as are those discussions with J. C. Garriott of Birdwell, Gildas Omnes and Charles Naville of Compagie Generale de Geophysique, and S. M. Bartz, Darrell A. Terry, Carl K. Poster, and Bruno Seeman of Schlumberger.

A. H. Balch and M. J. Lee of the United States Geological Survey expended considerable effort to help me understand their extensive work in vertical seismic profiling. Also, the Nigerian National Petroleum Corporation, AGIP Petroleum Company, Chevron, Total, Statoil, and Occidental Petroleum (Caledonia) Ltd. must be thanked for permitting examples of some of their VSP surveys to be used in various places in the text. Bruno Seeman, Francis Mons, and V. de Montmollin of Schlumberger offered an assortment of unpublished experimental measurements which helped illustrate several important principles of borehole geophone systems and also demonstrated some of the fundamental behavior of three-component VSP data. Dominique Michon of Compagie Generale de Geophysique donated several illustrations of P-wave and S-wave VSP analyses that were a great asset in preparing parts of the text.

I sent the first draft of several sections of this manuscript to people whose published work I chose to use either as a framework about which to construct parts of the text or as an illustration of an important VSP principle. The response of these people was most gratifying and serves as a testimony of the fraternal bond among those members of the geophysical community who are researching and developing VSP technology. I wish to express particular thanks to Roger Johnson and Hugh Riches of Occidental Petroleum, M. N. Toksoz of the Massachusetts Institute of Technology, Gildas Omnes of Compagie Generale de Geophysique, Kenneth R. Parrott of Wexpro, Peter Kennett of Seismograph Service Ltd., Carl K. Poster, Francis Mons, V. de Montmollin, and Bruno Seeman of Schlumberger, A. H. Balch of Mobil Research, Roger M. Turpening, Rob Stewart, and C. H. Cheng of the Earth Resources Laboratory (MIT), T. W. Spencer of Texas A&M University, and James P. DiSiena of Arco Research for the time they took to review the manuscript drafts I sent them, and for the advice and counsel they offered for improving the final text.

I have used data illustrations and copyrighted material from several technical journals in various sections of this text, and the editors and staff of these journals were most cooperative in granting permission to use these materials. I wish to acknowledge Jerry W. Henry of the Society of Exploration Geophysicists, G. F. Millington of the Canadian Society of Exploration Geophysicists, Klaus Helbig of the European Association of Exploration Geophysicists, Myra Martin of the Society of Professional Engineers, and W. W. Souder of the Society of Professional Well Log Analysts for their assistance in helping me obtain releases of copyrighted material published by, or belonging to, each of their respective organizations.

The majority of the VSP data that I had access to, and which were offered to me when preparing this text, were recorded and illustrated in English units of measure. I hoped that all data could be illustrated and described in the International System of Units

(SI), but sometimes it was just too awkward to convert seismic trace spacings, geophone depths, and other parameters into SI units. Consequently, either English or SI units (or both) occur in the text and illustrations. This mixture of units is not good but could not be avoided.

I extend a special thanks to Rosa Armitage and Joan Sprague for their drafting work and to Diana Beers and Judy Beets who converted my scribbling into the final manuscript. Several people in Phillips Petroleum reviewed the manuscript before it was released in order to confirm that it contained no technical, legal, or social blunders. This review was a time consuming task for these people, and they offered valuable suggestions that improved the final text. I want to thank Gary W. Crosby, Gary M. Hoover, John T. O'Brien, Ray J. Darveau, Allen Richmond, Steve B. Wyatt, Kay D. Wyatt, and Joe G. Gallagher for the effort they expended in this release process. I am particularly indebted to Steve Wyatt because he invested far more time and energy in reviewing the manuscript than did anyone, and his editing and suggestions were extremely valuable.

Bob A. Hardage
Bartlesville, Oklahoma
March, 1983

TABLE OF CONTENTS

CHAPTER 1

INTRODUCTION

The Concept of Vertical Seismic Profiling

A vertical seismic profile (VSP) is a measurement procedure in which a seismic signal generated at the surface of the earth is recorded by geophones secured at various depths to the wall of a drilled well. The direction that geophones are deployed during the data acquisition thus differs by 90 degrees relative to the lateral geophone placement used when recording seismic reflection data at the earth's surface. This basic difference between these two types of seismic measurements is illustrated in Figure 1-1. Because a

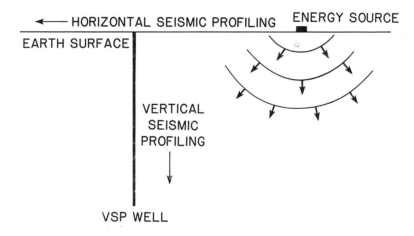

Figure 1-1   The contrast between geophone deployment in vertical and horizontal seismic profiling.

geophone is located far below the earth's surface when recording VSP data, it responds to both upgoing and downgoing seismic events. This type of geophone response is a second important difference between VSP data and surface-recorded reflection data because downward traveling events cannot be identified in data recorded by geophones positioned on the earth's surface.

A vertical seismic profile is closely related to a velocity survey since the source and receiver geometry is the same for both measurements. However, the following major differences between a vertical seismic profile and a velocity survey do exist:

1. The distance between geophone recording depths in a vertical seismic profile is much less than the distance between velocity check shot levels. VSP data are typically recorded every 10 to 25 meters along a borehole; velocity check shot levels can be separated by several hundred meters.

2. First break times are the critical information needed from the downhole signal in a velocity survey, but first breaks plus the upgoing and downgoing events that follow first breaks must be recorded in a vertical seismic profile.

The data recorded in a vertical seismic profile can, in concept, give insights into some of the fundamental properties of progagating seismic wavelets and assist the understanding of reflection and transmission processes in the earth. These insights, in turn, should improve the structural, stratigraphic, and lithological interpretation of surface seismic recordings. For example, a principal use now made of VSP data is to define upgoing and downgoing seismic events within the earth and thereby determine which events arriving at the surface are primary reflections and which are multiples. Other applications of VSP data include estimation of reflector dip, correlation of shear wave reflections with compressional wave reflections, location of fault planes, determination of lithological effects on propagating wavelets, looking for reflectors ahead of the drill bit, determining hydrocarbon effects on propagating wavelets, identification of intrabed multiples, measurement of both compressional and shear wave velocities, and estimation of the conversion of compressional to shear and shear to compressional energy modes within the earth. All of these applications, plus others, will be discussed in subsequent chapters.

The elements of an onshore vertical seismic profiling field experiment are shown in Figure 1-2. Presently, the only downhole recording equipment available to American and

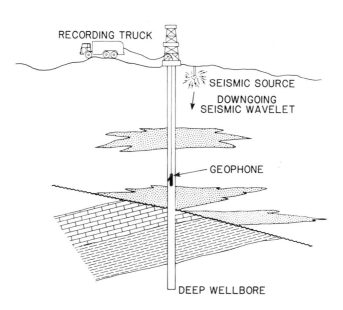

RECORDING TRUCK

SEISMIC SOURCE

DOWNGOING
SEISMIC WAVELET

GEOPHONE

DEEP WELLBORE

Figure 1-2   The basic components of a vertical seismic profile consist of a drilled well (cased or uncased), a standard seismic energy source at or near the earth's surface, a geophone package that can be clamped to the borehole wall, a digital data recording system, and a cable to connect the downhole geophone to a surface recorder (After Hardage, 1983).

European geophysicists consists of a single assembly of geophones which has to be positioned at each depth where one desires to record data. Consequently, many individual seismic shots have to be recorded during the course of a vertical seismic profile since there must be at least one shot for each geophone level. As already mentioned, the recording geometry shown in Figure 1-2 allows a geophone to record both downgoing and upgoing seismic events; whereas, downgoing events cannot be isolated and analyzed in standard surface seismic recordings. The fact that both downgoing and upgoing wavefields are recorded in a vertical seismic profile allows VSP data to be used in some unique exploration applications.

Vertical seismic profiling is a relatively new discipline for American petroleum companies even though borehole geophysics has been an active area of research and data acquisition by American geophysicists for several years. In the 1982 edition of INDEX OF WELLS SHOT FOR VELOCITY published by the Society of Exploration Geophysicists there are 10,200 onshore and offshore wells in the United States listed in which velocity check shot information has been recorded (Zwart and Baer, 1982). A vertical seismic profile was recorded in only one of these wells. However, it must be emphasized that the data in this index is voluntary information provided by petroleum companies operating in the United States. Even though a few U.S. based oil companies have recorded numerous vertical seismic profiles, they have not volunteered to include these data in this SEG publication.

Only a few investigations into the principles of vertical seismic profiling can be found in English language literature previous to 1981. On the other hand, vertical seismic profiling has been used for at least 25 years by Russian geophysicists. Geophysicists in the western world are the last in the geophysical community to become aggressively interested in using the technique in their seismic exploration programs. In the past two to three years in particular, vertical seismic profiling has captured the attention of American and European petroleum companies, and several geophysical service companies now offer vertical seismic profiling capability to the petroleum industry. Because of the present strong, widespread interest in vertical seismic profiling, it seems appropriate to offer the information in this book to all explorationists.

Historical Review of Vertical Seismic Profiling

Exploration seismology began as a surface based technique for imaging the earth's interior structure and stratigraphy. Today it is still largely restricted to surface positioned seismic energy sources and receivers. However, shortly after the initial proposals and attempts to develop surface reflection seismology, a few people began to think about using boreholes as locations for either seismic sources or geophones. Fessenden's patent (1917) appears to be the first documented seismic application involving buried sources and geophones. One illustration from this patent is shown in Figure 1-3. In this diagram, item 49 is a borehole acoustic energy source, and items 15 and 18 are borehole acoustic receivers. The dashed lines labeled 46 and 55 show possible travel paths of acoustic energy which interact with a buried ore body 44. In describing the method used to locate the ore body Fessenden states, "The vertical angle of reflection may be determined by hauling the transmitter 49 or the receivers 18, 15 up or down in the drill holes". Obviously, the rudiments of vertical seismic profiling can thus be traced back through time at least to the date of this patent. Consequently, Fessenden's work is

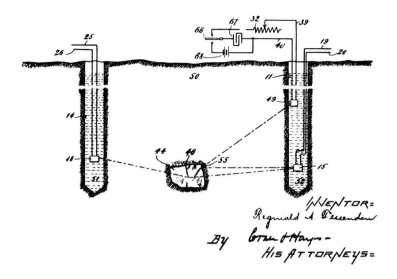

Figure 1-3   In 1917, Fessenden patented a method for detecting subsurface ore bodies by means of acoustic sources and receivers in water-filled drill holes. This illustration is Figure #2 from Fessenden's patent.

chosen as the starting point for constructing a chronological review of the development of borehole seismic measurements into what we now call vertical seismic profiling (VSP). A graphical representation of this chronological picture is portrayed in Figure 1-4.

In his analysis of how the newly developing seismic method could be used to map geological structures, Barton (1929) refers to Fessenden's earlier work and describes possible uses of seismic measurements in boreholes. McCollum and LaRue's (1931) paper is significant in that they stressed that existing wells were not optimally used for collecting exploration data.   Their proposal, that local geological structure could be determined by measuring the travel times from surface energy source positions to subsurface geophones located in wells, is now an active and valuable application of modern vertical seismic profiling.

After McCollum and LaRue's paper, enough of the essential elements and potential applications of vertical seismic profiling were captured in public geophysical literature so that VSP technology should have flourished and developed into a widely used procedure for

6

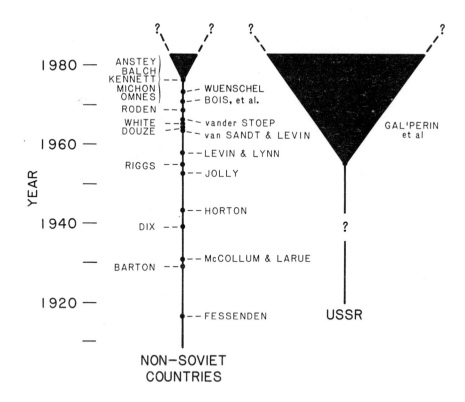

Figure 1-4   These two vertical lines provide an approximate chronological picture of the development of vertical seismic profiling and indicate some principal developers of the technology. The width of each line indicates the amount of interest focused on VSP. The extensive VSP work in the Soviet Union has not been chronologically studied. Gal'perin's (1974) book contains a comprehensive list of Soviet references.

hydrocarbon exploration. However, the geophysical community largely ignored the area of borehole seismology and continued to place its emphasis on surface reflection measurements. The major use of boreholes for seismic purposes was, for several years, limited primarily to the measurement of seismic wave propagation velocities (Dix, 1939). These investigations lead to the development of the valuable velocity check shot survey technology that is now widely used by the petroleum industry.

Although this emphasis on borehole velocity determination was invaluable for a rigorous interpretation of surface-recorded seismic data, it perhaps caused geophysicists to focus undue attention on just the information contained in the direct arrivals recorded by borehole geophones. It was not until the work of Jolly (1953), Riggs (1955), and Levin and Lynn (1958) that the potential of the total borehole geophone response was emphasized. They showed that rigorous wave propagation studies, interactions between primary and multiple reflections, and seismic wavelet attenuation could all be addressed by studying not just the first breaks but also the character of the seismic response that follows the direct arrivals. The present state of vertical seismic profiling is to a large extent the result of the work done by these men in the 1950's.

Even though the value and extensive applications of vertical seismic profiling were known and publicized by the mid-1950's, geophysicists in the Western hemisphere continued to use boreholes only for velocity measurements and not for rigorous vertical seismic profiling. Figure 1-4 shows two important historical facts about the development of VSP technology:

1. Documented studies of seismic measurements in boreholes began to appear with considerable frequency in the Soviet Union in the 1950's, and Soviet geophysicists aggressively developed VSP applications through the 1960's and 1970's. Gal'perin's (1974) book provides an excellent description of this Soviet work and contains an extensive list of Soviet papers written in this time period. According to Gal'perin (1974, p.2), VSP field work was initiated in the Soviet Union by the Institute of Physics of the Earth of the Academy of Sciences of the USSR in 1959.

2. During the 1950's, 1960's, and into the mid-1970's, only a thin thread of VSP interest continued outside of the Soviet Union.

The long lack of serious interest in vertical seismic profiling in the Western hemisphere cannot be easily explained or justified. Fortunately, in the late 1970's some non-Soviet geophysicists began to publicize the need for increased emphasis on vertical seismic profiling. Five of the earliest spokesmen in this VSP renewal are listed in Figure 1-4 (Anstey, 1977; Balch, 1980; Kennett, 1973, 1978; Michon, 1976; Omnes, 1978). The interest in vertical seismic profiling in non-Soviet countries now appears to be growing at a rate equal to or greater than that which was experienced in the early years of VSP development in the USSR.

Patent literature provides one source of information by which a detailed history of a technology developed in the 20th century can be constructed, as has already been illustrated by the discussion of Fessenden's work (Figure 1-3). For this reason, several patents subsequent to Fessenden's which cover various aspects of vertical seismic profiling, such as geophone designs, borehole energy sources, proposed field geometries,

8

and exploration ideas involving either geophones or sources in boreholes, are listed in the References section following Chapter 9. A chronological picture of this patent activity is shown in Figure 1-5.

This patent review is not offered as a complete and comprehensive collection of all VSP related patents. It is a rather exhaustive patent search, however, and is presented as a reasonably accurate picture of how VSP technology developed in the United States, since essentially all of the patents involved in this chronology were issued by the U.S. Patent Office. Only a few non-U.S. patents are included, but these were not issued until the early 1980's. No serious effort was made to search European patent literature. No doubt many important non-U.S. VSP patents have been omitted because of this approach.

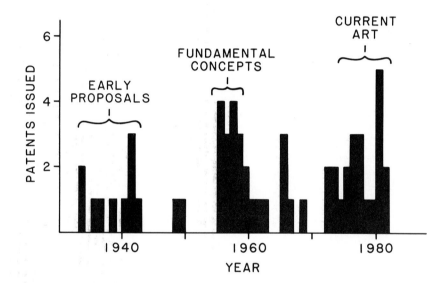

Figure 1-5   This profile of patent activity shows that most VSP technology was developed in three short time periods occurring about 1940, in the late 1950's, and in the late 1970's. Although the principles of vertical seismic profiling can be constructed from patents issued in the 1930's, the art was not actually realized until the 1950's.

An examination of Figure 1-5 shows that some principles of vertical seismic profiling were patented in the 1930's and early 1940's. Among these patents are those issued to McCollum (1933a, 1933b, 1935), Weatherby (1936), Salvatori (1938), Athy (1940), Beers (1941a, 1941b), and Slotnick (1941). VSP development then languished until the mid

1950's, when it made a dramatic resurgence. A rich source of VSP patents begins in 1955 and extends to 1960. In this brief period, patents describing most of the present fundamental concepts of vertical seismic profiling were issued to Weiss (1955), Bardeen (1955), Bazhaw (1955), Howes (1956a, 1956b, 1957), Ording (1955, 1956), Jolly (1957), Peterson (1957), Swan (1957), Clifford, et al. (1958), and Hildebrandt (1959). The patent claims made by McCollum, Howes, Clifford, Beers, and Slotnick are recommended reading. Clifford's patent shows one of the early wiggle trace plots of VSP data recorded over an extensive depth interval. VSP development in the U.S. then retreated into another dormant period until the late 1970's when interest again revived. Patents describing some of the current art in VSP technology would be those issued to Anstey (1980), Mons (1980a, 1980b), Hawkins (1980, 1981), Cowles (1981), and Fitch (1983a, 1983b, 1983c).

Two technical conferences and two continuing education courses dedicated to vertical seismic profiling particularly helped to build much of the current interest among non-Soviet geophysicists in VSP technology and deserve special mention. The dates of these meetings and the growth of interest in VSP technology manifested by attendance at them is shown in Figure 1-6.

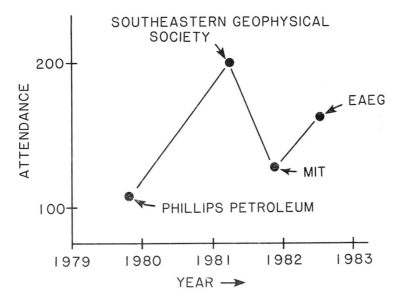

Figure 1-6   Conferences and short courses dedicated to vertical seismic profiling outside of the Soviet Union.

The first of these was a two-day seminar held at the Phillips Petroleum Company Research Center, Bartlesville, Oklahoma in October, 1979 at which Dr. E. I. Gal'perin was a featured lecturer. Over one hundred representatives of petroleum companies, seismic service companies, and academic institutions attended, and a dramatic increase in American VSP experiments immediately followed these discussions.

The first continuing education course was a one-day VSP school held in New Orleans, in March, 1981. This course was arranged by the Southeastern Geophysical Society because they saw a need to increase the industry's awareness of this "new" technology. Approximately 200 geophysicists, representing a cross-section of the American oil exploration industry and academia, attended and shared their experiences in developing VSP capabilities. Again, VSP experimentation increased in the American petroleum industry following this short course as it did following the Phillips seminar. The proceedings of the Phillips seminar were not published, but a limited quantity of the papers presented at the New Orleans course was distributed to attendees.

Following the New Orleans meeting, the Department of Earth and Planetary Sciences, Massachusetts Institute of Technology, held a one day symposium on vertical seismic profiling in December, 1981 in Houston, Texas as a part of their Industrial Liaison Program. Professor M. N. Toksoz guided the symposium. Papers given at this conference focused attention on three-component VSP data acquisition and interpretation with emphasis on fracture detection, seismic attenuation, and unified compressional and shear wave interpretations. A little more than one hundred representatives of companies involved in energy exploration attended the symposium.

The second continuing education course was a one-day school on vertical seismic profiling principles offered by the European Association of Exploration Geophysicists at their annual meeting in June, 1982 in Cannes. Approximately 75 explorationists and academic researchers attended this course which was taught by Peter Kennett and R. L. Ireson of Seismograph Service, Ltd. Likely more educational courses on all aspects of vertical seismic profiling will continue to be offered to explorationists worldwide by aggressive geophysical societies.

The number of papers dealing with various aspects of vertical seismic profiling which are presented at technical meetings of geophysical societies is increasing each year. The first technical session devoted exclusively to vertical seismic profiling at an international SEG convention occurred during the 51st annual meeting in October, 1981. These papers, particularly if presented by major petroleum exploration companies or by representatives of established geophysical service contractors, are a reliable indicator of the financial investment and degree of professional interest in vertical seismic profiling throughout the industry. Several unpublished, confidential VSP experiments and technical accomplishments usually exist for each VSP paper that is publicly presented.

CHAPTER 2

EQUIPMENT AND PHYSICAL ENVIRONMENT NEEDED FOR VERTICAL SEISMIC PROFILING

Subsequent chapters refer to certain features and characteristics of the equipment used in vertical seismic profiling; therefore, it seems essential to describe the basic hardware that must be assembled in order to record VSP data before discussing the various aspects and uses of the data. A simplified picture of the equipment and physical environment needed in a VSP land experiment is shown in Figure 1-2. The experimental components of a VSP survey which will be emphasized in this chapter are the:

- borehole,
- energy source,
- downhole geophone,
- recording cable,
- surface digital recorder.

Equipment and physical factors other than these items can be involved in vertical seismic profiling, but a discussion of the essential characteristics of these few components will provide an adequate background for the topics that will be presented in following chapters. Equipment descriptions will not be restricted to just those items needed for onshore vertical seismic profiling as implied by Figure 1-2; marine VSP equipment will also be illustrated and discussed.

The Borehole

An appropriate borehole must exist in order to execute a vertical seismic profile. This fact is so obvious that it is sometimes overlooked in an explorationist's rush to record VSP data in a hot prospect area. Several factors should be considered when selecting an "appropriate" borehole. Among these factors would be the following:

Vertical nature of hole - The collection and interpretation of VSP data in a deviated borehole can be more expensive and more difficult than the collection and interpretation of data in a straight vertical hole, primarily because there is more uncertainty concerning the position of the downhole geophone relative to the energy source and because the source may have to be moved to several different locations during the course of the data recording. Consequently, the vertical depths at which primary reflections are generated, as well as the vertical timing relationships between primary and multiple reflections, are more suspect in deviated hole VSP's. These interpretational problems are more common in marine vertical seismic profiles because offshore wells drilled from production platforms are often highly deviated. As will be emphasized in Chapter 8, there are some advantages to recording VSP data in a deviated borehole since this type of well allows the subsurface underneath the borehole to be imaged laterally with great resolution. Such seismic detail is invaluable when trying to interpret a structurally complex reservoir or when mapping subtle facies changes. The vertical nature of a borehole selected for vertical seismic profiling should be matched to the problem to be studied. For example, if horizontal imaging of the subsurface is required, a deviated borehole has several advantages. If the intent is to simply identify the depth and one-way time of primary reflectors, a vertical hole is a better choice. Given an option, VSP data can usually be recorded quicker, easier, less expensively, and more accurately in a vertical well than in a deviated well.

Casing and cementing conditions - Usable VSP data can be recorded only when seismic body waves in the earth can be transmitted across the borehole-formation interface to a downhole geophone with minimum alteration of waveform character. Several factors can affect this interfacial coupling. Foremost among these factors are the casing and cementing conditions that exist in the well. It is desirable to record VSP data in a cased well because the borehole is protected from sloughing and differential pressure sticking problems. Since both the borehole and the downhole equipment are in a secure, protected environment in a cased well, VSP experiments can be extended for many hours without having to re-enter the borehole and condition it by reaming or circulating mud. However, if a VSP well is cased, there must be some type of medium between the casing and the formation which is a good transmitter of seismic energy. The best medium is cement, but data shown in following chapters will demonstrate that, under some conditions, uncemented casing will also suffice. The one casing situation that invariably prevents usable VSP data from being recorded is multiple casing in which the casing strings are not rigidly cemented to one another and also to the formation. Examples of data recorded inside multiple casing will be shown in Chapter 3. The preferences that Soviet geophysicists have developed regarding VSP borehole casing and cementing conditions are shown in Figure 2-1 (Gal'perin, private communciation). The best

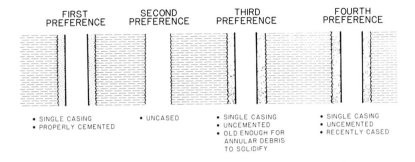

FIRST PREFERENCE   SECOND PREFERENCE   THIRD PREFERENCE   FOURTH PREFERENCE

- SINGLE CASING
- PROPERLY CEMENTED

- UNCASED

- SINGLE CASING
- UNCEMENTED
- OLD ENOUGH FOR ANNULAR DEBRIS TO SOLIDIFY

- SINGLE CASING
- UNCEMENTED
- RECENTLY CASED

Figure 2-1   Common borehole conditions shown in the order of their preference for VSP data recording.

borehole situation for recording VSP data is one where there is a single string of casing that is rigidly cemented to the formation over the complete depth interval where data are to be recorded. The second preference is an uncased, open borehole. In both of these situations, there is excellent coupling of body waves to a geophone clamped in the borehole. The third preference is a single casing string that is uncemented but which has been in place for several years. Experience has shown that the drilling mud, rock cuttings, and sloughing that fill the annulus between the casing and the formation tend to solidify as a well ages, causing a pseudo cement bond to be created between the casing and formation. The last choice is a cased, uncemented well that has only recently been cased. "Recently" is a relative term and can mean anything from a few hours to a few days. The longer one can wait after uncemented casing is set before commencing a VSP experiment, the better the chance that there will be improved seismic coupling between the formation and the casing.

Borehole diameter - A proposed VSP study well may be cased or uncased. If the hole is uncased, then borehole rugosity can affect the clamping of a geophone to the formation. In particular, large washouts often prohibit clamping by not allowing the geophone locking arm to touch the borehole wall; specific examples of data recorded with unclamped geophones will be given in Chapter 3. A caliper log of an uncased hole should be used to select geophone recording depths where no washouts occur and to determine the length of the locking arm that should be used with a geophone tool. On the other hand, the diameter of some wells may be too small to allow the clamping mechanism of a downhole geophone to extend far enough so that an adequate clamping force can be generated. The basic fact to keep in mind is that the diameter of a borehole may prevent good geophone coupling to the formation, thus borehole size must be carefully studied when selecting a VSP experimental well and the type of downhole geophone to use. From theoretical and experimental studies, Blair (1982) concludes that a seismic detector can be installed at any point on the circumference of a cylindrical borehole and still record the same particle motion as long as the wavelengths of interest in the seismic wavelet propagating past the borehole are greater than ten times the hole's circumference (not the hole's diameter). For shorter wavelengths, a detector should be coupled to the opposite side of the hole from that which is first displaced by the propagating signal in order to minimize the amplitude distortions and phase shifts created by the presence of the borehole. This preferred position for the detector is called the wave shadow side of the borehole. When recording VSP data in standard diameter oil and gas wells it should make no difference where a geophone is clamped around the circumference of the wellbore. However, in an extremely large diameter hole, such as that involved in one experiment conducted by Balch et al. (1980b) described in Chapter 8, and in high frequency borehole-to-borehole work, Blair's analysis should be considered.

Borehole obstructions - Production engineers sometimes set packers or frac rings inside a cased well before a VSP experiment is started. The depths at which these obstructions exist may not allow a geophone device to be lowered to those depth intervals critical to an experiment's objectives. An example of VSP data recorded in a well where a frac ring prevented data from being recorded in the bottom portion of the borehole is shown in Figure 3-22. The presence or absence of borehole obstructions must be confirmed before commencing a vertical seismic profile. If one is unsure whether or not borehole obstructions exist, then a cheap sacrificial tool, equal in size or larger than the VSP geophone package, should be carefully lowered to total depth (TD) and returned to the surface before starting a VSP survey.

Amount of borehole information - A complete rigorous interpretation of VSP data can be done only if a broad suite of independent data which specifies the physical characteristics of the formation around the borehole has been recorded in the VSP study well. A borehole in which caliper, sonic, density, electric, and radioactive logs are recorded, where drill cuttings are preserved, and where cores are taken would be much preferred over a borehole where only scant borehole information exists. Without this type of supporting information, one cannot determine if a numerical change exhibited by a propagating VSP wavelet is an indication of a real subsurface geological condition or an artifact of recording and data processing procedures. Also, cement bond logs and measurements of the depths of all casing joints are needed in cased wells in order to verify the nature of the coupling between a VSP geophone and the formation.

Surface seismic data at well site - Many objectives of vertical seismic profiling are directed toward improving or clarifying subsurface structural and stratigraphic interpretations made from surface-recorded seismic data. Consequently, when selecting a VSP study well, the quantity, quality and proximity of surface seismic data should be considered. Other factors being equal, a borehole which has a good quality surface seismic line crossing directly over the well location is the best choice for VSP data recording.

Principles of VSP Energy Sources

A major use of VSP data is to support and clarify interpretations of the subsurface made from surface-recorded seismic data. A unified interpretation of surface-recorded data and VSP data is best accomplished if the same seismic source wavelet is contained in both sets of data so that reflection character appears the same, regardless of which set of data is being analyzed. Consequently, it is preferred that the seismic energy source used in a vertical seismic profile be the same as that used when recording surface seismic data near a VSP well. In many instances this recommendation cannot be followed, and wavelet equivalence between the two sets of data must be achieved by numerical data processing procedures such as deconvolution or frequency and phase shaping of wavelet spectra.

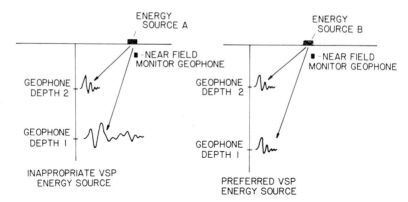

Figure 2-2   The energy source used in vertical seismic profiling must generate many highly repeatable shot wavelets. A single VSP experiment may require that several hundred seismic shots be recorded. An energy source that functions like the one on the right is preferred because the near-field geophone which captures the outgoing shot wavelet indicates that this wavelet remains almost constant as data are recorded between depths 1 and 2.

With the exception of equipment used in the Soviet Union, the downhole geophone devices used in vertical seismic profiling record data at only one depth level for a single seismic shot. In order to record seismic responses at finely spaced intervals over an appreciable vertical distance, a VSP seismic energy source must be activated many times and usually must do this while remaining at a fixed surface location. It is therefore essential that VSP energy sources produce a highly consistent and repeatable shot wavelet time after time; otherwise, it is difficult to correlate equivalent features of upgoing and downgoing wavelets throughout the vertical section over which data are recorded. An illustration of this desirable feature of a VSP energy source is shown in Figure 2-2.

The output strength of a VSP energy source must be carefully chosen at each experimental well site. Many geophysicists have discovered that the attitude, "bigger energy sources are better", is not always a good philosophy when recording surface reflection data, and this caution is perhaps even more important in vertical seismic profiling. Figure 2-3 shows a simplified picture of the ray paths recorded by VSP data. In all vertical seismic profiles, downgoing wavefields are much stronger than upgoing wavefields. Most VSP applications involve analyses and interpretations of these weak upgoing wavefields. As the output strength of a VSP energy source increases, more downgoing events are usually created due to numerous reverberating layers in the shallowest part of a stratigraphic section. This increase in the number and amplitude of

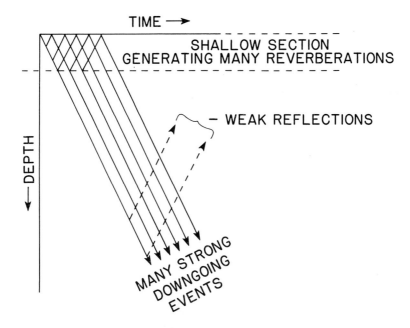

Figure 2-3   Increasing the strength of a VSP energy source may make it more difficult to retrieve upgoing events, depending on the nature of the shallow earth section near the source position.

downgoing events may be a greater contamination of the data than is the gain in amplitude of upgoing events. It is not unusual to record better VSP data when using an energy source of modest strength than when using a source of considerable strength.

Dynamite

Chemical explosives (the generic name, dynamite, will be used hereafter) are traditional seismic energy sources which have been used in hydrocarbon exploration for many years. Buried dynamite charges are particularly effective producers of seismic body wave energy, and are widely used as an energy source in vertical seismic profiling in the Soviet Union (Gal'perin, 1974). Levin and Lynn (1958) recorded VSP data with dynamite shots suspended in air and found that the downhole signal amplitude was reduced by a factor of 30 compared to the signal generated by a buried shot, even though the air charge size was reduced by only a factor of 2. Outside of the Soviet Union numerous velocity check shot surveys have been made using dynamite as an energy source, but little publicly documented VSP data exist in which dynamite shots were used to generate the recorded wavelets. Papers by Wuenschel (1976), Levin and Lynn (1958), and Lash (1980, 1982) are easily available non-Soviet investigations showing the type of VSP data generated by dynamite shots.

One objection commonly voiced by VSP explorationists concerning dynamite energy sources is that it is too difficult to shoot several tens of dynamite shots, as you have to do in vertical seismic profiling, and maintain consistent shot wavelets. Consequently, any VSP data application which demands invariant wavelets can be thwarted if dynamite is used. However, if great care is exercised in the field, reasonably invariant shot wavelets can be obtained. People who have successfully used dynamite as a VSP energy source find that two field conditions must be satisfied in order to obtain such wavelets.

(1)   The shot hole must be carefully prepared so that it remains reasonably constant in diameter and depth during the course of the shooting. Optimal conditions would be to drill a hole below the weathered zone (or as deep as feasible), case and cement it back to the surface, and keep it filled with water.

(2)   Explosive charges should be as small as possible. Charges of one to three pounds (0.5 to 1.5 kg) are common. Sometimes charges as small as one-quarter pound (0.1 kg) are satisfactory. On-site testing in which reflection signals from a variety of charge sizes are recorded at several borehole depths is the only rigorous way to define proper charge size at a well location.

An illustration of a carefully prepared dynamite shot hole suitable for recording VSP data is shown in Figure 2-4. The depth, H, should be at least twelve to fifteen meters, or should extend below the weathered zone, whichever is the greater depth. The height, h, of the charge above the bottom of the hole should be great enough to prevent enlargement and deepening of the hole; a distance of three meters is usually suitable for a small charge of dynamite. Some investigators privately claim that highly repeatable, wide bandwidth wavelets are best achieved if the diameter of the hole is about one meter. A mobile drill rig such as is sometimes used to prepare the initial surface casing hole at a drill rig site can often be used to drill a shot hole with a diameter this large. Thirty or thirty-six inch casing is used to protect the formation. Smaller holes and casing are used if the weathered layer is deep or if the near-surface is competent material. Wuenschel

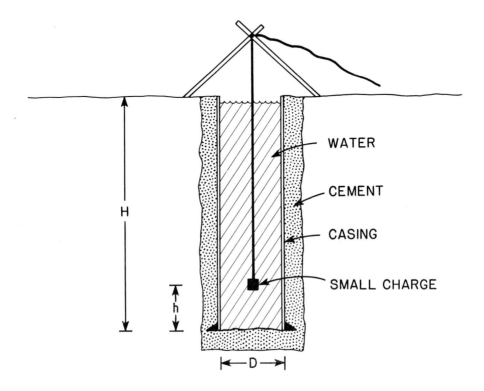

Figure 2-4   Dynamite shot hole designed to produce many invariant VSP shot wavelets. The charge size should be as small as possible, and the diameter, D, as large as possible.

(1976) describes a carefully prepared shot hole for vertical seismic profiling in which 8 inch diameter, water-filled casing was cemented to a formation, and charges were kept away from the casing by centralizers. More than 200 charges, each weighing $2\frac{1}{2}$ pounds (1.1 kg), were shot before this casing ruptured. Most field personnel emphasize that a significantly increased signal results when a dynamite charge is mass-loaded with a column of water as was done in this experiment. Additional information describing how to construct reproducible shot holes can be found in Wuenschel's patents (1972, 1974, 1975).

Shown in Figure 2-5 are some experimental data resulting from Wuenschel's field work. The patents referred to above provide specific information about the type of casing and cementing used in these tests. The data show that if many highly repeatable shot wavelets are desired, then a large shot hole diameter is required when the surrounding formation is soft. The hole size can be reduced in hard formations. As these data imply, the appropriate diameter for a cased shot hole which will allow a large number of highly repeatible shot wavelets to be generated is a rather complicated function of the strength of the explosive used, the number of reproducible wavelets required, and the elastic properties of the surrounding earth medium. Field tests of this nature should precede any VSP experiment involving dynamite as an energy source in order to establish how to properly construct a shot hole.

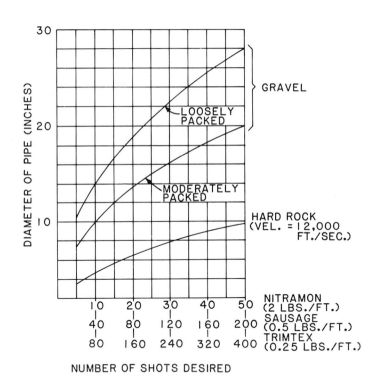

Figure 2-5   Experimental data showing the relationship between shot hole casing size, elastic properties of medium around the shot hole, and number of shots before casing rupture (After Wuenschel, 1972). Charge weights are as large as 2.5 pounds (Wuenschel, 1976). The explosive names are registered DuPont trademarks.

Dynamite energy sources will be prohibited at many VSP well sites because of environmental conditions, cultural restrictions, and federal or state regulations. Dynamite is generally never considered as a marine energy source for environmental reasons. In addition, any VSP experiment which requires several different source locations in order to image a desired geological feature is generally more difficult and more expensive to execute with drilled shot holes than it is with mobile surface energy sources.

Impulsive Surface Sources

A variety of seismic sources exist that can apply a large impulsive vertical force to the surface of the ground, and all of these devices are viable energy sources for onshore VSP work. Included in this category are weight droppers and devices that use explosive gases or compressed air to drive a heavy pad vertically downward with great force. These types of sources will not be illustrated nor discussed because they are adequately described in other texts (Dobrin, 1976; Telford, et al., 1976; Cholet and Pauc, 1980), but they do need to be acknowledged as acceptable VSP energy sources for many situations. Actual field tests should be made at each onshore VSP well site in order to determine if it is inadvisable to use an impulsive surface source because of the severe shallow reverberations which they tend to generate, or because of inadequate energy input, restricted bandwidth, or other reasons.

Vibrators

A wide variety of Vibroseis® type energy sources are used in hydrocarbon exploration. The generic term, vibrator, will be used to refer to these types of seismic sources in this text. Vibrators have several features that make them attractive as energy sources in onshore VSP work. They are quite mobile, allowing an explorationist to execute VSP experiments employing many different shot point locations in an efficient and ex- peditious manner. Also, the frequency content of a vibrator signal can often be adjusted

---

® Registered Trademark of Conoco, Inc.

to meet resolution requirements needed in a particular VSP recording. The magnitude of the input energy can be tailored for optimal signal-to-noise conditions by varying the size or number of vibrators or by altering the output drive of individual vibrators. All in all, vibrators are probably one of the most versatile onshore VSP energy sources. Most of the land-recorded VSP data shown in this text were generated by vibrators, and both positive and negative features of these sources will be discussed as the data are presented. As a general observation, if a VSP experimental site is plagued by random noise, vibrators are an excellent energy source because the correlation process used to reduce the data to an interpretable form discriminates against noise outside the sweep range, and summing several sweeps cancels noise within the sweep range. However, if coherent noise having frequencies within the vibrator sweep frequency range are present, then the correlation process accentuates these noise modes. Much literature exists describing the operating characteristics of vibrators, and that information will not be rephrased in this text. The following references are only a portion of the available literature. (Crawford, et al., 1960; Geyer, 1969, 1971; Seriff and Kim, 1970; Dobrin, 1976; Goupillaud, 1976; Telford, et al., 1976; Bernhardt and Peacock, 1978; Waters, 1978; Cunningham, 1979; Edelmann and Werner, 1982).

Physics of Airguns Pertinent to VSP Usage

A fundamental equation describing the physical parameters of a high pressure oscillating air bubble in a fluid is the Rayleigh-Willis formula:

$$T = 1.14 \ (\rho)^{0.5} \ (KQ)^{0.333} \ (d+33)^{-0.833} \tag{1}$$

where T is the bubble oscillation period in seconds, ρ is the density of the surrounding fluid in gm/cc, Q is the potential energy of the expanded bubble, K is a constant whose value depends on the units chosen to express Q, and d is the depth in feet of the center of the bubble in the fluid. A mathematical development of this equation together with a discussion of the assumptions involved and a table of K values are provided by Kramer, et al. (1968).

A key insight into the physics of airgun operation provided by this equation is that the bubble oscillation period can be altered by changing the potential energy, Q, of the fully expanded bubble or by varying the depth, d, where the bubble is created. The potential energy can be altered by varying either the working volume or operating pressure of the airgun which creates the bubble. Of these two choices, airgun users usually choose to vary the working volume of airguns in order to change the oscillation period.

A typical far-field pressure response created by a single airgun is shown as the top trace of Figure 2-6. Positive pressures are plotted as downward excursions, so the initial pressure pulse occurs at time T_0, and the first bubble oscillation at time T_1. Subsequent bubble oscillations occur but are ignored since they have smaller amplitudes. If a second, larger airgun is simultaneously fired at the same depth and operating pressure as is Gun 1 but far enough away so that its bubble does not interact with that of Gun 1, then its response would be as shown in the second trace. Its initial pressure pulse also occurs at time T_0, but the first bubble oscillation occurs at time T_2. In constructing airgun arrays, one design concept is to adjust airgun sizes so that the pressure increase of the first bubble oscillation from one gun occurs at the same time as a pressure rarefaction from a second gun, such as shown by the pressure relationships between Gun 1 and Gun 2 at time T_2. The net result is that these two pressure effects of the first bubble oscillation tend to cancel each other.

Adding a third, smaller airgun can yield a pressure response such as shown in the third trace with still a different bubble period. Summing all three responses yields the array response in the bottom trace. This composite pressure response exhibits much less bubble oscillation than exists for any single airgun. Two factors contribute to this result:

(1)   All initial pressure pulses at time T_0 sum in phase so that the amplitude of the first pressure pulse of the composite trace is much larger than the amplitude of any individual bubble oscillation.

(2)   Pressure increases and rarefactions from individual bubble oscillations tend to cancel each other.

A modified version of actual measurements of bubble period oscillations made by Kramer, et. al. (1968) is given in Figure 2-7. These data describe the oscillation of a 300 in^3 (4914 cm^3) volume of air having an initial pressure of 2000 psi (13788 kPa) as a function of the depth of the bubble. The dashed line shows the theoretical response expected from the Rayleigh-Willis formula described by Equation 1. The measured oscillation period agrees with the theoretical prediction except at shallow depths where the loss of water mass above the bubble invalidates some of the assumptions used to derive the Rayleigh-Willis equation. It should be noted that these shallow water depths are the type of environment in which marine airguns are used in pits or

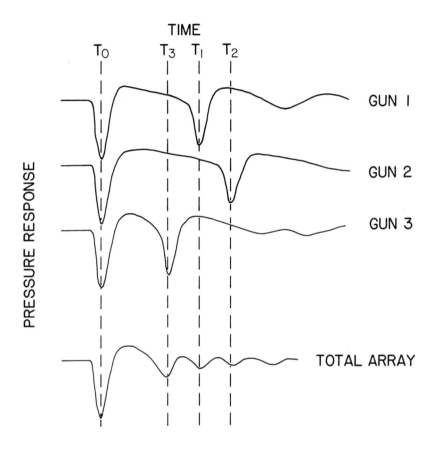

Figure 2-6   Attenuating bubble oscillations by arraying airguns of various sizes. Each trace is a measure of the far-field pressure variation plotted so that a pressure increase appears as a trough.

shallow shot holes in onshore VSP work. The Rayleigh-Willis equation cannot be used to design airgun arrays having efficient attenuation of bubble oscillations in these onshore situations. Kramer, et al. (1968) show that placing a massive steel plate above a shallow airgun effectively increases the mass-loading above its air bubble, and as a result, the measured bubble oscillation curve moves much closer to the Rayleigh-Willis prediction.

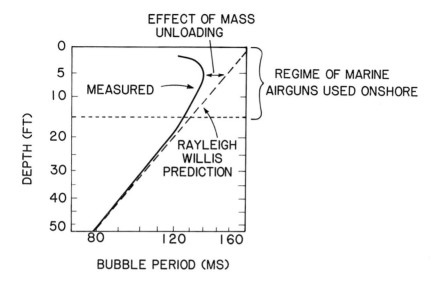

Figure 2-7   Bubble oscillation period versus depth for 300 in^3 of air at 2000 psi showing effects of mass-unloading above an airgun discharge (Modified from Kramer, Peterson, Walter, 1968).

Two concepts seem to prevail as to how an offshore array of airguns should be designed in order to generate a shot wavelet having optimum resolving power.   One approach is to arrange two or more airguns so that their respective air bubble oscillations tend to cancel each other as already discussed.   The second concept is that one should forget trying to cancel individual airgun bubble oscillations and simply create an arrangement of guns having different peak frequencies so that the composite frequency spectrum of the array is as broadband as possible.

This second approach to airgun array design is shown in Figure 2-8.   The peak frequency of an airgun is determined primarily by its chamber volume, working pressure, and operating depth.   From Safer's (1976) analysis, the dominant frequency, f_0, of an isolated airgun discharge can be written as

$$f_0 = 51.5(V^{-0.33})(P^{-0.31})(1+d/10)^{0.81} \tag{2}$$

where V is the chamber volume in litres, P is the discharge pressure in bars, and d is the firing depth in meters.   In developing this equation, it has been assumed that the ratio of

26

specific heats C_p/C_v for air is 1.08 (C_p is the specific heat at constant pressure and C_v is the specific heat at constant volume); that the airgun occupies 80 percent of the equilibrium volume of the bubble; and that no physical boundaries are near the bubble. If N airguns are fired simultaneously at the same depth, then this third assumption is violated because reflecting boundaries are created by the (N-1) bubble cavities surrounding any given airgun, and a correction term must be added to Equation 2. Safer's equation 48 defines this correction term as a function of average bubble radius and the airgun separation distances. For a single gun operating at a fixed depth, the spectrum of the shot wavelet may peak as shown by curve 1 in Figure 2-8. A second gun of different size or at a different depth can yield an output spectrum such as curve 2. By adjusting the size, discharge pressure, and firing depth of two or more guns, the composite output energy spectrum can often be extended over a broad frequency range and thus create a short, compact wavelet. Even if bubble oscillations occur, a wavelet with a broad frequency spectrum will not be as long nor have as many persistent peaks and troughs as a wavelet having a narrow frequency spectrum. In addition to Safer's analysis, comprehensive reviews of airgun operation and airgun array behavior are also given by Giles (1968), Schulze-Gattermann (1972), Vaage, et al. (1982), and Ziolkowski (1982a, 1982b).

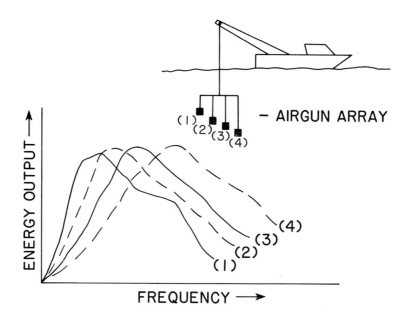

Figure 2-8   Airgun array designed for broadband VSP data recording.

Ziolkowski (1982a) and Larner, et al. (1982) have presented arguments showing that the philosophy of designing elaborate airgun arrays to suppress bubble oscillations has perhaps been over emphasized. This conclusion is, to a large extent, based on the concept that one should sacrifice wavelet tuning in order to achieve higher output energy which will, in turn, improve the signal-to-noise characteristics of deep reflectors. For this reason, some marine VSP experimenters are now preferring to use a single, large airgun rather than a tuned array so that better signal-to-noise conditions are created in the upgoing wavefield and also because a single gun tends to generate more consistent wavelets over the course of several hours of VSP recording than does an array of guns.

Onshore Use of Marine Airguns

Marine airguns have several features that make them attractive as onshore VSP energy sources; namely, they are small and portable, they can be fired at intervals of a few seconds, and they usually create highly repeatable wavelets. Airguns must be submerged in water in order to function properly, thus some type of reservoir capable of holding a rather large volume of water must be constructed when airguns are used in onshore VSP work.

Probably the most common technique for creating a water environment on land for marine airgun operation is to dig a pit and fill it with water, as shown in Figure 2-9. Typical pits have a width, L, of seven or eight meters and a depth, D, of five or six meters.

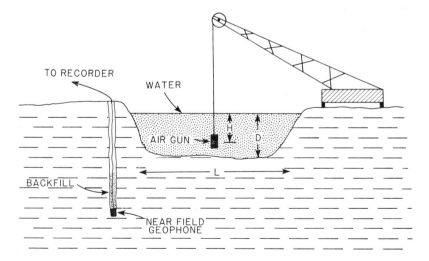

Figure 2-9   One scheme for using marine airguns as an onshore VSP energy source.

Figure 2-10 (A)   Marine airgun (120 in^3 size) and water-filled pit such as could be used as an energy source in onshore VSP work. This pit should probably be larger. Some field people have noted that more energy enters the earth as the base area of an airgun pit is enlarged.

(B)   Air bubble created by submerged airgun erupting through water surface.

(Courtesy D. E. Lauffer, Phillips Petroleum Company)

The airgun should be set at a depth, H, of two to four meters. Photographs of one experimental field site where a water-filled pit and an airgun were used as an energy source are shown in Figure 2-10. One common problem of airguns operating in pits is that sediment is continually dislodged from the bottom and sides of the pit, and if the mud content becomes too high, the gun mechanisms can become plugged. Plastic pit liners reduce this erosion and also prevent excessive water loss if a pit must be dug in porous soil. A near-field monitoring geophone should be planted at a depth of at least ten meters near the airgun pit.

A second possible onshore airgun configuration is shown in Figure 2-11. This water volume is much smaller than that of a dug pit; consequently, the hole has to be cased and cemented to prevent it from being destroyed. The top of the cased hole should be sealed

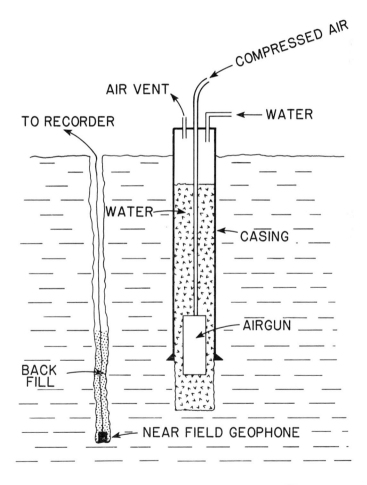

Figure 2-11   Technique for using a marine airgun as an onshore VSP energy source.

except for an air vent and a supply line for maintaining the water level. This seal helps to mass-load the air bubble as well as to prevent a high rate of water loss from the hole.

One shortcoming of airguns used in water-filled pits or boreholes is the lack of mobility of such sources. If appropriate VSP data can be obtained with a stationary source, then airguns deployed as in Figures 2-9 and 2-11 are satisfactory. Mobile sources are necessary for many VSP experiments, and in such cases an airgun confined to a relatively small water reservoir which, in turn, is attached to a truck or tractor chassis can be used. The elements of such a device are shown in Figure 2-12. The total weight of the water and its container is on the order of 2000 pounds (8896 Newtons), and the reservoir can be quickly raised so that its supporting vehicle can be driven to a new location. Bolt land airguns operate in this manner, but the internal design and operating principles of these devices differ in specific details from what is shown here. The mechanical principles involved in a device of this type are described in patents issued to Chelminski (1974, 1978). A downhole energy source, based on a variation of this concept, is also described by Chelminski (1976).

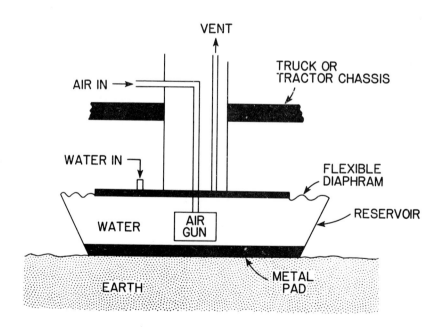

Figure 2-12   Mobile airgun system for onshore vertical seismic profiling.

Airguns used in onshore VSP data recording range in size from 40 in^3 to 200 in^3. Seldom are more than two guns used, although there is no reason for not using more if there is a sufficient volume of water for them to operate in and an adequate supply of compressed air to keep them firing at a desired time interval.

Offshore Energy Sources

Airguns are by far the dominant energy source used in offshore vertical seismic profiling. Implosive types of marine sources show promise as reliable VSP energy sources, but little or no testing of these devices for vertical seismic profiling has been done. If an offshore well is vertical, then many VSP objectives can be attained by keeping the airgun at a fixed location near the wellhead. This shooting geometry is one of the most convenient because it allows the airgun to be suspended from a work crane such as shown in Figure 2-13. Also, large capacity air compressors, which are standard equipment on these drill rigs, are available for operating the guns. Although guns placed close to a drill rig may vibrate the rig when they discharge, they do not cause structural damage and pose no safety hazard to offshore rigs.

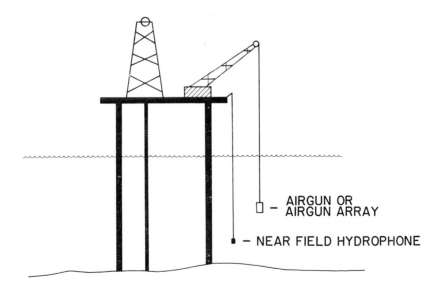

Figure 2-13 Airgun used as a stationary energy source in marine vertical seismic profiling.

32

There are several types of marine VSP exploration applications which require that VSP data be recorded from various source positions. In addition, many offshore wells are deliberately deviated from vertical, and recording VSP data from a fixed source location near the wellhead is an unsatisfactory procedure for these wells. In these situations a shooting boat must be employed to position airguns at desired shotpoint locations. Illustrations and discussions of possible shooting boat positions needed while recording data in a deviated well are included in Chapter 4 (Figures 4-20 and 4-21). Most of these same shooting boat concepts apply if a borehole is vertical but a variety of source locations must be used to accomplish certain objectives.

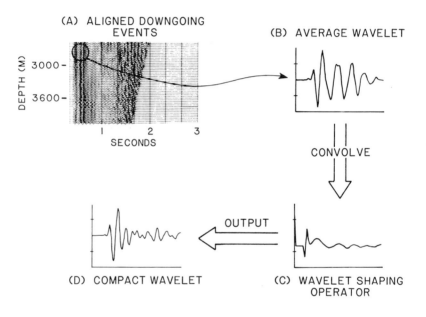

Figure 2-14 An example of the numerical conversion of a long, oscillating airgun wavelet to a shorter wavelet better suited for stratigraphic interpretation. This test shows that operator (C) can be successfully applied to the total VSP data set (A).

Either a single airgun or an array of several airguns can be used in marine VSP work. A single gun often creates more repeatable wavelets than does an array of guns, and some explorationists prefer single gun sources for this reason. However, the shot wavelet created by a single airgun is long and has several high amplitude peaks and troughs due to oscillations of the air bubble created by the gun discharge, as has already been pointed out. VSP data created by a single airgun are shown in Part A of Figure 2-14. The data processing procedures which create vertically aligned downgoing events as shown here are described in Chapter 5. The average wavelet, B, recorded in this survey prohibits any detailed stratigraphic resolution of the subsurface because of its long, resonating appearance. Usually, numerical wavelet shaping procedures, such as shown in Part C of this illustration, can remove the bubble oscillations and create a short, compact wavelet which allows improved resolution of subsurface geology.

Shear Wave Sources

All of the onshore VSP energy sources previously described generate both compressional and shear body waves. In order to study the physics and exploration applications of shear waves, it is often necessary to deliberately increase the amount of shear wave energy in the downgoing wavefield. This objective is accomplished in vertical seismic profiling by using surface sources which apply horizontally directed impulses to the earth or by vibrators which oscillate horizontally rather than vertically. In either case, a heavy pad is usually used to impart horizontal movement to the earth by means of projections on its bottom side which extend into the earth, as shown in Figure 2-15. A specific design for a horizontal vibrator can be found in a patent issued to Fair (1964).

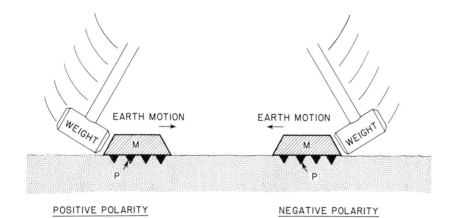

POSITIVE POLARITY   NEGATIVE POLARITY

Figure 2-15   Concept of surface shear wave generator used in vertical seismic profiling. A heavy mass, M, is coupled to the earth by means of projections, P, that are pressed into the ground. An impulsive source is indicated. A shear wave vibrator employs a similar type of earth coupling.

A horizontal movement of the mass M forces the projections P to shear the earth, and a large amount of SH energy is created as well as compressional wave modes. This shearing action results in considerable surface damage. One field example is shown in Figure 2-16. Some landowners and government agencies prohibit the use of such sources in some areas because of this fact.

Figure 2-16   An example of surface damage caused by a shear wave energy source. This pyramid shaped hole is approximately 15 inches (38 cm) deep, 18 inches (46 cm) wide, and 24 inches (61 cm) long. A man's cap is shown in the hole.

After a certain number of horizontal impulses or vibrations, the hole created by each projection is larger than the projection itself, as shown in Figure 2-17. The number of impulses or vibrator sweeps necessary to create this situation varies according to soil type and surface conditions. Once the hole reaches a certain size, a horizontal vibrator can no longer properly couple to the earth since the magnitude of its horizontal motion may be less than the indicated void space. Even an impulsive horizontal source will begin to encounter some coupling problems if the hole becomes too large. The only recourse is to move the pad to a new location. Each time the pad is moved, the elastic parameters of the earth at the new location are possibly different. Consequently, different mechanical coupling exists, and the wavelet created at the new location differs from any previously generated. The fact that surface shear wave sources have to be continuously moved and recoupled to the earth increases the likelihood that variable shot wavelets are created during the course of VSP data recording. Most VSP studies assume invariant input wavelets. There appears to be no easy solution to this shear wave source dilemma.

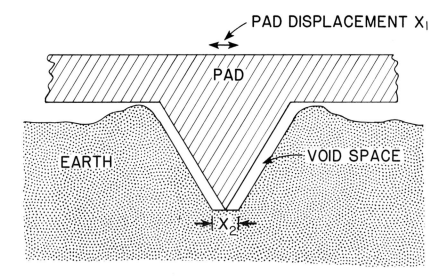

Figure 2-17   Horizontal vibrators have to be moved often because coupling between the earth and the pad cannot be maintained if numerous sweeps are made in one location so that hole dimension X_2 exceeds pad movement X_1.

Downhole Geophone Tools

There is a significant difference between the physical appearance of a borehole geophone used to record VSP data and a geophone used to record surface seismic data. It is not uncommon for a VSP geophone assembly to be 25 to 30 times longer and perhaps 500 times heavier than a typical velocity-sensitive surface geophone, as shown by the comparison in Figure 2-18. The massive size of the downhole geophone is due in part to the fact that the device must withstand the high pressure environment of a deep well. Any metal container capable of withstanding several hundred kilobars of fluid pressure must necessarily be thick-walled and have mating pieces large enough to contain carefully machined O-ring glands. Also, a VSP geophone must be rigidly clamped to the borehole

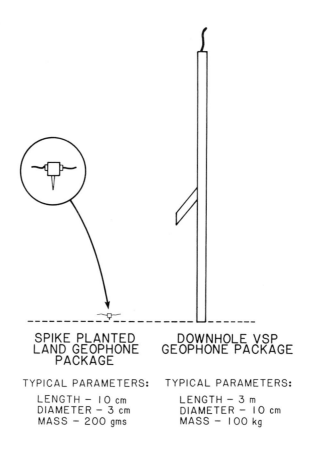

SPIKE PLANTED   DOWNHOLE VSP
LAND GEOPHONE   GEOPHONE PACKAGE
PACKAGE

TYPICAL PARAMETERS: TYPICAL PARAMETERS:

LENGTH – 10 cm   LENGTH – 3 m
DIAMETER – 3 cm   DIAMETER – 10 cm
MASS – 200 gms   MASS – 100 kg

Figure 2-18   Comparison of the physical sizes of geophone devices used to record surface seismic data and VSP data. Both devices are drawn at the same scale.

wall in order to move in phase with the borehole particle displacements created by seismic disturbances, and this requirement means that some type of bulky locking mechanism has to be incorporated into the geophone package. Some examples of common geophone clamping designs are shown in Figure 2-19. The retractable type of locking device is manipulated by electrical or hydraulic motors inside the geophone package, and these motors also add to the size and weight of the downhole tool. Some designs of

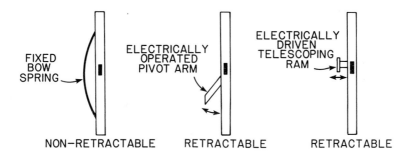

Figure 2-19   Three general types of locking mechanisms used to clamp a VSP geophone to a borehole wall.

downhole geophones are provided by Hildebrandt (1949, 1959), Bardeen (1955), Ording (1955, 1956), Howes (1956, 1957), Jolly (1957), Evans (1962), Rademacher (1965), Malmberg (1965), and Fair (1973).

Non-retractable types of locking devices were used in early designs of borehole geophones, but only a few tools using this locking concept still exist and are in use. This type of locking mechanism consists of a strong bow spring which is held closely to the barrel of the geophone tool by means of a mechanical latch as the geophone is lowered downhole. This latch can be connected to either a trip lever on the bottom of the tool or to a telescoping cable head at the top of the tool. Examples are shown in Figure 2-20.

In the telescoping cable head design, the cable head can be extended only when a large upward impulse is applied to the cable, thus the geophone must be moved slowly and gently on its downward trip into the wellbore. Once the geophone is at the depth selected for recording data, the cable head can be lengthened by dropping the geophone tool a few feet in free fall and then braking hard. An upward extension of the cable head removes the latch restraining the spring, and the expanded spring forces the geophone assembly against the wall of the borehole. If a trip lever on the bottom of the tool barrel is used to release the bow spring, the tool must be dropped against the bottom of the hole with enough force to depress the lever catch. This latter release system has been the more popular, but neither technique for releasing the bow spring is completely foolproof. Unexpected acceleration or deceleration of the tool can cause premature release of the bow spring.

VSP geophones having non-retractable locking arms have limitations that have prevented their widespread acceptance by explorationists. Since the spring arm cannot be retracted once it is released, there is always the danger that the tool can become jammed

38

below a ledge as it is being raised and result in an expensive fishing job. The bow spring can extend to only a limited radial distance, thus the geophone cannot be coupled to the borehole wall in any washed out interval whose diameter exceeds the maximum expansion diameter of the bow spring. Because of these limitations, these types of geophones should be used only in cased holes having no obstructions. Even in a cased hole, once the bow spring is released the geophone can be moved only in an upward direction. It cannot be lowered back to a deeper depth to repeat a measurement. The ability to move a geophone either up or down a borehole is often a great advantage in conducting an efficient VSP experiment. However, for some VSP applications, and in proper boreholes, this type of geophone device is still a viable equipment option to consider.

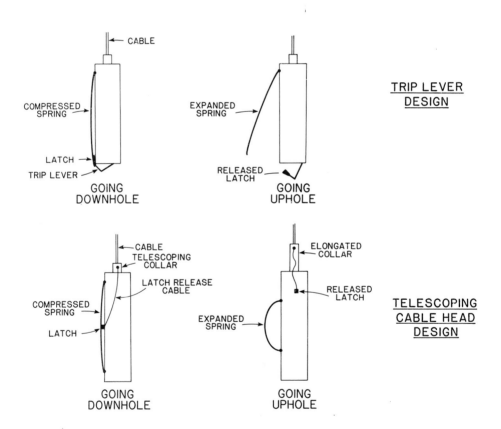

Figure 2-20   The basic design concepts and operating principles of a non-retractable locking device for VSP geophones.

Retractable geophone locking devices employing motor driven pivot arms are the most common type of VSP geophone design now in use. Levin and Lynn (1958) show comparisons between downhole signals recorded with a bow spring geophone and a locking arm geophone. Specific locking arm geophone systems differ primarily with regard to how many pivot arms are used and how much lateral locking force the arms can generate. Some versions of these types of tools have a single locking arm which pivots outward until it contacts resistance, and the drive motor then applies only a small additional pivoting force before shutting off. The lateral force pressing this type of geophone against the borehole wall is usually less than the weight of the tool. The lateral force is increased by slacking the cable so that the tool's weight causes an additional pivoting force about the fulcrum point of the extended arm. This locking principle is called "locking by vertical wedging" and is illustrated in Figure 2-21. The exact amount of lateral force, F, created

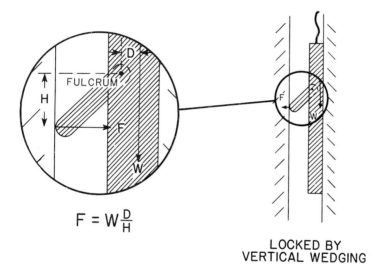

$$F = W\frac{D}{H}$$

LOCKED BY
VERTICAL WEDGING

Figure 2-21   The concept of coupling VSP geophones to a formation by vertical wedging.

by vertical wedging depends on W, the bouyant weight of the tool suspended in the borehole fluid; on the mechanical design of the tool which establishes the distance, D, between the fulcrum point of the arm and the line passing through the center of gravity of the tool; and on the combination of arm length and borehole diameter which defines the magnitude of moment arm H. The moment equation which determines the magnitude of F is shown in the illustration.

VSP geophone tools which depend upon vertical wedging in order to couple moving coil geophone elements to a borehole wall record good quality data if the geophone elements are vertically oriented, but sometimes the magnitude of the laterally directed force is not large enough to properly couple horizontally oriented geophone elements with the formation. Consequently, recent geophone designs have increased the amount of force generated by the pivot arm drive motors so that the lateral force pressing the tool to the formation before the cable is slacked is two to three times the tool's weight. This increased locking force allows high quality data to be recorded by both vertical and horizontal geophones.

Only a few borehole geophones have been built which use a laterally directed telescoping ram such as shown in Figure 2-19 to press the geophone package to a borehole wall. The concept seems practical and may gain wider acceptance. A tool using this locking technique is described by Wuenschel (1976) and Gustavson et al. (1973). One advantage of this locking design is that the locking force is the same for any borehole diameter smaller than the maximum extension length of the ram. The locking force generated by a pivoting arm type of locking mechanism depends on the angle of contact between the arm and the borehole wall (i.e., the locking force is a function of borehole diameter).

Geometrical Arrangement of Geophone Transducer Elements

Geophone transducer elements are positioned in a VSP geophone package in one of three geometrical arrangements; these being (1) vertically oriented, (2) three-component XYZ, or (3) three-components tilted at 54 degrees. The most common situation is to orient all transducer elements vertically along the longitudinal axis of the geophone tool. In vertical boreholes, this geophone geometry measures only the vertical component of particle motion. Some VSP tools may have as many as six geophone elements stacked on top of each other along the longitudinal axis of the tool. These transducers are electrically connected in parallel in order to generate a strong downhole voltage output and also in order to provide geophone backup in case one or more of them should fail. Some vertical geophones are gimbaled so that they always point upward, even in deviated boreholes.

A large amount of valuable VSP data has been recorded with vertically oriented geophones; however, some valuable applications of vertical seismic profiling can be made only if one measures the complete X, Y, and Z components of particle motion created by downgoing and upgoing seismic events. For example, the reflection and transmission properties of shear body waves, the energy mode conversions that occur at impedance boundaries, and the determination of reflector orientation by the direction that compressional energy reflected from the interface arrives at downhole geophones, can all be estimated by recording subsurface particle movements in three mutually orthogonal directions. In order to record three-component particle motion, the downhole VSP geophone instrumentation must contain non-vertical geophone elements. A convenient geophone geometry which responds to three-component particle movement can be created by orienting one or more geophone transducer elements along each of the three orthogonal Cartesian coordinate axes, i, j, and k, shown in Part A of Figure 2-22. In this configuration, the k axis coincides with the longitudinal axis of the downhole geophone

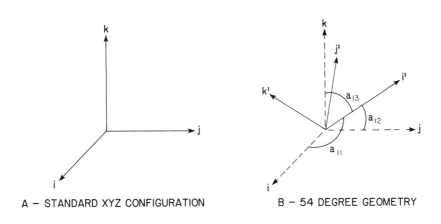

A — STANDARD XYZ CONFIGURATION   B — 54 DEGREE GEOMETRY

Figure 2-22   Geophone transducer elements which measure three-component particle motion can be positioned inside a VSP geophone package along the solid line axes shown in one of these two geometrical configurations. The k axis points vertically along the longitudinal axis of the geophone package. a_{mn} is the direction cosine angle specifying the direction of rotated axis m relative to axis n of the original ijk system. In B, the angle between the k axis and any of the three rotated prime axes is 54.74 degrees.

tool. Geophones oriented along the k axis measure vertical particle motion in vertical boreholes, and geophones oriented along the i and j axes measure horizontal particle motion. This geophone geometry is probably the most common type now used outside of the Soviet Union to record three-component VSP data. Care must be used when fabricating this type of geophone assembly so that the following objectives are achieved:

(1)   The horizontal geophones must exhibit amplitude and phase responses identical to that of the vertical geophone. The three geophones used in a VSP tool should have identical seismic response characteristics since the components of a linearly polarized signal arriving at each of the three orthogonal geophones exhibit identical phase and frequency behavior. If the phase relationships or frequency content of the horizontal signal components are altered differently than are the phase and frequency of the vertical component, then precise three-component data analyses cannot be performed.

(2)   The wall locking device that presses the geophone tool to the borehole wall must create a geophone-to-formation bond that results in the horizontal geophones being mechanically coupled to the formation in the same way that the vertical geophone is. Otherwise, anomalous phase shifts and amplitude distortions are created in the horizontal component data. It is difficult to fabricate wall locking devices that mechanically couple vertical and horizontal geophones to a formation in the same way. The resultant force pressing the geophone package to the formation does not, in general, have equal components along the three orthogonal geophone axes.

These design objectives can often be better achieved by rotating the i, j, k positions of the geophone transducer elements to new orthogonal directions i', j', k' shown in Part B of Figure 2-22. This rotated geophone configuration is the third type of geophone geometry used in vertical seismic profiling, particularly in the Soviet Union. In this figure, a_{11} is the direction cosine between the i' and i axes, a_{12} is the direction cosine between the i' and j axes, and a_{13} is the direction cosine between the i' and k axes. If a_{mn} is the direction cosine specifying the direction of rotated axis, m, relative to axis n of the original ijk system, then for a fixed n, a fundamental law of direction cosines is (Spiegel, 1959):

$$\sum_{m=1}^{3} (a_{mn})^2 = 1 \cdot \tag{3}$$

Setting n to a value of 3 will specify the direction of each rotated axis relative to the original vertical k axis which points along the longitudinal axis of the geophone tool. For n=3, the direction cosine law becomes

$$a_{13}^2 + a_{23}^2 + a_{33}^2 = 1. \tag{4}$$

If the three rotated geophone elements are symmetrically located about the k axis then

$$a_{13} = a_{23} = a_{33}. \tag{5}$$

This symmetry requirement, combined with Equation 4, sets each of these three direction cosine angles at 54.74 degrees; consequently, this type of VSP geophone transducer geometry is commonly referred to as the "54 degree" geometry. The geophone elements in their final rotated positions are oriented at an angle of 54.74 degrees away from the original vertical k axis, and there is an azimuth angle (measured in the original i-j plane) of 120 degrees between them. Since the transducers are symmetrically oriented about the longitudinal axis of the geophone tool, a laterally directed locking force tends to create a geophone-to-formation coupling that is more equally distributed among the three geophones than would be the case if they were arranged in an XYZ geometry. In addition, geophone elements with identical spring constants and coil suspensions can be used for all three coordinate axes in the 54 degree geometry. Because of their geometrical symmetry about the longitudinal axis of a downhole tool, all three geophones in a 54 degree configuration can be simultaneously impulse tested by a single impulse directed along the k axis. These facts mean that data recorded by transducers positioned in a 54 degree geometry usually exhibit better calibration and more consistent amplitude, frequency, and phase behavior among the three components of motion than do data recorded by geophone elements positioned in an ijk (XYZ) geometry.

   Three-component VSP data recorded with 54 degree geophones can be converted to three-component data in an XYZ coordinate system by appropriate orthogonal transformation equations. This transformation is important because an XYZ coordinate system is easier to visualize since the XYZ axes can be related to north, east, and vertical directions. This transformation is easier to visualize when one views the geometrical relationships between the ijk (XYZ) axes and the i'j'k' (54 degree) axes in the manner shown in Figure 2-23. In this vertical view of the i-j plane, the i' axis is positioned directly above the i axis. The angle between any of the primed axes and the i-j plane is 35.26 degrees, so all three of the 54 degree axes are projected onto the i-j plane by multiplying by cos(35.26°). The 60 degree angles shown in the figure are measured in the i-j plane. Let $A_1(t)$, $A_2(t)$, $A_3(t)$ represent the components of a VSP signal recorded by 54 degree geophones oriented along the i'j'k' axes. The equivalent response $B_1(t)$ which would be measured by a geophone oriented along the i axis is then

$$B_1(t) = \cos(35.26°)A_1(t) - \cos(35.26°)\cos(60°)A_2(t)$$
$$- \cos(35.26°)\cos(60°)A_3(t) \tag{6}$$

or

$$B_1(t) = \sqrt{\frac{2}{3}}\, A_1(t) - \sqrt{\frac{1}{6}}\, A_2(t) - \sqrt{\frac{1}{6}}\, A_3(t) \ . \tag{7}$$

If a geophone were oriented along the j axis, its response would be

$$B_2(t) = \cos(35.26^\circ)\sin(60^\circ)A_2(t) - \cos(35.26^\circ)\sin(60^\circ)A_3(t) \tag{8}$$

or

$$B_2(t) = \sqrt{\frac{1}{2}}\, A_2(t) - \sqrt{\frac{1}{2}}\, A_3(t) \ . \tag{9}$$

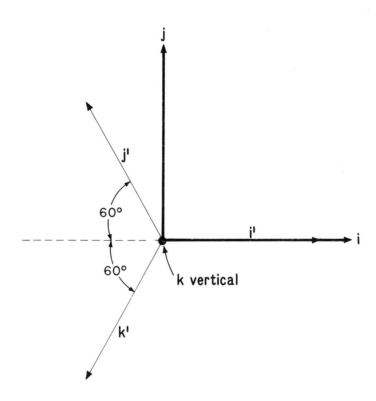

Figure 2-23   This view looks down the k axis onto the i-j plane. The light lines indicate the projections of the three 54 degree axes, i'j'k', onto the i-j plane assuming that the i' axis is positioned directly above the i axis.

A vertical geophone oriented along the k axis would have an output given by

$$B_3(t) = \cos(54.74°)A_1(t) + \cos(54.74°)A_2(t) + \cos(54.74°)A_3(t) \qquad (10)$$

or

$$B_3(t) = \sqrt{\frac{1}{3}} A_1(t) + \sqrt{\frac{1}{3}} A_2(t) + \sqrt{\frac{1}{3}} A_3(t) . \qquad (11)$$

The transformation from 54 degree geophone responses to XYZ geophone responses can thus be written in matrix form as

$$\begin{pmatrix} B_1(t) \\ B_2(t) \\ B_3(t) \end{pmatrix} = \begin{pmatrix} \sqrt{\frac{2}{3}} & -\sqrt{\frac{1}{6}} & -\sqrt{\frac{1}{6}} \\ 0 & \sqrt{\frac{1}{2}} & -\sqrt{\frac{1}{2}} \\ \sqrt{\frac{1}{3}} & \sqrt{\frac{1}{3}} & \sqrt{\frac{1}{3}} \end{pmatrix} \begin{pmatrix} A_1(t) \\ A_2(t) \\ A_3(t) \end{pmatrix} . \qquad (12)$$

The unit vectors i', j', and k' defining the 54 degree coordinate system are related to the unit vectors i, j, and k of the cartesian coordinate system by this same matrix equation; i.e.,

$$\begin{pmatrix} i \\ j \\ k \end{pmatrix} = \begin{pmatrix} \sqrt{\frac{2}{3}} & -\sqrt{\frac{1}{6}} & -\sqrt{\frac{1}{6}} \\ 0 & \sqrt{\frac{1}{2}} & -\sqrt{\frac{1}{2}} \\ \sqrt{\frac{1}{3}} & \sqrt{\frac{1}{3}} & \sqrt{\frac{1}{3}} \end{pmatrix} \begin{pmatrix} i' \\ j' \\ k' \end{pmatrix} . \qquad (13)$$

The cartesian ijk vector system is orthonormal, and it is important to verify that its vector properties are maintained in the rotated i'j'k' vector system. Inverting Equation 13 defines the i'j'k' vectors as

$$\begin{pmatrix} i \\ j \\ k \end{pmatrix} = \begin{pmatrix} \sqrt{\frac{2}{3}} & 0 & \sqrt{\frac{1}{3}} \\ -\sqrt{\frac{1}{6}} & \sqrt{\frac{1}{2}} & \sqrt{\frac{1}{3}} \\ -\sqrt{\frac{1}{6}} & -\sqrt{\frac{1}{2}} & \sqrt{\frac{1}{3}} \end{pmatrix} \begin{pmatrix} i' \\ j' \\ k' \end{pmatrix} . \qquad (14)$$

Algebraic manipulation of these vector components shows that

$$i' \cdot i' = 1 \qquad\qquad i' \times i' = 0$$
$$j' \cdot j' = 1 \qquad\qquad j' \times j' = 0$$
$$k' \cdot k' = 1 \qquad\qquad k' \times k' = 0 \qquad\qquad (15)$$
$$i' \cdot j' = 0 \qquad\qquad i' \times j' = k'$$
$$j' \cdot k' = 0 \qquad\qquad j' \times k' = i'$$
$$k' \cdot i' = 0 \qquad\qquad k' \times i' = j' \quad .$$

These relations confirm that the i'j'k' system is orthornormal. A similar mathematical analysis of a 54 degree geophone system is presented by Goff and O'Brien (1981). Der (1970) has published an example of earthquake data which are recorded in a deep well with a vertical array of 54 degree triaxial geophones and mathematically converted to horizontal and vertical components of motion by a numerical coordinate transformation.

A 54 degree VSP geophone assembly is shown in Figure 2-24. In order to keep the overall tool diameter as small as possible, the geophones are stacked vertically at their appropriate orientation angles rather than grouped in the same horizontal plane with ends butted together as implied by the picture of the i' j' k' axes in Figure 2-22.

Summary of VSP Geophone Design Concepts

Very little information can be found in public geophysical literature that documents how, and to what extent, specific VSP geophone design parameters affect geophone performance. Most experienced VSP data users have observed that some geophone systems record good data while others do not, and internally within their own exploration or research groups, these people have decided what features of a borehole geophone are desirable and which should be avoided. However, at this date, no one has published a thorough study of VSP geophone design concepts for the industry at large to assimilate.

After numerous private communications with groups who either build borehole geophones or have used a wide variety of geophone systems to record VSP data, a general pattern of preferred VSP geophone design characteristics seems to emerge. The following discussion presents this generalized view of an idealized borehole geophone, but no experimental data will be offered to substantiate that these common industry opinions are indeed true. There is a critical need for good experimental studies to be published showing in detail how specific geophone design parameters, such as length, weight, locking force, etc., affect VSP data quality. Some recent unpublished investigations of this nature will be shown in Chapter 9 (Figures 9-6, 9-7, and 9-8).

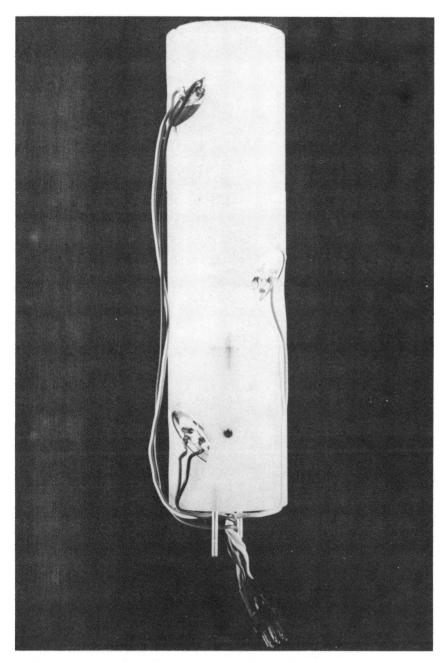

Figure 2-24   Geophone elements assembled in a 54 degree geometry arrangement.

What appears to be the industry's "idealized" VSP geophone for oil and gas exploration is shown in Figure 2-25. At least nine design characteristics are commonly mentioned by VSP users as highly desirable in such a tool. These features are listed in the figure and described in the following numbered paragraphs.

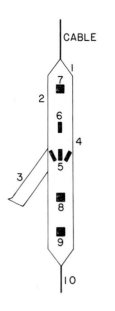

CABLE

1 — POINTED END

2 — SMALL DIAMETER

3 — RETRACTABLE LOCKING ARM

4 — SHORT LENGTH AND LIGHT WEIGHT

5 — THREE—COMPONENT GEOPHONES

6 — VERTICAL GEOPHONE FOR CALIBRATION

7 — ORIENTATION MEASURING SYSTEM

8 — MECHANISM TO DETERMINE GEOPHONE— TO—FORMATION COUPLING

9 — DOWNHOLE DIGITIZING OF ALL DATA SENT UPHOLE

10 — CONNECTION TO OTHER VSP GEOPHONE SYSTEMS

Figure 2-25   Desired characteristics of a VSP borehole geophone.

1. Pointed ends - Tube waves are a common coherent noise problem in vertical seismic profiling and will be discussed in detail in Chapter 3. A tube wave travels primarily in the fluid column in a wellbore and will reflect from any abrupt change in fluid column cross-section, such as occurs at the end of a VSP geophone assembly. If borehole diameter changes are tapered rather than abrupt, less tube wave energy is reflected, and the recorded coherent noise pattern created by tube waves is less complicated. Consequently, tapered geophone ends help reduce the number of upgoing and downgoing tube wave modes in a vertical seismic profile. Also, tapered ends reduce the possibility of the tool becoming jammed underneath ledges and borehole obstructions.

2.   Small diameter - The diameter of a geophone package also affects how much of a tube wave's amplitude is reflected as a tube wave passes by a borehole tool. The smaller the diameter of the tool, the less the change in the cross-section of the fluid column in the wellbore, and consequently, a tube wave traveling along the fluid column is less sensitive to the presence of a slim tool. In addition, narrow geophone tools can be used in both small and large boreholes and are less likely to be stopped by borehole obstructions.

3.   Retractable locking arm - It is essential that a VSP geophone be capable of either uphole or downhole movement during a VSP experiment. The ability to position a geophone at any depth in a borehole for particular seismic shots allows tremendous versatility in setting up optimum field geometry for specific subsurface imaging. Complete freedom of geophone movement is guaranteed only if the locking mechanism is a retractable type. Retractable arms are particularly essential when recording VSP data in uncased boreholes since tools with non-retracting arms are too prone to becoming jammed against obstructions. Only one locking arm is indicated in Figure 2-25, but two arms may be needed if tool length exceeds two meters. The locking mechanism should create a lateral force which is at least two to three times the weight of the tool.

4.   Short length and light weight - Experimental measurements show that surface geophones should have (a) minimum mass and (b) maximum contact area with the earth in order to move in phase with the earth without introducing extraneous coupling resonances (Hoover and O'Brien, 1980). It is assumed that these two design criteria should carry over to the design of borehole geophones. Contact area can be increased by lengthening a VSP geophone tool, but this increase in tool size also increases geophone mass. The general concensus of VSP researchers is that a VSP geophone should be short and have minimum mass. The ratio of geophone length to geophone mass is obviously an important design parameter that should be studied. Some effects of geophone mass on VSP data quality will be illustrated in Chapter 9. In wells containing heavy muds, there will often be times when weight may have to be added to the bottom end of the recording cable in some manner in order to get a light weight geophone to go downhole.

5.   Three-component geophones - In order to study shear body waves and to pursue any VSP data application which focuses on particle motion behavior, it is necessary to record three-component VSP data. A 54-degree geophone geometry is indicated in Figure 2-25. Ideally, a VSP downhole tool should have easily interchangeable

geophone modules which allow an operator to use 54-degree geophones, an XYZ geophone arrangement, vertical geophones, or gimbaled geophones, depending on the application intended for the recorded data.

6. Vertical geophone for calibration - This design feature is applicable only if the geophone configuration used in the VSP tool is the 54-degree type. For a 54-degree geophone geometry, each of the three geophones is oriented at 54.74 degrees away from the longitudinal axis of the tool. If the outputs, $X_i(t)$, of these three identical geophones are arithmetically summed, all vertical components combine so that, except for a multiplicative constant, K, the three component sum, $\sum X_i(t)$, is the same as the output, $y(t)$, of a single vertical geophone. Placing a fourth geophone along the longitudinal axis of the tool allows the output of the 54-degree geophones to be checked in situ for accuracy via this arithmetic sum technique for each recorded trace. During a field experiment, if a camera playout or CRT display of the three-component geophone sum differs by a preselected amount, C, from the playout of the output of the vertical geophone; i.e., when

$$\sum_{i=1}^{3} X_i(t) > Ky(t) + C, \tag{19}$$

then an on-site VSP quality control person has an indication that one or more of the 54-degree geophones has become uncalibrated. The tool can be brought to the surface to correct the problem, or the experiment can continue with the knowledge that accurate three-component data may not be obtained.

7. Orientation measuring system - Three-component VSP data can be fully utilized only if one knows which directions the geophone elements were pointing when data were recorded at each depth. Consequently, some type of orientation measuring scheme needs to be included in the geophone tool. If VSP data are recorded only in uncased borehole intervals, then reliable orientation systems comprised of magnetometers that measure azimuth from magnetic north and gravity sensitive accelerometers that measure deviations from vertical are commercially available in small packages suitable for a borehole geophone assembly. Camera systems that photograph the position of a magnetically sensitive needle against a fixed compass grid or the position of a bubble floating in a graduated scale are too inefficient for VSP usage. If data must be recorded inside casing then magnetometers are ineffective, and gyroscopes must be considered. No one has publicly described a successful

mating of a gyroscopic package with a VSP velocity sensitive geophone assembly, so it is unknown if this combination will work. Some researchers think that a gyroscope will introduce mechanical noise that will be picked up by geophones. Also, the length of time required to record VSP data and the mechanical abuse imposed on a VSP geophone package while lowering, locking, and raising it in a well will challenge the durability of a mechanical gyroscope system. The new concept of a laser gyroscope seems attractive, but no one has yet tried to package that technology into a borehole geophone tool. Geophone orientation measurements will be discussed in more detail in Chapter 9.

8. <u>Mechanism to determine geophone-to-formation coupling</u> - When analyzing the amplitude behavior of VSP data in order to calculate reflection coefficients, or when determining the data's frequency content in order to estimate attenuation effects, it is usually assumed that the phase and amplitude properties of the in situ wavelets are precisely captured at each geophone level. This assumption is valid only if the geophone-to-formation coupling at each recording level is identical. If this coupling varies, then amplitude and phase variations are introduced into the data that have nothing to do with the in situ behavior of a propagating body wave. There is thus a need to have within, or attached to, the geophone assembly a means by which the geophone elements can be mechanically impulsed, or mechanically vibrated over a wide frequency range, after the geophone is locked in place at each recording depth. Recording the output of each geophone element's response to this deliberately generated mechanical disturbance allows a data processor to mathematically adjust the amplitude and phase spectra of the data recorded at each depth so that the transmission properties of the geophone-to-formation interface is depth-invariant. The mechanically generated disturbance must, of course, be identical for each coupling test, which is a difficult design objective to accomplish.

9. <u>Downhole digitizing system</u> - Standard logging cable contains only seven wires which allow only three simultaneous analog signals to be sent uphole if signal crossfeed is to be minimized. This limitation prohibits the simultaneous recording of the fourth vertical calibration geophone output proposed earlier. Downhole digitizing would allow a large number of simultaneous and independent measurements to be sent uphole on a single wire in multiplex form.

10. <u>Connection to other VSP geophone systems</u> - One common request among VSP users is for the capability of recording data simultaneously at more than one depth. Multi-level VSP recording would greatly reduce the length of time required to

complete VSP field exercises and would be a significant economic benefit. Less expensive VSP data acquisition means more VSP data will be recorded and used. The goal of multi-level recording can be achieved only if two or more geophone assemblies are connected to comprise a single recording system. No such system commercially exists outside of the Soviet Union.

Temperature and Tilt Angle Calibration of Geophones

A VSP geophone can be subjected to extreme changes in temperature while recording data. Temperatures at the bottom of a VSP well can be 200°C or more and reduce to 0°C or less at the surface. It is possible that some moving-coil geophones can change their resonance behavior over such large temperature ranges, and this geophone behavior must be known if important rock properties, such as the intrinsic attenuation of compressional and shear body waves, are to be accurately calculated from recorded borehole data. Specifically, the mechanical suspension and magnetic field strength of the geophone coil and the resistance of the coil and damping resistor all change with temperature. Oven tests of VSP geophone resonance and phase behavior should therefore be made over wide temperature ranges and appropriately documented. Calibration tests of each individual geophone in the total geophone package should be made before the tool is sent to the field to determine the effects of temperature and tilt angle on each individual geophone's signal. High frequency geophones (e.g. 15 Hz resonance) are less sensitive to tilt angle than are low frequency geophones (e.g. 5 Hz resonance) and are preferred by some VSP experimenters for this reason. In situ measurements of tilt angle (via an orientation measuring device) and internal tool temperature should be made immediately after recording each data trace. These field measurements, combined with the laboratory derived calibration curves, allow a propagating seismic wavelet to be more accurately reconstructed throughout a drilled stratigraphic section.

Recording Cable

With the exception of a few special research systems, the recording cable used to transmit the voltage output of a downhole geophone to the surface is standard logging cable used worldwide by all major well logging service companies. These cables contain only seven wires, and this construction limits the number of simultaneous analog signals that can be transmitted. Special downhole circuits have to be built to transmit the analog output of even three different geophones simultaneously so that there is no significant crossfeed between them. Some VSP development groups are considering systems that digitize all pertinent information downhole, and this capability will allow a large number of distinct signals to be transmitted in multiplex form to a surface recorder. The next generation of well logging systems will likely digitize all log data downhole, and the logging industry should gradually convert to single wire logging cable once these systems are built. This change in logging cable construction will force VSP data to be transmitted uphole in digital rather than analog form.

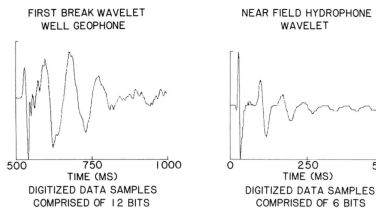

FIRST BREAK WAVELET
WELL GEOPHONE

TIME (MS)
DIGITIZED DATA SAMPLES
COMPRISED OF 12 BITS

NEAR FIELD HYDROPHONE
WAVELET

TIME (MS)
DIGITIZED DATA SAMPLES
COMPRISED OF 6 BITS

Figure 2-26   Effect of insufficient digitizing on the resolution of a near-field wavelet. Energy source was a single 80 in^3 airgun in a shallow marine environment.

Recording System

Seismic recording equipment used for VSP surveys should meet rigid standards regarding resolution, dynamic gain, and recording format. Both the downhole geophone data and the near-field monitor geophone response should be recorded with a resolution of at least 12 bits (including sign) in order to capture high quality waveforms. It is important that a good quality near-field waveform be recorded in all vertical seismic profiles, but especially so when performing source signature deconvolution of marine or land data shot with an energy source that creates a long wavelet such as untuned airguns do.

An example of data recorded with a system which digitized the downhole geophone output with 12 bits of resolution, but digitized the near-field monitor geophone channel with only 6 bits of resolution is given in Figure 2-26. The near-field wavelet exhibits a clipped, stair step behavior because of the small number of bits used to describe its amplitude behavior. Attempts to use the near-field wavelets recorded in this experiment to create a consistent input wavelet for every recording depth by numerical wavelet shaping algorithms were only marginally successful because of the shot-to-shot variations created by the limited 6-bit digital data samples. This type of recording equipment is satisfactory for recording velocity check shot data but is being replaced throughout the industry with systems that digitize all channels with at least 12 bits of resolution so that rigorous wavelet processing of VSP data can be performed.

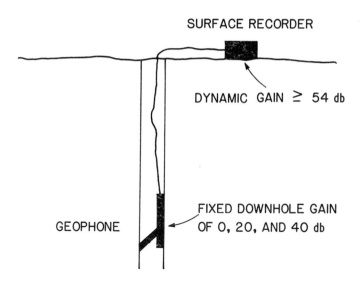

Figure 2-27 Two types of signal amplification/attenuation conditions are involved in a VSP recording system. The downhole geophone package should be capable of either amplifying or attenuating the signal that it transmits up the recording cable. A fixed gain that can be incremented over a 40 or 50 db range is adequate for handling the downhole signal level. The surface recorder should have a dynamic gain range of at least 54 db.

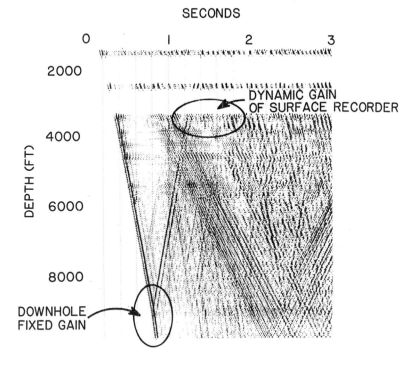

SECONDS

Figure 2-28   Regimes where the two basic VSP data recording gains play dominant roles. Downhole fixed gain insures that the first arrival signal feeding into the recording cable is neither too weak nor too large. The dynamic gain of the surface recorder resolves weak events after the correct downhole gain is determined.

VSP recording equipment should have a proper dynamic gain in order to avoid overscaling and clipping the output of the downhole geophone, especially as the geophone approaches the surface where the downgoing wavefield has its largest amplitudes. To avoid these problems, the downhole geophone data should be recorded with a minimum of 54 db of dynamic gain and with a fixed downhole gain adjustable up to a maximum of 40 to 50 db as shown in Figure 2-27. This fixed downhole gain must be adjustable from a surface control panel in order to have complete recording flexibility. Basically, the downhole gain sets the amplitude of the strong direct arrival to a numerical range that will not overdrive the surface amplifiers, and the surface dynamic gain capability preserves the weak events that follow the direct arrival. The relative domains of influence of these two gain controls on VSP data are shown in Figure 2-28. A numerical dynamic gain function has been applied to these data during data processing so that all events have approximately the same amplitude. Only a few geophone levels were recorded above a depth of 4000 feet. No high amplitude wavelets are clipped because the downhole gain could be adjusted over an appropriate range. Likewise, weak reflection events approaching the surface are recovered by the dynamic recording gain of the field system used in this experiment.

CHAPTER 3

NOISE PROBLEMS ENCOUNTERED IN VERTICAL
SEISMIC PROFILING

Numerous noise sources contaminate seismic measurements made on the earth's surface. Considerable care must be exercised both in the field and during data processing in order to overcome the undesirable effects of ground roll, air waves, wind, electrical power transmission lines, and inconsistent geophone-to-ground coupling. Such noise contamination is generally reduced in a marine environment, but there too, unwanted noise is still a problem that must be combated. On land it is generally recognized that the

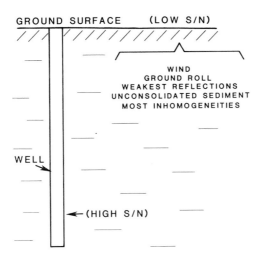

Figure 3-1   Signal-to-Noise (S/N) regimes encountered when recording seismic data.

noise level is decreased and improved signal fidelity is achieved if one buries the geophones. Sometimes a geophone buried only a few inches demonstrates a considerable reduction in noise level relative to one positioned directly on the earth's surface. This fact is one reason for the strong interest in vertical seismic profiling; i.e., a geophone can be rigidly locked at great depths in a VSP measurement and thus avoid several of the common surface noise problems which are shown in Figure 3-1. However, there are noise problems that are unique to vertical seismic profiling, and it is essential that the sources of these noise modes be identified so that they can be recognized and avoided.

Prominent noise modes that are frequently observed in VSP recordings will be discussed and illustrated in this chapter. Real data examples will be used to show the effects that cable waves, geophone clamping, multiple strings of well casing, unbonded casing, tube waves and other noise sources have on VSP data. The importance of other factors, such as field geometry, accurate depth control, recording system gain, and proper depth sampling, on data quality will be discussed in other chapters. In some wells, formation sloughing, fluid movement behind casing, gas bubbles, and borehole mud movement may also create a random noise background if borehole environmental conditions are not stable, but these noises will not be considered.

Geophone Coupling

Some of the seismic noise observed in surface geophones is caused by the quality of the geophone plant; i.e., a loosely planted geophone is noisier than one which is rigidly bonded to the earth. The same principle applies for borehole geophones, except that a downhole geophone cannot be spike-planted into the borehole wall. Instead, some type of locking mechanism must press the downhole geophone assembly against the borehole or casing wall with sufficient force to achieve good seismic coupling between the geophone and the formation. The downhole VSP geophone tools presently available in the United States and Europe are typically 3 to 4 meters long and weigh 75 kilograms or more. Thus, the mechanism that bonds them to a borehole wall must be capable of generating a large radially directed force. For convenience, and in order to avoid fishing for jammed tools, the locking mechanism should have a reliable control system by which it can be extended and retracted remotely from the surface of the ground by electrical commands. Since one cannot visually observe that the locking mechanism has functioned properly in the borehole, indirect indications of the quality of the contact between the geophone and the borehole wall are necessary. During a VSP field exercise, one indirect indicator of the quality of the downhole geophone-to-formation coupling can be provided by a seismic camera playout system that displays a time record of the downhole geophone's voltage output. The movement of the light beam which draws the geophone's voltage fluctuations on camera playout paper, or on a cathode ray tube, should be monitored before each shot to judge the mechanical stability and noise environment of the downhole geophone. Post-shot inspection of a paper playout of the recorded data should be done to confirm that the pre-shot assumption of proper geophone clamping was correct. Wuenschel (1976) describes a VSP geophone system which has small internal geophone shakers that measure the frequency response of triaxial geophone elements in their downhole planted condition. The determination of these "plant resonances" defines the consistency of the geophone-to-formation coupling at all recording depths. This concept has considerable merit but has not been aggressively followed by most subsequent VSP geophone designers. A second technique for achieving uniform geophone coupling at each depth level which has been suggested by some VSP researchers is to record several shots at each depth, unlock and relock the geophone (and perhaps even move it a small distance) after each shot, and average the responses. This procedure is time-consuming and thus is seldom followed in practice.

The necessity for achieving a good downhole geophone-to-formation coupling is illustrated by the data in Figure 3-2. Even though the geophone in this example is 1295 meters below ground level and far removed from common surface noise sources, the geophone cannot simply hang free in the fluid column and record meaningful data. The response of the unlocked geophone before the high amplitude direct arrival represents noise that is transmitted down the cable. This topic will be further discussed in the next

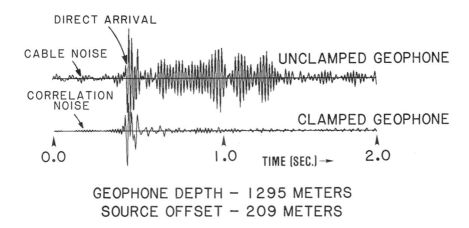

GEOPHONE DEPTH — 1295 METERS
SOURCE OFFSET — 209 METERS

Figure 3-2   Effect of geophone clamping on signal response.

section. Not only is the basic character of the direct arrival altered by the response of the loose geophone, but the strong noise level and resonance occurring after the direct arrival completely mask any upgoing or downgoing events that occur. The energy source used in this test was a surface compressional vibrator, and the low amplitude oscillations preceding the direct arrival of the trace labeled "Clamped Geophone" are correlation noise. Tests of this nature show that any downhole geophone tool used to record VSP data must be capable of rigidly and faithfully bonding the geophone case to the borehole wall; otherwise, weak upgoing reflections cannot be resolved. Since an unclamped geophone (or hydrophone) sometimes produces data in which high amplitude first arrivals can be identified, it might be possible to use such data for any VSP analysis which depends only on measuring the travel times of first arrivals. However, as illustrated by the data in Figure 3-3, there is no guarantee that even first arrivals can be resolved if the recording transducer is not bonded to the formation. As a specific example of the data deterioration that can be introduced by an improperly clamped geophone, van Sandt and Levin (1963) have noted that the noise level increases by 40 db when downhole data are recorded with a free-hanging geophone rather than with a wall-locked geophone.

A downhole geophone device can have an excellent locking mechanism and yet not be capable of clamping the geophone to the formation if the borehole diameter is larger than the extension length of the locking arm. Consequently, if a vertical seismic profile is executed in an uncased well, it is important to record a caliper log before beginning the

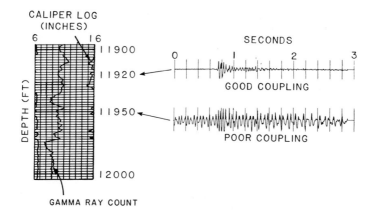

Figure 3-3   The locking arm of the downhole geophone assembly used in this well could extend only 16 inches. Consequently, the geophone could not be coupled to the formation in those depth intervals where large washouts occurred. If a caliper log were not available, a field observer would have no idea where the nearest depth would be where he could achieve good geophone-formation coupling. (Courtesy J. G. Gallagher, Phillips Petroleum Company).

VSP experiment so that large washout zones can be identified. An example of the undesirable geophone response that results when a borehole is too large to allow a geophone locking device to function properly is shown in Figure 3-3. If a caliper log of this borehole were available, an operator could quickly reposition the geophone to the nearest depth where the tool could be properly clamped.

Most VSP data recording systems allow an operator to determine when the locking arm is exerting an acceptable force on the borehole wall by monitoring the current requested by the downhole d-c motor which extends the locking mechanism. However, a caution should be expressed about using this motor current behavior alone to define when a downhole geophone has been properly locked in place in uncased boreholes because a strong locking arm pressure does not necessarily imply a good formation-to-geophone coupling. See, for example, the hypothetical situation in Figure 3-4. Here, the assumed

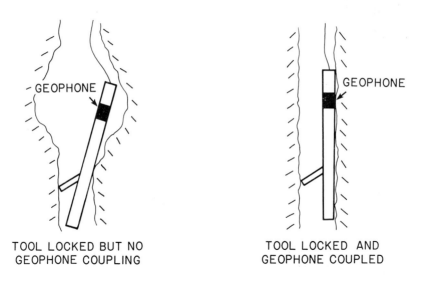

TOOL LOCKED BUT NO
GEOPHONE COUPLING

TOOL LOCKED AND
GEOPHONE COUPLED

Figure 3-4   A good tool lock indication does not necessarily imply adequate geophone-to-formation coupling in an uncased, highly rugose borehole. A caliper log made shortly before a VSP experiment would help one to recognize potential problem intervals.

downhole locking concept is the common wedge-action type. In this situation, the downhole locking arm is firmly planted, but the geophone is still not coupled to the formation because of the borehole enlargement. The equipment operator will likely observe a quiet geophone output on the surface monitor camera and decide to activate the seismic energy source. However, once seismically disturbed by the first arrival, the downhole tool can pivot and rotate about its clamping point and create an erroneous seismic response after the first break. This example again emphasizes the need to incorporate caliper log data into the planning of a field experiment so that optimum geophone depths are selected for data recording.

Cable Waves

Even if a geophone is rigidly locked to a borehole wall, it can still exhibit a high noise level because vibrations can be transmitted from the surface down the recording cable to the geophone assembly. The propagation velocity of waves along multistrand armored cable, such as used in vertical seismic profiling, is variable and depends on specific cable construction. Gal'perin (1974) quotes propagation velocities in the range 2500 m/sec to 3500 m/sec. Dix (1945) measured velocities between 8500 and 9500 ft/sec (2591 to 2896 m/sec). In shallow holes and in slow velocity stratigraphic sections, a cable wave can be the first arrival measured by a borehole geophone, and if not correctly recognized, these waves can cause erroneous formation velocity analyses. Also, spurious, late arriving cable wave events may overlay seismic reflections occurring after first breaks and prohibit one from making correct seismic interpretations of subsurface geological conditions.

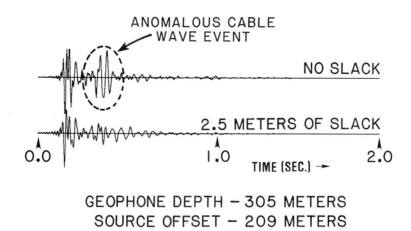

Figure 3-5   Effect of cable slack on geophone signal.

Cable-borne events are generally caused by wind or machinery vibrating the recording cable above ground level. Thus, this type of noise can be reduced by slacking the cable after the tool is locked downhole, and thereby allowing the vibrations to damp themselves out before reaching the geophone. The amount of slack necesssary for noise suppression is small; 3 or 4 meters of slack is often sufficient even at depths of 3000 to 4000 meters. Most wireline operators are reluctant to allow more than 4 or 5 meters of slack for fear that the cable may wrap around itself or the downhole tool and result in a fishing job.

An example of the dramatic change in the seismic response that can result from cable slack is shown in Figure 3-5. These data were recorded in a situation where there was negligible wind and mechanical vibration at the surface to create cable waves. Consequently, no erroneous events occur before the first breaks, as was the case in the top trace of Figure 3-2. The absence of noise before the first breaks is also due to the fact that the geophone was locked when both data traces in Figure 3-5 were recorded. However, when the Rayleigh wave created by the seismic surface source swept past the wellhead, the ground roll it created caused the mast supporting the recording cable to sway. The anomalous event marked in the figure is interpreted to be the cable wave resulting from this motion of the surface support for the recording cable. Each trace is plotted at a fixed gain. Note that the amplitude of the cable wave is as large as the amplitude of the first arrival. The cable wave disappears when the cable is slacked.

Wind velocities of 50 km/hr and higher have sometimes occurred during the recording of VSP data in the Mid-Continent areas of the United States. There have been times when wind generated cable motion above the ground prohibited data recording even after the cable was slacked several meters. In such cases, securing the cable at the wellhead with a clamp after slacking it 3 or 4 meters usually reduces the downhole noise to an acceptable level. Air waves created by some seismic sources can also create undesirable cable motion. In severe wind conditions, the cable may have to be slacked, clamped at the wellhead, and then slacked again so that it lays on the ground.

If a vertical seismic profile is executed after the drill rig moves away from a study well, a mast truck or workover rig is often used for wireline support for the downhole tool. These masts are typically 10 to 15 meters high and are quite susceptible to wind and ground roll motions. A compact pulley system secured directly to the wellhead will keep the logging cable between the wireline truck and the wellhead closer to the ground. This type of low elevation wireline support is a better way to transfer a recording cable to a borehole geophone in order to minimize wind-induced cable motion.

64

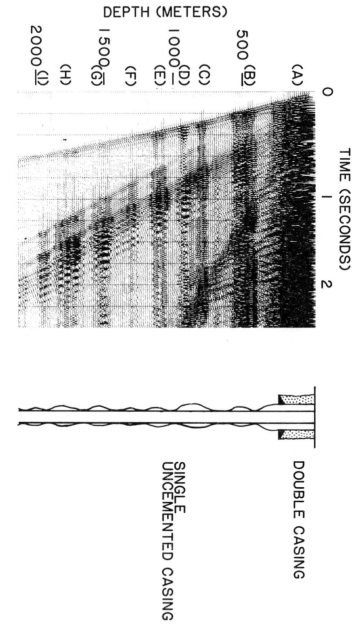

Figure 3-6 Examples of casing noise - (A) Resonance in multiple, unbonded casing. (B) through (I) - Resonance in single, uncemented casing. The borehole diameter on the right is drawn deliberately exaggerated in the intervals labeled (B) through (I). Below 250 meters, the highest amplitude resonances occur in the tube wave mode. These data were recorded with a vertically oriented geophone. (Modified from Hardage, 1981b).

Resonance in Multiple Casing Strings

Usable VSP data are difficult to record inside multiple casing strings because one or more of the casings may not be bonded to the formation. Consequently, poor quality data are usually recorded near the surface where multiple casing strings are encountered. For example, the data in Figure 3-6 were recorded by a vertically oriented, moving-coil geophone in a well which was cased with a single string of casing after a surface casing was set at 225 meters below ground level. These data are numerically processed to preserve both downgoing and upgoing compressional body wave events in their correct relative amplitudes. This processing technique will be described in Chapter 5. The procedure basically readjusts the magnitude of every data point by an amount equal to the geometrical attenuation that a spherically spreading compressional wavefront in a horizontally stratified earth would experience as it travels along either a direct or a reflected path from the surface to the VSP geophone. Although the amplitudes of non-compressional events will not be exactly correct in this display, some diagnostic interpretations of the amplitude behavior of both compressional and non-compressional events can be made. By visual inspection of the VSP data, one can see that the data character in the top 250 meters or so of the well differs dramatically from the data character in the interval where there is only a single string of casing. The shallow data in the double-cased interval exhibit a high amplitude resonating behavior that persists for the entire record time. Upgoing reflections will be virtually impossible to track through this resonating interval.

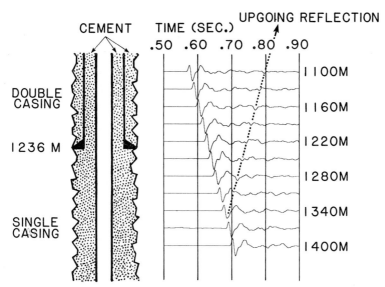

Figure 3-7   Comparison of waveshapes in properly cemented single and double casings.

Soviet researchers have noted that the particle motion recorded by a borehole geophone inside a resonating, multiply-cased, poorly-cemented interval is highly polarized in the vertical direction. Consequently, they claim that vertically oriented geophones record a dominating coherent type of noise but non-vertical geophones may or may not do so (Gal'perin, 1974). The physics of particle motion inside unbonded concentric pipes is not well understood, but these Russian field measurements suggest that horizontally polarized wave modes may be successfully recorded in uncemented multiple casings even though vertically polarized events cannot. Theoretical calculations of the particle motion behavior inside and immediately outside of a fluid filled, uncased borehole are shown in Figures 3-16 and 3-17.

Russian geophysicists combat the problem of recording data in the shallow, multiply-cased section of a well by drilling a second shallow hole near the VSP well. They leave this hole uncased, or protect it with a single string of cemented casing, and record shallow subsurface data in it rather than in the VSP well.

In order to record usable VSP data inside multiply-cased intervals of a well, all casing strings must be rigidly cemented to each other and also to the borehole formation. An example of good quality data recorded in a properly cemented, double-cased section of a wellbore is shown in Figure 3-7. There is no significant change in the character of the downgoing first arrival as it propagates from the double-cased to the single-cased interval, and a reflection that originates from an interface at a depth of 1350 meters in the single-cased interval can be traced upward through the double-cased section. No resonance behavior like that observed in the double-cased portion of the well shown in Figure 3-6 occurs in this well. Good quality cementing of all casing strings to each other and to the formation seems to be a key requirement for recording VSP data inside multiple casings.

Bonded and Unbonded Single Casing

If VSP data are recorded in an interval of a well which contains only one casing string, there must be a medium between the casing and the borehole wall capable of faithfully transmitting seismic energy from the formation to the geophone. The best medium is cement because field observations by van Sandt and Levin (1963) and Gal'perin (1974, p. 17) show that there is no significant difference in the shape of a seismic wavelet recorded in an open, uncased hole and the same wavelet recorded in a properly cemented single casing. As additional evidence for this statement, consider the data in Figure 3-8 which show the downgoing first arrivals measured in an uncased interval and in a

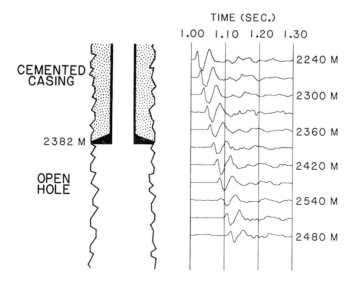

Figure 3-8   Comparison of waveshapes in open hole and in a properly cemented single casing.

cemented, single cased interval of an experimental well. The uniform shape of the downgoing wavelets, combined with the consistent signal-to-noise character of the data in the cased and uncased intervals, confirms that seismic measurements made inside a rigidly cemented single casing are equivalent to those made in an uncased borehole. This is an important insight because one wishes to capture a seismic wavelet in its true in situ behavior in a stratigraphic unit, and many wells must be cased in order to preserve the borehole before a VSP experiment can be executed.

All of the traces in Figure 3-8 are plotted with the same fixed gain, so amplitude differences are diagnostic of subsurface rock conditions, as long as the geophone is properly locked to the casing or formation wall. Most of the amplitude changes in these first arrivals can be correlated with acoustic impedance changes in the formation at the depths where the VSP data were recorded. The acoustic impedances are estimated from a sonic log (which is not shown) that was recorded in the well before it was cased. The amplitude of a seismic first arrival wavelet measured by a velocity-sensitive borehole geophone is inversely proportional to the acoustic impedance at the depth where the VSP data are recorded; i.e., a weak amplitude first arrival is measured inside a high impedance interval, and a high amplitude first arrival is recorded in a low impedance interval. This seismic wave amplitude behavior will be further discussed in Chapter 7.

68

The seismic response recorded in a single-cased interval that is not definitely known to be bonded to a formation by cement is difficult to predict before actually recording some downhole data in that interval. The wavelet character and the signal-to-noise behavior of the data may be equivalent to that recorded in a properly cemented interval, or they may be considerably inferior. The data in Figure 3-6 are evidence of this fact. Below a depth of 250 meters there are several intervals (labeled B, C, D,...I) where the amplitudes of some recorded events are anomalously high. Between these intervals, however, are regions where the data behave normally, yet none of the casing in the entire interval B to I is cemented. Note that much of the anomalous amplitude behavior in Figure 3-6 occurs because tube waves are present. If field techniques are used that prevent tube waves from being created, the data would not appear to be so resonating. For example, excessive tube wave amplitudes occur in all intervals B, C, D,...I, but anomalously high compressional first arrivals do not occur in intervals F, G, H, and I.

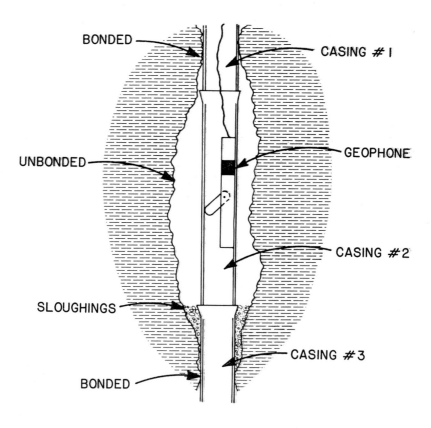

Figure 3-9   Resonance environment inside single, uncemented casing in intervals with large washouts.

A commonly accepted, but unproven, explanation for the VSP data behavior shown in Figure 3-6 is that single uncemented casing can be seismically bonded to a formation when random borehole ledges and protrusions rigidly bind a casing to a formation over appreciable depth intervals, or when cutting chips, sloughings, and solidified drilling mud form a type of pseudo cement in the annulus around the casing. In long intervals of extreme borehole washouts, there can be no bondings of this nature, and a geophone inside a casing cannot couple to the formation in such washout zones. These types of borehole conditions are shown in Figure 3-9. Casing 1 is seismically bonded to the earth by an annular ledge which presses against the casing. Casing 3 is coupled to the formation by a similar borehole restriction and also by an accumulation of borehole sloughings. Good quality data can thus be recorded inside each of these casings. Casing 2 makes absolutely no contact with the formation, and anomalous data will likely be recorded inside it. Field observations show that the seismic bond between an uncemented single casing and a formation is usually quite good in highly deviated boreholes. This behavior is thought to be true because gravity causes the casing to touch more of the formation on its bottom side than does an identical casing which is hanging in a similarly sized vertical borehole. Interestingly too, some VSP field crews claim that usable data can be recorded in some deviated cased wells without even deploying a geophone's locking arm. This operating practice is certainly not recommended; however, the fact that acceptable data can occasionally be recorded in a deviated well with an unclamped geophone is further evidence that gravity assists the geophone-to-formation coupling in non-vertical boreholes.

A field example of the detrimental effect that unbonded casing can have on VSP data is shown in Figure 3-10. These data were recorded in a borehole interval spanned by a single casing that was supposedly cemented to the formation. The distance between successive traces is 15 meters. The character of the data, particularly the direct arrivals, between depths A and C is different from the data appearance below this interval. The propagation velocity of the first black peak in the 180 meters between A and B is 5200 m/sec, which is the velocity of sound in steel casing. However, the propagation velocity of the first black peak in the 300 meters between C and D is 3430 m/sec, which is the formation compressional velocity in this interval, as defined by sonic logs. The interpretation of this data behavior is that the casing is adequately bonded to the formation between C and D, but that none of the casing joints between A and B contact the formation. Thus, very little seismic energy can transfer from the formation to the casing between A and B. Instead, the seismic compressional body wave disturbance arriving at A travels down the casing with a velocity of 5200 m/sec to depth B, where the casing again becomes bonded to the formation. A significant amount of the seismic displacement pulse arriving in the casing at B is reflected upward and travels back to A with a time-depth slope opposite to that of the downgoing first black peak. This reflection event is the second black peak in the depth interval AB, and it also travels in the casing with a velocity of 5200 m/sec.

Note that the first breaks of the correctly recorded data between depths C and D are downgoing troughs. In the top half of interval AB, the first breaks are weak troughs, which implies that a small amount of compressional energy is transferred directly from the formation to the borehole geophone at each recording level. However, the casing disturbance at depth A (the first black peak) travels with a velocity greater than the formation compressional velocity, so the black peaks overtake these weak first break troughs in the lower third of interval AB.

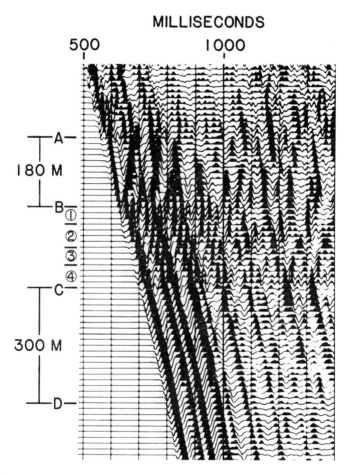

Figure 3-10   Deterioration of VSP data in severely unbonded single casing.

Between depths B and C there are four intervals, labeled 1, 2, 3, and 4, where a different type of data deterioration occurs. In each of these intervals, the first break troughs have a variable shape, and the first black peaks tend to align vertically at the same recording time. In each one of these intervals, all casing joints appear to move as a single rigid body whenever any part of the casing string in that interval is seismically impulsed. Note that the vertical alignment of black peaks persists for several hundred milliseconds. This phase alignment implies that the seismic propagation velocity in each of these four intervals is infinitely large. Continued investigations are warranted in order to construct satisfactory physical explanations of this data behavior.

In summary, in many instances the data recorded inside single, uncemented casing are equal in quality to data recorded in an uncased section of the same well. This fortunate circumstance is apparently due to the fact that a single, uncemented casing is still seismically bonded to the formation because of the cavings, drill cuttings, and solidified drilling mud which remain in the annulus between the casing and the formation and also because there are numerous rigid contact points between the casing and the formation. The degree and quality of this bonding increases as a well ages so that it is more attractive to record VSP data in old, cased, uncemented wells than it is in newly cased, uncemented wells. This logic leads to the preferred order of VSP borehole recording conditions which are illustrated in Figure 2-1. However, as documented by the data in Figures 3-6 and 3-10, one can almost guarantee that inferior VSP data will be recorded in some depth intervals of a well spanned by a single casing string which is not properly cemented to the formation. In most wells, these noisy intervals comprise only a small part of the total cased interval so that they do not completely prohibit one from recording VSP data in uncemented single casing strings.

Tube Waves
_____

Exploration seismology deals primarily with the generation, recording, and analysis of compressional and shear body waves because these wave modes reach to, and reflect from, deep subsurface points where exploration targets exist. In onshore exploration, Rayleigh and Love waves propagate along the earth's surface in all directions away from a seismic energy source and imprint over these deeply reflected body waves as they are recorded by a geophone located at or near the surface. Most explorationists view these surface waves as undesirable noise because they are superimposed on reflection signals and prevent optimum imaging of stratigraphic and structural conditions in the subsurface.

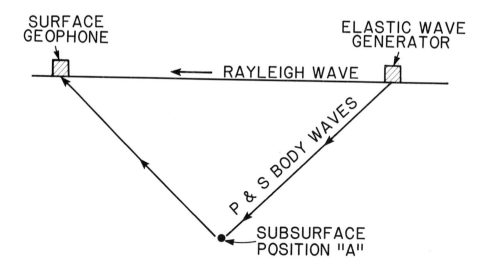

Figure 3-11   Elastic wave modes involved in surface seismic recording.

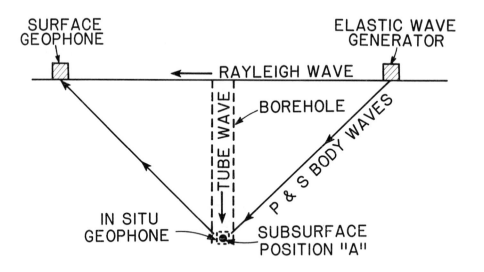

Figure 3-12   Elastic wave modes involved in subsurface seismic recording.

A simplified picture of the surface wave and body wave propagation involved in surface reflection seismology is shown in Figure 3-11.

The purpose of vertical seismic profiling is to implant a geophone at depth and record seismic wavefields in the interior of the earth rather than at the surface. This concept is shown in Figure 3-12. Rayleigh and Love waves traveling along the surface of the earth do not interfere with the recording of body wave events by the deeply buried geophone used in this type of recording geometry. Unfortunately, the fluid-filled borehole by which one gains access to the earth's interior introduces a cylindrical discontinuity which serves as a medium for propagating undesirable interfacial waves, just as does the planar earth-air interface involved in surface recordings. These fluid-borne borehole wave modes are commonly called tube waves in American geophysical literature (White, 1965; Cheng and Toksoz, 1981a, 1982c), and that terminology will be used in this text. (In Soviet geophysical papers this same wave mode is called a fluid wave, whereas a tube wave is defined as a disturbance that propagates along steel casing with a velocity of approximately 5.5 km/sec (Gal'perin, 1974, p. 19)). Thus, in the seismic recording of elastic body wave behavior in the earth, one must contend with the problem of some type of unwanted interfacial wave mode, irrespective of geophone location. Surface geophones are affected by Rayleigh and Love waves, and borehole geophones are influenced by tube waves.

Tube waves are one of the most damaging types of noise that can exist in vertical seismic profiling because they represent a coherent noise mode that repeats itself time after time for every seismic shot. Random seismic noise can be reduced by repeating seismic shots and summing several geophone responses into a single, composited record. Coherent tube wave noise cannot be reduced by summing repeated shots; in fact, it is usually amplified since its waveform character is consistent for all records being summed.

Mathematical Descriptions and Physical Properties of Tube Waves

The problem of wave propagation in a fluid filled cylinder surrounded by an elastic medium has been analyzed by several people. The geometry describing the general problem is shown in Figure 3-13. Solutions are published for situations where the cylinder is empty and when it is filled with fluid (Biot, 1952; Cheng and Toksoz, 1981a; Peterson, 1974; Roever, et al., 1974; Tongtaow, 1980; Tsang and Rader, 1979; White and Zechman, 1968, Wyatt, 1979). These solutions show that potential functions satisfying the wave equation in the cylindrical coordinate system illustrated in Figure 3-13 can be written as:

$$\emptyset = \left(AK_0(nr) + BI_0(nr)\right) \exp\left(ik(z-vt)\right) \tag{1}$$

$$\psi = \left(CK_1(mr) + DI_1(mr)\right) \exp\left(ik(z-vt)\right) \tag{2}$$

where A, B, C, D are constants, K_0, K_1, I_0 and I_1 are modified Bessel functions of zero and first orders, v is the phase velocity of the propagating wave, k is the wavenumber in the Z direction, and n and m are wavenumbers in the radial direction defined as:

$$n^2 = k^2 (1 - (v/\alpha)^2) \tag{3}$$

$$m^2 = k^2 (1 - (v/\beta)^2) \ . \tag{4}$$

The parameters α and β designate the compressional and shear wave velocities in the elastic medium around the cylinder.

BOREHOLE WAVE

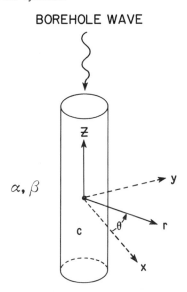

Figure 3-13   The cylindrical coordinate system (r, θ, Z) used to describe wave propagation in a borehole surrounded by an elastic medium. α and β are the compressional and shear wave velocities in the elastic material, and c is the compressional wave velocity in the borehole fluid.

Because of the mathematical behavior of the modified Bessel functions, K_i and I_i, the amplitudes of the particle displacements calculated from these potential functions decrease exponentially with radial distance, r. Thus, tube wave propagation is confined to the fluid column shown in Figure 3-13 and to a relatively thin shell of the formation around the borehole. Note also that, as long as the wavenumber k is a real number, Equations 1 and 2 allow no amplitude attenuation in the Z direction. Consequently, tube waves are sometimes called "guided waves" because they tend to focus energy along the axis of a fluid-filled borehole and allow only small amounts of energy to leak out into the formation. Thus, tube wave amplitudes do not diminish with travel distance as dramatically as do compressional or shear body wave amplitudes since tube waves cannot expand spherically in all directions as a body wave does.

Some of the more complete studies of borehole wave propagation have concentrated on solutions describing sonic logging measurements in which both the energy source and the receiver are located in the same wellbore. The frequencies involved in sonic logging are considerably higher than the frequencies recorded in vertical seismic profiling, and the source-receiver geometry of a sonic logging tool is certainly different from the source-receiver geometry used in a VSP experiment. Nonetheless, mathematical solutions describing sonic log waveforms recorded in a fluid-filled borehole are a good foundation on which to establish fundamental properties of tube waves.

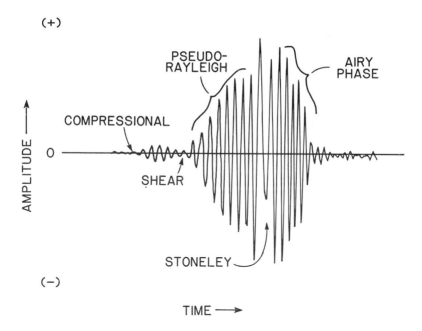

Figure 3-14   A common appearance of the wave modes propagating in a fluid-filled borehole surrounded by an elastic medium. (After Cheng & Toksoz, 1981a)

An example of a typical waveform that can be generated and recorded by a sonic logging tool is shown in Figure 3-14 (Cheng and Toksoz, 1981a). The compressional and shear body waves, which are the critical measurements in sonic logging, are followed by a long, high amplitude waveform composed of several interfacial waves that propagate along the fluid-rock boundary. An analysis of this packet of interfacial waves is essential for understanding the physics of tube waves.

The waves that travel with the greatest velocity along the cylindrical fluid-rock boundary are the pseudo-Rayleigh waves. They are followed by a second interfacial wave called the Stoneley mode (Stoneley, 1949). The behavior of the propagation velocities for these wave modes is shown in Figure 3-15. For short wavelengths (i.e., large wave-numbers), the group velocity for all of these modes is essentially the same as the com-

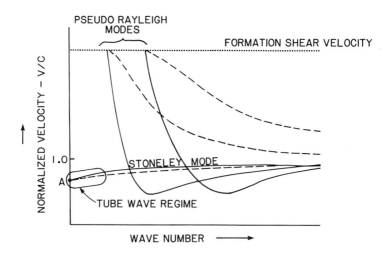

Figure 3-15   General behavior of the phase (---) and group (——) velocities for wave modes propagating in a fluid-filled borehole.   C is the compressional velocity in the borehole fluid. (After Cheng & Toksoz, 1981a)

pressional wave velocity in the borehole fluid.   As the wavelength increases (i.e., wavenumber decreases), the group velocities become smaller with the group velocity of the pseudo-Rayleigh modes decreasing more rapidly than does the velocity of the Stoneley mode.   At a certain wavelength, whose value depends upon borehole diameter and the wave velocities, c, α, and β, the group velocity of each pseudo-Rayleigh mode reaches a minimum.   That portion of the pseudo-Rayleigh wave energy which travels with this minimum velocity is called the Airy phase and is shown as the latest part of the wavelet packet in Figure 3-14. For wavelengths longer than that associated with this velocity minimum, the group velocity of a pseudo-Rayleigh mode increases rapidly. The group velocity increases until it equals the formation shear wave velocity, and then the pseudo-Rayleigh mode no longer exists as an interfacial wave.

Pseudo-Rayleigh modes are not recorded in typical vertical seismic profiles because the cutoff wavelength beyond which these modes cannot exist, although large when viewed as a component of a sonic logging signal, is shorter than the wavelengths contained in the propagating seismic wavelets used for oil and gas exploration. The interfacial wave that is of most interest in VSP data analysis is the Stoneley mode shown in Figures 3-14 and 3-15. The Stoneley mode can exist in normal sized boreholes when the wavenumber values are extremely small (even zero). Thus, this mode can contain wavelengths corresponding to those measured in vertical seismic profiling. In the long wavelength limit, a Stoneley mode propagating in a fluid-filled borehole is what is called a "tube wave" in this text.

The frequency attenuation factor, $1/Q$, of a borehole guided wave is a weighted combination of the compressional and shear wave attenuations in the formation around the borehole and the attenuation in the borehole fluid (Anderson and Archambeau, 1964). The three weighting factors used to sum these attenuation values are partial derivatives of the phase velocity of the guided wave evaluated at a constant frequency and are called partition coefficients. Cheng and Toksoz (1981b, 1981c) show that the partition coefficient associated with acoustic wave attenuation in a borehole fluid is typically four times larger than the partition coefficient associated with a formation's shear wave attenuation, and that the partition coefficient for a formation's compressional wave attenuation is so small that it can be ignored. Thus, tube wave attentation is controlled by fluid wave and shear wave attenuation values. The tube wave attenuation contributed by the fluid in a borehole is the product of the fluid's partition coefficient and the actual seismic attenuation coefficient of the fluid. Unless an unusually high attenuating mud is present in a borehole, tube wave attenuation tracks the behavior of the shear wave attenuation in the surrounding formation, but is smaller in magnitude.

A mathematical analysis by Cheng and Toksoz (1982c) of the particle displacements created by a propagating tube wave is summarized by the curves plotted in Figure 3-16. These displacement values were calculated by assuming compressional and shear propagation velocities appropriate for either competent rock or unconsolidated sediment. In each calculation, the borehole diameter is 8 inches. These results demonstrate the following facts about tube wave propagation:

1. The size of a particle's displacement orbit in the fluid column of a borehole is much larger than a particle's displacement orbit in the surrounding formation.

2. The radial component of a particle orbit in the fluid column increases linearly from a value of zero at the center of the borehole to a small value at the fluid-formation interface.

3. In the fluid column, the axial component of motion is much larger than the radial component of motion.

4. The radial displacement is continuous across the fluid-formation interface.

5. The axial component of motion decreases by a factor of approximately 20 across the fluid-formation interface when the tube wave frequency is 400 to 425 Hz and by a factor of 100 or more when the tube wave frequency reduces to 75 to 80 Hz.

6. At high frequencies, the particle motion in a soft formation is more circular than is a particle orbit in a hard formation.

7. In hard formations, the major axis of a particle orbit is vertical (axial).

8. In soft formations, the major axis of a particle orbit is radial at low frequencies.

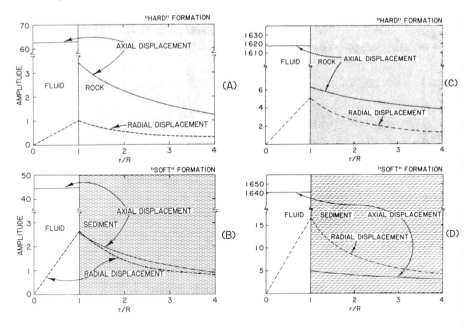

Figure 3-16   Axial and radial particle displacements created by VSP tube waves for boreholes penetrating a hard formation and a soft sediment. R is the borehole radius. Only a single frequency is analyzed in each plot. These frequencies are (A)-409 Hz, (B)-427 Hz, (C)-82 Hz, and (D)-74 Hz. ((A) and (B) modified from Cheng and Toksoz, 1982c. (C) and (D) from Cheng, private communication.)

Within any vertical plane containing the Z axis (i.e. set θ = constant in Figure 3-13), the particle motions created by tube waves would thus appear like those shown in

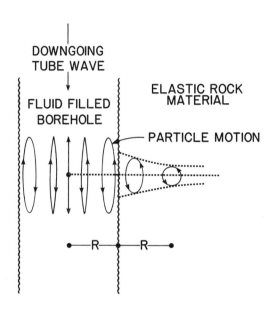

Figure 3-17 A prograde elliptical particle motion is created by a propagating tube wave. The ellipticity in the fluid is greater than drawn here. Pure rectilinear motion occurs at the center of the fluid column, and the ellipticity decreases with radial distance from the axis. The axial component of motion is discontinuous at the boundary; the radial component is continuous. The amplitude of the orbit decays exponentially away from the boundary.

Figure 3-17 if a hard formation surrounds the borehole. The fluid motion is strictly linear along the Z axis (r = 0) but elliptical elsewhere. The shape of the elliptical orbit depends on the radial distance r. The ellipticity decreases with increasing radial distance, so the orbits become slightly more circular toward the circumference of the fluid column. There is a dramatic reduction in tube wave amplitude across the borehole boundary, and the motion in the elastic material approaches a circular orbit at a rate dependent upon frequency content, radial distance, and the material's elastic constants. The axial (Z) component of tube wave particle motion is discontinuous at the fluid-rock boundary, but the radial (r) component is continuous. The amplitude of the particle orbit decays exponentially with increasing distance from the fluid-formation interface. At low

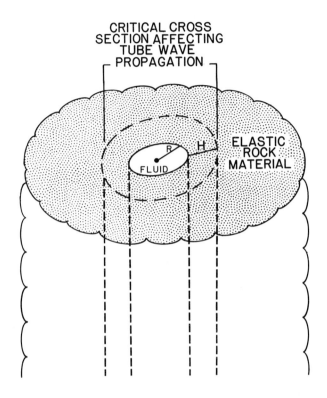

Figure 3-18   The energy carried by a propagating tube wave in a fluid-filled borehole is confined primarily to the fluid column. The particle motion amplitude in the surrounding rock formation is at least a factor of 10 less than that in the fluid. The distance, H, at which the particle displacement reduces to less than half that which exists in the rock immediately next to the fluid interface is of the order of 1 to 3 borehole radii.

frequencies in a soft formation, the particle motion in the rock material would be elongated in the radial direction rather than in the vertical direction as drawn.

This particle displacement behavior shows why it is important to use a velocity-sensitive VSP geophone that clamps rigidly to the wall of a borehole. If tube waves exist in a well, then a hydrophone or geophone hanging free in the fluid column would be subjected to the strongest tube wave effects. In fact, tube wave particle motion plots are one way to judge geophone-to-formation coupling in a qualitative way. If a borehole geophone is not coupled to a formation, the vertical component of tube wave motion will be much larger than the radial component. When a geophone is rigidly coupled to a formation, the tube wave motion will be smaller and much less elliptical because a velocity-sensitive geophone (or accelerometer) clamped to the borehole wall will move only as much as do the rock particles in the wall. This observation explains the high amplitude appearance of the tube wave in intervals B through I of Figure 3-6; i.e., the geophone was not coupled through the casing to the formation in these intervals so the recorded tube wave response was the motion in the fluid and not the motion of the formation. Figure 3-17 shows that, as far as tube wave propagation is concerned, particles in the formation move less than do the fluid particles. Thus, a wall-clamped geophone will filter out some tube wave energy; a free hanging transducer will not. This claim is further substantiated by Russian research which has found that the amplitudes of recorded tube waves are inversely proportional to the amount of locking force pressing a geophone to a formation wall (Gal'perin, 1974, p. 21). This same reference states that tube wave amplitude is inversely proportional to the density of the fluid in a borehole and that tube waves tend to attenuate quickly if a borehole contains heavy mud.

The limited volume of the earth about a borehole which affects tube wave propagation is illustrated in Figure 3-18. Tube wave particle motion is generally confined to the fluid column, as shown in Figure 3-16. The exponential decay of tube wave particle displacements in the radial direction away from a well means that little tube wave energy exists beyond a distance of a few borehole radii from the fluid interface. Thus, the distance H in Figure 3-18 at which tube wave particle displacement reduces to one-half of its value at the fluid interface is on the order of one to three borehole radii. Because the volume of earth supporting the majority of tube wave propagation has a small circular cross-section about the axis of the wellbore, tube waves are greatly affected by changes in borehole diameter, casing conditions, cement bonding, or any feature influencing acoustic impedance within this small circular cross-section. This behavior means that a tube wave having a wavelength in the Z direction of several tens of meters can be altered by a physical borehole discontinuity spanning only a few centimeters in the radial direction.

Although the majority of tube wave energy is confined to a cylinder of radius (R+H), such as shown in Figure 3-18, tube wave disturbances propagating in one well can still be seismically detected as weak signals in other wells a considerable distance away. The geometry of one field experiment that demonstrates this behavior is shown in Figure 3-19. One pound of dynamite was detonated at a depth of 515 feet in the water-filled borehole labeled "Source Well". Twelve hydrophones were suspended in a second water-filled borehole offset 40 feet from this source well. The hydrophones were vertically separated by 20 feet, with the deepest hydrophone positioned at a depth of 300 feet and the shallowest at a depth of 80 feet. None of the hydrophones were clamped to the borehole wall. Three triaxial geophones were cemented in a third well which was also offset 40 feet from the source well. These 3-component geophones were spaced 50 feet apart, with the deepest one placed at a depth of 294 feet and the shallowest at a depth of 194 feet.

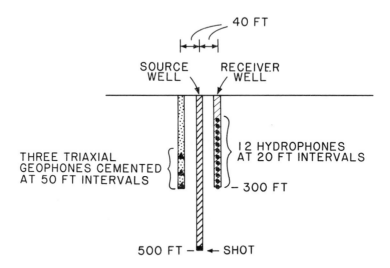

Figure 3-19   Field geometry used in an unpublished experiment conducted by Spencer. The source well and the receiver well are fluid-filled. Hydrophones hang free in the receiver well. (By permission of T. W. Spencer, Texas A & M University and Chevron U.S.A., Inc.)

The responses recorded by the free-hanging hydrophones are shown in Figure 3-20. Three distinct upgoing wave modes can be identified. The first arrival is the compressional body wave which propagates past the hydrophones with a velocity of 8200 ft/sec. The second and third modes, represented by the upgoing black peaks, travel vertically with a velocity of 3000 ft/sec. These modes are assumed to be tube waves for the following reasons:

1.   They reflect downward when they encounter changes in borehole impedance, as a tube wave would. This tube wave reflection behavior is further documented by the data shown in Figure 3-22.

2.   The responses measured by the cemented 3-component geophones are not shown, but they indicate that the shear body wave travels with a velocity of 4600 ft/sec in this depth interval. Since the modes represented by the

black peaks travel with a velocity of 3000 ft/sec, neither of them could be a converted mode created by a shear wave propagating along the borehole wall.

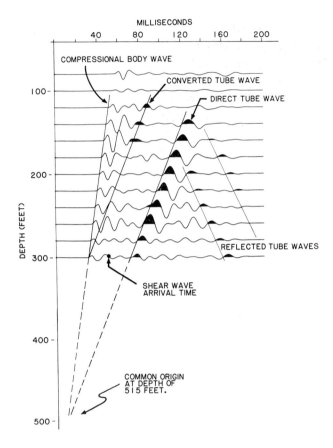

Figure 3-20 Data recorded by hydrophones hanging in the receiver well shown in Figure 3-19. (Courtesy T. W. Spencer, Texas A & M University and Chevron U.S.A. Inc.)

The event labeled "Converted Tube Wave" is created by the compressional body wave at the bottom of the receiver well, as shown by the time-space coincidence of the trajectories of these two events at a depth of 300 feet. The predicted arrival time of the shear wave mode at the base of the receiver well is also indicated on the bottom trace. The event labeled "Direct Tube Wave" is not coincident with either the upgoing compressional or the upgoing shear body wave arrival on this deepest trace.

Consequently, this tube wave mode has no relationship with body waves arriving at the bottom of the receiver well. Because of its low velocity, neither can it be created by compressional or shear body waves propagating up the wall of the receiver well.

Extrapolating the time-depth stepout trends of the compressional first arrival and the Direct Tube Wave mode downward shows that they intersect at a depth of 515 feet, which is the depth of the shot in the source well. For this reason, the Direct Tube Wave in the receiver well is interpreted as being created by a stronger tube wave which propagates up the fluid column in the source well. The amplitude of the tube wave in the source well must therefore be great enough to exert an influence on the fluid column in the receiver well 40 feet away. The frequency content of the Direct Tube Wave is concentrated at 50 Hz, so its dominant wavelength is 60 feet. If the assumption that the Direct Tube Wave is actually a continuation of a stronger tube wave located in the source well is correct, then measurable tube wave effects can extend out at least one wavelength from the fluid column in which a tube wave is propagating.

No hydrophones were placed in the source well. Thus, it cannot be confirmed that the upgoing tube wave in that well travels at a velocity of 3000 ft/sec, as does the Direct Tube Wave. The dynamic recording gain has not been removed from the hydrophone data in Figure 3-20, so the amplitude of the Direct Tube Wave should not be directly compared with the amplitude of the compressional arrival.

Reflection and Transmission of Tube Waves

Tube waves propagate by displacing fluid particles in the fluid column of a wellbore and rock particles in a small annulus around the borehole. Vertical cross-sections of simple-shaped, fluid-filled wellbores are shown in Figure 3-21. A downgoing tube wave, with displacement amplitude U_I, is shown approaching a borehole impedance change at depth Z_1, where the borehole cross-sectional area changes from a_1 to a_2, except in example C. An upgoing reflected tube wave of displacement amplitude U_R and a downgoing transmitted tube wave of displacement amplitude U_T are created at Z_1 in all three instances.

Since the borehole diameter is much less than the wavelength of a seismic tube wave, it will be assumed that the tube wave can be represented mathematically as a plane wave traveling in the axial direction. This assumption is a reasonable approximation of the tube wave behavior in the fluid column where the majority of the tube wave disturbance occurs.

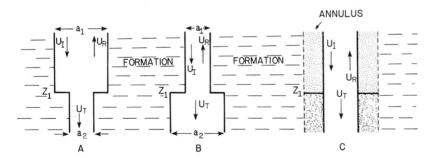

Figure 3-21   Borehole impedance changes that affect tube waves. Tube waves propagate
by displacing fluid particles in a wellbore and rock particles in a small
annulus around the borehole. In A and B, the change in borehole diameter
means that either more or less fluid has to be displaced at depth Z_1,
depending on the direction of tube wave propagation. In C, fluid flow is not
impeded at Z_1, but an acoustic impedance change in the annulus around the
fluid column means that more (or less) tube wave energy is needed to
displace rock particles in the annulus when the tube wave reaches Z_1.

A plane wave assumption may not be completely valid in the elastic medium around
the borehole. Likewise, the change in the cross-sectional area of a fluid column will not
always be as abrupt as those shown in Parts A and B of Figure 3-21. Thus, the following
expressions for tube wave reflection and transmission coefficients should be viewed only
as good estimates and not as precise determinations. However, the errors involved in a
plane wave approximation and in a simple borehole geometry assumption are probably
acceptable, considering the actual caliper log measurement errors that would be involved
in determining the correct shape and cross-sectional area of a typical uncased hole. Any
frequency component, ω, of the incident tube wave particle displacement disturbance can
be represented as

$$U_I(Z,t) = A_I \, e^{j(\omega t - kZ)} \quad , \tag{5}$$

and the corresponding reflected and transmitted particle displacement as

$$U_R(Z,t) = A_R \, e^{j(\omega t + kZ)} \tag{6}$$

and

$$U_T(Z,t) = A_T \, e^{j(\omega t - kZ)} \quad . \tag{7}$$

The amplitude weights A_I, A_R, and A_T are constant for a fixed frequency, ω, but assume new values for each frequency component.

Two boundary conditions must be imposed in order to calculate the reflection and transmission coefficients of the incident tube wave. The first boundary condition is that at depth Z_1 the volume of fluid leaving cross-sectional area a_1 must be the same as the volume of fluid entering cross-sectional area a_2. This requirement means that

$$a_1 U_I(Z_1,t) + a_1 U_R(Z_1,t) = a_2 U_T(Z_1,t). \tag{8}$$

For convenience, the origin of the Z axis will be assumed to be at Z_1, so that $Z_1=0$. This conservation of mass requirement then becomes

$$a_1(A_I + A_R) = a_2 A_T . \tag{9}$$

The second boundary condition is that the pressure, P, must be continuous at Z_1; i.e.,

$$P_I(Z_1,t) + P_R(Z_1,t) = P_T(Z_1,t) . \tag{10}$$

The relationship between pressure and particle displacement is (Kinsler and Frey, 1950, p. 129; Ingard and Kraushaar, 1960, p. 616)

$$P_I = \rho c U_I$$
$$P_R = - \rho c U_R \tag{11}$$
$$P_T = \rho c U_T$$

where ρ and c are the density and compressional wave velocity in the fluid. The conservation of pressure at $Z_1 = 0$ is thus

$$A_I - A_R = A_T , \tag{12}$$

and the reflection and transmission coefficients for the incident tube wave amplitude become

$$R = (A_R/A_I) = \frac{(a_2/a_1) - 1}{(a_2/a_1) + 1} \tag{13}$$

and

$$T = (A_T/A_I) = \frac{2}{(a_2/a_1) + 1} \cdot \tag{14}$$

Note that the reflection coefficient for tube wave amplitude is negative if the downgoing tube wave is approaching a reduction in the cross-sectional area of the fluid column; i.e., if a_2 is less than a_1. This situation would exist at the top of a blunt ended geophone tool hanging in a wellbore for a downgoing tube wave and at the bottom of this geophone tool for an upgoing tube wave.

At the bottom of a fluid-filled well, $R = -1.0$ for a downgoing wave since the area a_2 in Part A of Figure 3-21 would be zero in such a case. At the top of a fluid-filled well, an upgoing tube wave encounters a free surface that can expand with no resistance. The effect is similar to that which a downgoing tube wave in Part B of Figure 3-21 would encounter if it enters an area a_2 that is infinitely large. In this case, and also at the top of a fluid column, the tube wave reflection coefficient is $R = +1.0$.

Significant impedance changes in the annulus around a borehole, such as shown in Part C of Figure 3-21, can also conceivably create reflected tube waves, but the reflection coefficient analysis developed here does not address this situation. The approximation in Equation 13 will suffice for most tube wave analyses encountered in oil and gas exploration uses of VSP data.

Sources of Tube Waves

The mechanisms that create tube waves can often be identified by determining the depth and time where tube wave modes are generated in a vertical seismic profile. Some valuable diagnostic VSP data which contain several strong tube waves are shown in Figure 3-22, and these will be used to examine tube wave source mechanisms. These data were recorded onshore using surface compressional vibrators, and four different tube wave modes are indicated. The casing and cementing conditions that existed in the well when the data were recorded are shown at the right. A numerical automatic gain control function is applied to the data to enhance the visibility of all events. These data were recorded with vertically oriented geophones and thus represent only the vertical component of subsurface particle motion.

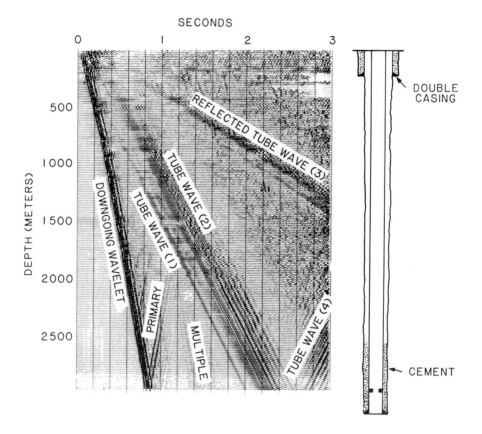

SECONDS

Figure 3-22 A VSP data set dominated by tube waves. The casing and cementing conditions in the well at the time the VSP data were recorded are shown at the right. The VSP geophone could not pass by a frac ring set at 2819 meters below ground level. The borehole caliper is not accurately drawn because the casing and formation were in frequent physical contact with each other in the uncemented interval (Modified from Hardage, 1981b).

One objective of this vertical seismic profile was to identify primary seismic reflectors and correlate their reflection signals with surface seismic reflection data recorded near the well. This objective is not completely achieveable because the strong downgoing tube waves obliterate so many upgoing events. Consequently, tube waves are justifiably regarded as undesirable VSP noise in this experiment.

Detailed examination of these data shows that the four labeled tube wave modes are created by the following processes:

Tube Wave Mode 1 - The source of this mode can be identified by tracing the event upward to find its point of origin. The decrease in the amplitude of tube wave mode 1 above 1000 meters is caused by the size of the data window used to calculate the numerical gain function that is applied to the first 200 ms of data following the first arrival. Even though the tube wave amplitude fades, the linear slope of this mode can still be followed upward to intersect the compressional first arrival at a depth of 225 meters. This depth coincides with the termination of the surface casing, which causes a significant change in borehole impedance. Thus, one mechanism that creates tube waves is the disturbance created in the mud column when a propagating compressional body wave interacts with a strong impedance change occurring within a small cylindrical volume centered about the wellbore. There should also be an upgoing tube wave created at this same depth (225 meters), but it cannot be identified in these data. The change from uncemented to cemented casing at a depth of 2500 meters evidently causes only a small change in borehole impedance in this well because the compressional direct arrival does not create a tube wave there, nor do any of the downgoing tube waves reflect upward with a large amplitude at that depth. In some wells, the top of a cemented interval can be a large enough change in borehole impedance to create significant reflected tube waves.

Tube Wave Mode 2 - This is the dominant tube wave mode in this data set for two reasons:

1. It is the strongest of the four modes.
2. It creates modes 3 and 4.

Consequently, it is essential to identify the mechanism that creates this mode so that appropriate preventative measures can be designed to eliminate it. Tracing the event upward to determine its point of origin shows that this mode is generated at the earth's surface by the first arrival. Again, the amplitude of this mode decreases near the surface because of the numerical gain technique used to dynamically adjust the magnitude of the data. When the geophone is at the surface, the first arrival is the Rayleigh wave ground roll and not a compressional body wave as was the case for mode 1. Thus, the dominant source of tube waves in this well, and in most onshore wells, is the ground roll wave which sweeps across the wellhead and creates vertical motion in the mud column in the well.

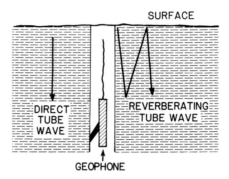

SURFACE

DIRECT TUBE WAVE

REVERBERATING TUBE WAVE

GEOPHONE

Figure 3-23   Tube wave mode 2 shown in Figure 3-22 results from the first passage of the surface-generated tube wave past the downhole geophone. Some of the energy in mode 2 reflects from the top of the downhole geophone package and returns to the surface. The second passage of this energy past the geophone creates tube wave mode 3. Mode 3 could be falsely interpreted as a surface generated event that travels down the wellbore with a velocity one-third that of the direct tube wave. These raypaths are drawn outside of the fluid column only for clarity.

Tube Wave Mode 3 - This mode is a reverberation of mode 2 between the earth's surface and the top of the downhole VSP geophone package. The diameter of the downhole geophone device is 4 inches, so the instrument almost fills the $5\frac{1}{2}$ inch diameter casing used in this well. The reduction in effective borehole cross-section created by the geophone is a dramatic impedance change to the downgoing tube wave and almost 40 percent of its amplitude is reflected upward in accordance with the relationship shown in Equation 13. The open end termination of the casing at the surface creates a reflection coefficient of almost +1.0 for this reflected upgoing tube wave, so essentially all of its energy is reflected downward. As this energy passes by the geophone a second time, it is recorded as tube wave mode 3. This event appears to travel downward with a velocity 1/3 that of modes 1 and 2 because it has traveled three times as far as these modes before it is recorded, as illustrated in Figure 3-23. A second tube wave reverberation between the surface and the geophone package would appear to travel downward with a velocity 1/5 that of modes 1 and 2, but the recording time in Figure 3-22 is not long enough to reliably confirm that a second reverberation was recorded. It should be emphasized that mode 3 would not exist if tube wave mode 2 were eliminated.

Tube Wave Mode 4 - This tube wave mode is a reflection of mode 2 from the bottom of this VSP well. A tube wave reflection coefficient of -1.0 exists at the closed end of the wellbore, so essentially all of the energy in mode 2 which reaches the bottom of the borehole is reflected upward. Mode 4, like mode 3, would not exist if tube wave mode 2 could be eliminated.

Tube waves are thus created when fluid particles in any part of the mud column in a wellbore are displaced. The dominant source of this fluid displacement is the surface Rayleigh wave which rolls across a wellhead and shakes the uppermost part of the liquid column filling the wellbore. Once a tube wave is generated in a well, it propagates up and down the entire hole several times and creates secondary tube waves at prominant borehole impedance contrasts. It is important to note that, even though tube waves have wavelengths that are several tens of meters long, these wave modes are reflected by borehole impedance contrasts that are less than one meter in diameter. It should also be noted that each tube wave mode has a waveshape identical to that of the source wavelet which created it. Mode 1 in Figure 3-22 is a short symmetrically shaped wavelet like the compressional direct arrival that created it. Mode 2 is a long, high amplitude, ringy event, as is the Rayleigh ground roll wave that created it. The following three data characteristics are particularly helpful in identifying tube wave noise sources:

1. The depth at which a tube wave originates indicates the physical borehole anomaly that contributes to the formation of the tube wave.

2. If a second propagating energy mode coincides in depth and time with the origin point of a tube wave, then that wave mode is almost always the energy source that creates the tube wave.

3. The waveshape of a tube wave mode tends to be the same as the waveshape of the energy mode that creates it, and this fact can often distinguish which of several possible energy modes is the generating source.

A borehole anomaly and a generating energy source are both required in order to create the tube waves observed in vertical seismic profiling. (The termination of a wellbore at the earth's surface can be considered a "borehole anomaly".) If these two necessary ingredients for tube wave formation can be identified by any of the characteristics listed above, then preventative measures should be implemented to minimize the effect of one or both of them. Some of these procedures will be discussed in Chapter 4.

92

Tube waves are more of a problem in onshore vertical seismic profiling than they are in marine work. However, tube waves can occasionally be created during offshore vertical seismic profiling. An example of such a situation is shown in Figure 3-24. These data are a velocity check shot survey, not a vertical seismic profile, nonetheless tube waves are obvious in the data. No attempt was made to determine the source of these tube waves since they did not affect the compressional first breaks, which was the only information desired in the velocity survey. The airgun source used in this survey obviously created some type of disturbance in the fluid column at or near the surface in order to cause the downgoing tube wave. The upgoing tube waves are created either at casing points or at severe borehole washouts. An example of similar tube waves observed in a marine VSP is shown in Figure 3-27.

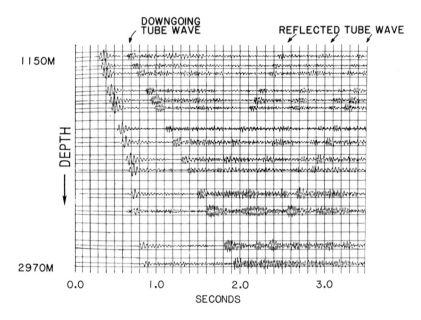

Figure 3-24   Offshore velocity survey data showing tube waves created in a marine environment.

Frequency Characteristics of Tube Waves

Hopefully, when undesirable noise contaminates VSP data, the frequency bandwidth of the noise lies outside the frequency bandwidth of seismic body wave signals so that the noise can be eliminated by digital filtering. Since tube wave noise can greatly reduce the interpretive value of VSP data, it is particularly important to compare the frequency characteristics of these fluid-borne wave modes with the frequency content of downgoing and upgoing body wave events recorded in a VSP well.

A comparative frequency analysis of the tube wave and body wave events contained in the vertical seismic profile shown in Figure 3-22 is shown by the power spectra plotted in Figure 3-25 (Hardage, 1981b). These spectra are calculated in restricted time and depth windows containing just the downgoing compressional first arrival, or just downgoing tube wave modes 1 and 2. (These wave modes are labeled in Figure 3-22.) The depth intervals where the analyses were performed are labeled beside each spectrum below. The spectra are necesssarily restricted to the vibrator sweep limits of 8 to 64 Hertz.

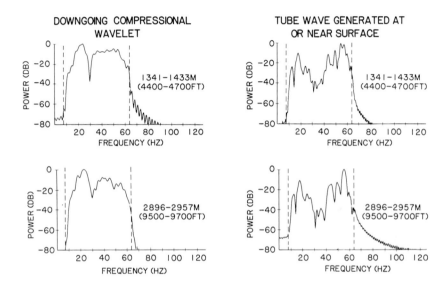

Figure 3-25   Comparison of the frequency content of compressional body wave signals and tube wave events recorded in the vertical seismic profile shown in Figure 3-22 (After Hardage, 1981b).

Although the tube wave spectra differ in some respects from the compressional body wave spectra, it is obvious that these tube waves span the same frequency range as the compressional wavelets do. This fact is not surprising since it has been emphasized that the tube waves in this experiment have the same waveshape as the wavelet that created the original disturbance in the fluid column, and one of these tube waves (mode 1) was created by the compressional first arrival. Since these tube waves have the same frequency components as the compressional wavelets, they cannot be eliminated by digital filtering without also severely attenuating downgoing and upgoing compressional wave signals.

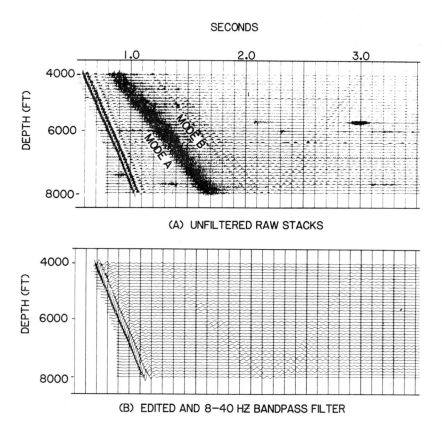

SECONDS

(A) UNFILTERED RAW STACKS

(B) EDITED AND 8–40 HZ BANDPASS FILTER

Figure 3-26   Example of high frequency tube wave eliminated by bandpass filtering. Energy source was an airgun in a water-filled pit positioned 200 feet from the wellhead.   The geophone was vertically oriented when recording the data.   The filter does not remove the low frequency tube wave mode B. (After Poster, 1982).

A second onshore vertical seismic profile containing strong tube waves is shown in Figure 3-26. These data were recorded in a vertical borehole with a vertically oriented geophone, as were the data analyzed in Figure 3-25. In this instance, two wave modes, labeled A and B, are created and propagate down the fluid-filled well with a velocity of approximately 4800 ft/sec (1463 m/sec).   Because of their propagation velocity, both modes are assumed to be tube waves. By visual inspection, it is obvious that mode A is of much higher frequency than either the compressional first arrival or tube wave mode B. Consequently, a simple bandpass filter (8-40 Hz) eliminates this noise mode.   Mode B remains after filtering. No VSP data are available above 4000 feet, so the sources of these tube waves cannot be rigorously determined. Mode B appears to have been generated at the surface by Rayleigh ground roll, but mode A is apparently created below the surface by some other mechanism.

An example of a tube wave generated in a marine VSP environment is shown in Figure 3-27. The energy source in this instance was a single 200 in^3 airgun. The water depth was 245 meters. Compressional events contained in this display are adjusted to true relative amplitudes by multiplying each trace value by a function of the form AT^n, where T is the record time corresponding to each sample point, and A and n are constants appropriate for a compressional wavefront. (See Equation 12, Chapter 5.) This gain function removes the effects of amplitude decay caused by spherical divergence. Thus, tube wave amplitudes are excessively amplified since they do not decay as much as does a spherically spreading compressional wavefield.

The amplitude of this tube wave is not as large as that of some tube waves created in onshore experiments where ground roll exists. The relatively weak third bubble oscillation is still strong enough to be identified as it crosses the tube wave on its downward path. However, these tube wave amplitudes are much larger than the amplitudes of upgoing reflections in this stratigraphic interval, and so the tube wave events are a serious detriment to a thorough interpretation of the upgoing wavefield.

By visual inspection, it is apparent that this tube wave has a distinctly higher frequency content than does the compressional wave. Because of the frequency difference between the tube wave and the compressional body wave signals, the tube wave events are effectively eliminated by applying a 0-65 Hz bandpass filter to the data. This high frequency behavior of tube waves appears to be a common phenomenon in marine environments. Note that the tube waves recorded in the offshore velocity survey shown in Figure 3-24 also have a higher frequency content than do the compressional first arrivals observed in that well. The mechanism that creates this high frequency type of tube wave is difficult to ascertain because VSP data and log data are seldom recorded in the shallow part of a well where the tube wave event is usually generated.

These VSP experiments confirm that there is a wide range of relationships between the frequency characteristics of tube waves and the frequency content of body waves arriving at a wellbore. Tube waves may span the same frequency band that body waves do (Figures 3-25 and 3-22), lie in a distinctly different frequency band (Figure 3-27), or contain modes whose frequencies lie both within and outside of the seismic band (Figure 3-26). Each vertical seismic profile containing tube wave noise has to be analyzed and processed on an individual basis to determine if bandpass filters can reduce tube wave contamination.

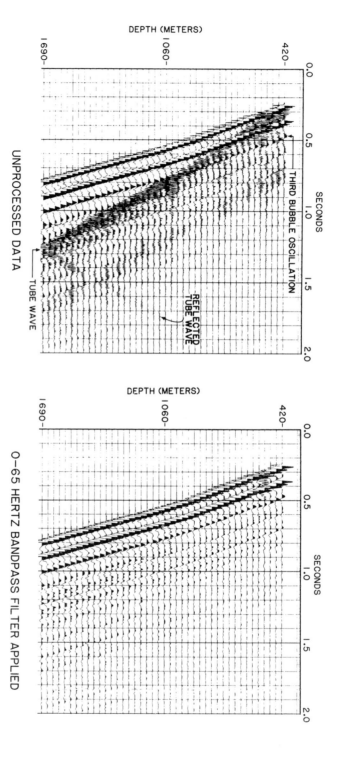

Figure 3-27   An example of a tube wave created in a marine environment. In this instance, the tube wave is dominated by high frequencies and can be effectively eliminated by simple bandpass filtering. (Courtesy of Statoil)

Surface Cultural Noise

The claim made at the beginning of this chapter, that a deep borehole geophone records seismic data where optimum signal-to-noise conditions exist, is usually true with, perhaps, the exception of cultural noise, which tends to concentrate at a VSP well site. Cultural noise is not included in the list of noise sources shown in Figure 3-1 for this reason.

Cultural seismic noise occurring during vertical seismic profiling is defined as mechanical and electrical noise which exists because of the presence of people or machinery near a VSP study well. Examples of such noise sources are diesel engines, air compressors, electrical generators, and drill site work activities such as welding, stacking pipe and metal goods, and general rig site maintenance. The amount of this type of noise depends on whether VSP data are recorded while a drill rig is still on-site at the well or after the drill rig has been disassembled and moved away. More cultural noise exists when a drill rig is on-site simply because more people, more machinery, and a higher level of work activity are present, as is illustrated in Figure 3-28.

It must be kept in mind that a fluid-filled borehole is an efficient wave guide that transmits any seismic event that disturbs the top of the fluid column, rotary table, or surface casing to considerable depths. Consequently, even low level noise sources located

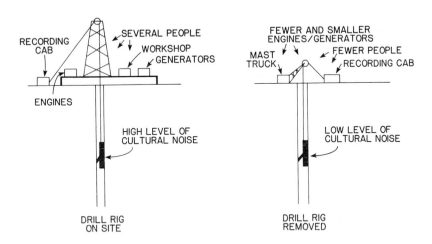

Figure 3-28   The amount of cultural noise that occurs during a vertical seismic profile is often a function of whether or not the drill rig is still on-site.

close to a well head can significantly increase the downhole noise level. A borehole is, in effect, an ear trumpet or stethoscope that directs surface cultural noise at the well head to a downhole geophone.

Some typical examples of well site cultural noise are shown in Figure 3-29. Mechanical noises, such as shown in A and B, travel down the fluid column. Even though the propagation velocity in the fluid is typically less than formation propagation velocities, these noise events can occur either before or after the body wave first arrival created by a VSP seismic source because the noise generation may have no specific time relationship relative to a seismic shot. For example, a diesel engine which vibrates a drill rig floor will create a continuous noise background that exists both before and after a seismic shot, as shown in Part A. Random impulsive noise, such as indicated in B, may occur before the body wave first arrival or overlay an upgoing reflection following the direct arrival. Drill rigs always have large capacity electrical generators that operate many lights and electrical devices. Consequently, electrical crossfeed into a VSP recording cable is common. This crossfeed also appears as a continuous noise background, as shown in Part C. Usually this type of noise can be reduced in the field by electronic filters or by digital filtering during data processing.

The only way to avoid severe cultural noise problems during VSP data recording is to create as quiet an environment at the well site as possible. If a drill rig is on-site, an optimum noise environment can often not be created because all equipment needed for human life support and for the security of drill rig equipment must continue to operate, and some work activity not associated with VSP data recording must necessarily occur. This cultural situation is particularly true in marine VSP work, where a drill rig or production platform will always be on-site during data recording. Only in onshore VSP activity can one think in terms of recording data after a drill rig has left, and a quieter cultural situation exists.

When recording VSP data after a drill rig is moved away, some level of cultural noise will still exist, and in some instances this noise may even exceed that observed in a drill rig environment. In general though, less cultural noise will exist if the drill rig is absent, as shown in Figure 3-28, and onshore vertical seismic profiles should be conducted at this time, if possible. This objective must be weighed against the fact that data will have to be recorded inside casing, and depth intervals where multiple, unbonded casings exist will likely have to be skipped.

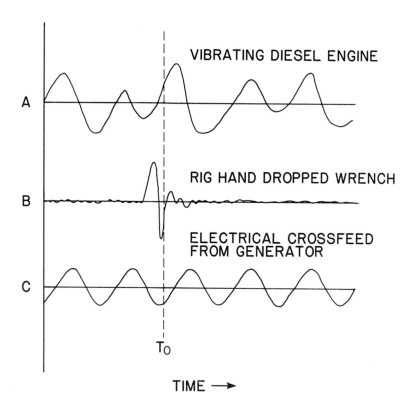

Figure 3-29   Typical cultural noises observed in vertical seismic profiling. These noises
may be continuous or occur at random times relative to the recording zero-
time which might, for example, be at T_O.

CHAPTER 4

VSP FIELD PROCEDURES

The overall quality of VSP data is probably more affected by procedures followed in the field while recording the data than by any other factor. Some field techniques cancel noise, while others result in noise contamination. One must be able to recognize inappropriate field operations because even elegant and sophisticated data processing techniques are of limited value if poor field practices have introduced excessive noise or unwarranted wavelet variations into the measurements. Some field operations are efficient, but others are time consuming and expensive. Efficient field operations are essential because the longer a VSP experiment is in progress, the more likelihood there is for mechanical failures of equipment or for borehole deterioration. Likewise, procedures that work to yield good data in one area may not work in another area. For these reasons, people responsible for recording VSP data must know and use a wide variety of field techniques in order to consistently record good quality data.

Correct VSP field procedures should begin and end with a stringent set of instrument tests which confirm that the complete recording system does not unknowingly attenuate some signal frequencies, that no significant crossfeed exists between data channels, that amplifier gains are correct, and that extraneous electrical noise is not introduced into the data. Other tests must be done to verify that sufficient signal energy is received by the geophone at all depth levels, that recording depths are accurately known, and that data are recorded at appropriate stratigraphic positions in the wellbore.

Geophone Tap Tests

A geophone tap test measures what polarity a geophone's output will be when the geophone case moves in a specified direction. These tests are vital when making interpretations of VSP measurements that are contingent upon wavelet polarity, or when studying particle motion behavior recorded by orthogonal, three-component geophone systems. A tap test also confirms that the geophone elements inside a VSP geophone package are functioning before lowering the instrument into a wellbore. The fact that the signal polarity measured by a moving-coil geophone depends upon its direction of motion is one reason why velocity sensitive transducer elements are used in recording VSP data. For a given compressional event, the polarity of the output of a pressure sensitive device, such as a hydrophone, is the same regardless of the direction of arrival of that event at the hydrophone's location. Consequently, a triaxial hydrophone system cannot be used to determine the direction of arrival of subsurface events at each recording depth nor to analyze polarized seismic wave motions. In addition, the sensitivity of most hydrophones is reduced in the high hydrostatic pressures found in deep wells.

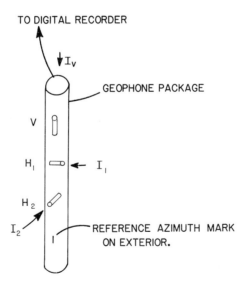

Figure 4-1   Geophone tap tests are performed by applying a small impulse, I, to the exterior of a VSP geophone package so that the impulse is directed along the longitudinal axis of one of the geophone elements inside the package. This diagram shows an orthogonal 3-component geophone system having a vertical geophone V and two horizontal geophones H_1 and H_2. As a minimum, three impulses, I_v, I_1, and I_2, should be applied along the axis of each element, and all three geophone outputs digitally recorded for each of these impulses.

A simplified picture of a tap test is shown in Figure 4-1. The impulses I_v, I_1, and I_2 shown in this figure can be created by lightly tapping the exterior of the geophone package with a small hand tool. The direction the hand tool is moving when it strikes the exterior surface of the geophone assembly represents the direction the earth would be moving when the geophone is clamped in the borehole. The output of all geophone channels should be digitally recorded for each tap in order to determine if an impulse in one direction generates responses in geophone units oriented orthogonally to the impulse direction. The principal objective of the test is to answer the question, "If the geophone case moves in direction A, is the output of the geophone element oriented in direction A represented by positive or negative numbers on the field data tape?"

When determining the polarity of geophones activated by horizontally directed impulses, it is helpful to have a reference mark on the exterior of the geophone package whose position is known relative to each of the internal geophone elements. Geophone responses can be recorded for horizontal impulses that begin at this reference mark and repeat at angular increments of 90 degrees around the device. Again, the objective is to determine what direction the earth has to move in order for geophones H_1 and H_2 shown in Figure 4-1 to generate outputs that appear as positive numbers on a field tape.

Multiple Recordings for Constant-Depth Stacking

It is advisable to record several seismic shots at each recording depth and then sum these records into a single VSP trace. This shooting procedure usually allows a relatively weak seismic energy source to be used in vertical seismic profiling, and the summing process is also an effective way to cancel random noise. The number of records that should be made per depth level must be a reasonable balance between economic considerations and the signal-to-noise ratio desired in the data. It is essential to record these shots as individual records and then sum them during data processing rather than sum them in a field recorder and lose the individual shot records. This procedure allows a skilled interpreter to edit traces having spurious noise and thereby create a high quality composited trace. An example of editing and compositing individual seismic shots recorded at a fixed depth is illustrated and discussed in Chapter 5. Many VSP energy sources create highly repeatable wavelets so that this summation process usually does not induce undesirable noise. The number of shots per depth level can vary from 3 to 10 or more, depending on the penetrating power of the seismic source.

Energy Input as a Function of Recording Depth

A fundamental difference between a vertical seismic profile and a velocity survey is that a velocity survey focuses only on first breaks; whereas, both first breaks and subsequent upgoing and downgoing events must be recorded in a VSP survey. The philosophy of energy input can be completely opposite for these two measurements. In a velocity survey, more energy may be input to the downgoing wavelet as the geophone depth increases so that first break amplitudes are maintained. The opposite approach is sometimes needed in a VSP measurement because more input energy is needed when the geophone is at a shallow depth so that weak, late arriving deep reflections can be detected. This energy requirement is shown by the example in Figure 4-2. If the shallow VSP data are recorded using only the criterion that first breaks have adequate amplitude,

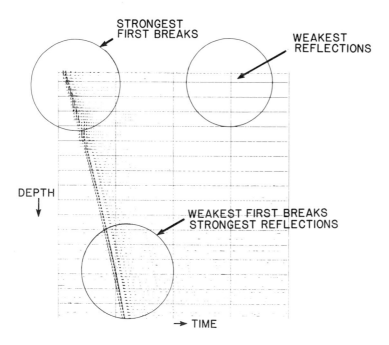

Figure 4-2   The behavior of first break amplitudes and reflection amplitudes as a function of recording depth. These data are numerically adjusted so that all first break amplitudes are equal. Good quality upgoing reflections exist in these data but only faint hints of them can be seen near the first breaks when the data are not displayed with an automatic gain adjustment as they are in Figure 4-6.

then deep reflection events may be lost because of their weak signal strength. In some situations, the energy input when recording VSP data at shallow depths may need to be two or three times that used when the geophone is at the bottom of a deep well. This energy increase can be accomplished by increasing the strength of each seismic shot when the borehole geophone is in the shallow part of a stratigraphic section, or by keeping the energy strength constant and increasing the number of shots per depth level. The latter option is better since increasing the strength of a seismic shot often creates input wavelets that have fundamentally different character than wavelets already recorded. Also, this need for increased energy input when recording shallow VSP data must be balanced with the caution expressed in Chapter 2; i.e., too much energy input can damage overall VSP data quality by creating a more dominating downgoing wavefield.

Since VSP data are recorded as a borehole geophone is raised from the bottom of a well to the surface, a recommended field procedure is to make exploratory data recordings at intervals of 300 to 500 meters as a VSP geophone is lowered to the bottom of a study well, as shown in Figure 4-3. These measurements allow quality control person-

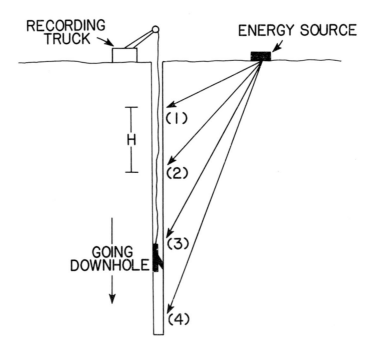

Figure 4-3   Appropriate recording gain and input energy strength should be determined at increments, H, of 300 to 500 meters as a geophone is lowered to the bottom of a test well. VSP data are then recorded as the geophone is brought uphole.

nel to determine appropriate recording gains, the amount of source energy output needed, the number of shots to sum for required signal-to-noise quality, and hopefully, they reveal some reflection events. These insights expedite the recording of VSP data on the return to the surface. One also obtains two independent seismic measurements, one made going downhole and one made coming uphole, of the VSP response at several depth levels which can be compared for data consistency. A further benefit is that the accuracy of the recording depths can often be checked by determining if physical or magnetic marks on the logging cable are located at the same reference point when the cable odometer indicates that the geophone is at the same depth on its downward and upward trips.

Time Sampling Interval

The appropriate time sampling rate, Δt, to use when recording digital time variant data is usually defined as

$$\Delta t \leq \frac{1}{2f_a} \tag{1}$$

where f_a is the highest frequency that is contained in the recorded data. Sample rates of 1 and 2 ms are commonly used when recording typical seismic exploration data. Even if a sample rate of 4 ms is satisfactory for recording VSP data from a frequency bandwidth consideration, it is recommended from a data processing point of view that a sample rate of 1 or 2 ms be used. Experienced data processors have observed that some numerical procedures, such as the design of deconvolution operators, work better if more sample points are used to describe wavelet character. Consequently, data recorded at a 2 ms sample rate sometime yield processing results superior to those obtained with data recorded at 4 ms, even though the frequency bandwidth of the data is the same in both instances. If precise numerical behavior of wavelet character is to be studied, for example if one wishes to estimate attenuation, then a 1 ms sample rate should be used regardless of whether or not the wavelet bandwidth justifies that small a sample rate.

Depth Sampling Interval

When data are digitally sampled, the sample interval must be selected so that at least two sample points fall within the shortest wavelength to be preserved in the data in order to avoid digital aliasing. In a vertical seismic profile, data have to be sampled not only in time so that all frequencies are preserved, but also in space so that all spatial wavelengths can be reconstructed. Proper depth sampling is crucial because some attributes of a vertical seismic profile can be revealed only after two-dimensional digital frequency filtering and velocity filtering have been performed on the data. In particular, velocity filters are often needed in order to satisfactorily separate upgoing and downgoing

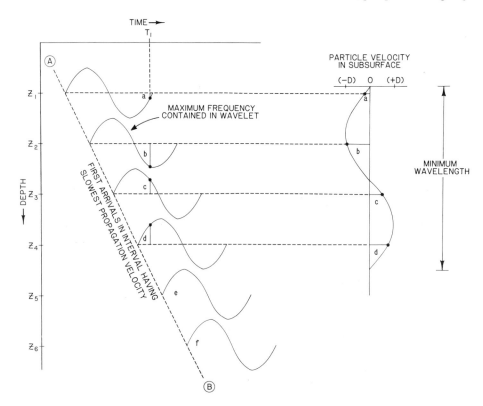

Figure 4-4   The relationship between wavelet frequency, seismic propagation velocity, and spatial wavelength. The minimum wavelength occurs when the maximum frequency component of a wavelet travels through a depth interval where the propagation velocity has its minimum value. This minimum wavelength, λ, defines the spatial sampling interval as $\Delta Z \leq \lambda/2$. The data shown here are properly sampled in space since there are four sample points within the minimum wavelength.

events in VSP data. If the data are not sampled at sufficiently small increments in both space and time, then digital aliasing exists, and velocity filters cannot separate energy modes with optimum resolution.

There is usually no problem in sampling VSP data properly in time, but improper depth sampling is a common operational error in VSP surveys. The magnitude of the correct spatial sampling interval can be determined if one knows both the velocity profile through the formations penetrated by the well and the frequency spectrum of the propagating VSP source wavelet. These two physical properties define the depth sampling interval, ΔZ, as

$$\Delta Z \leq \frac{V_{min}}{2f_{max}} \tag{2}$$

where V_{min} is the minimum velocity in the formations penetrated by the well, and f_{max} is the maximum frequency in the propagating wavelet. This Nyquist sampling requirement is illustrated in Figure 4-4. This illustration shows only the highest frequency component contained in some hypothetical VSP data. The line AB indicates the time-depth stepout behavior in that part of the stratigraphic section where the seismic propagation velocity is a minimum; consequently, each maximum frequency component is positioned at its proper arrival time for the recording depths Z_1 to Z_6. These traces can be sampled as a function of depth at a fixed recording time in order to see if spatial aliasing occurs. In this example, the data are sampled along record time, T_1, to obtain the spatial wavelet values a, b, c, and d. Since more than two depth sample values fall within the minimum wavelength shown at the right, no spatial aliasing occurs.

A sonic log recorded in a VSP study well is a valuable physical measurement that can be used prior to beginning VSP field work to select the spatial sampling interval at which VSP data should be recorded. A sonic log can also be used after VSP data are recorded to determine which depth intervals, if any, were not properly sampled in a spatial sense. One sonic log from a VSP study well is shown in Figure 4-5. To completely avoid spatial aliasing in the VSP data recorded in this well, no part of the sonic log in the depth interval where VSP recordings are made should fall below the line

$$\Delta t = \frac{1}{(2 \Delta Z)(f_{max})} = \frac{1}{V_{min}} \tag{3}$$

where ΔZ is the actual depth sampling interval used in the VSP survey. The position of this velocity cutoff line is controlled by the choices assigned to ΔZ, the spatial sampling increment, and f_{max}, the maximum frequency component in the propagating wavelet. If

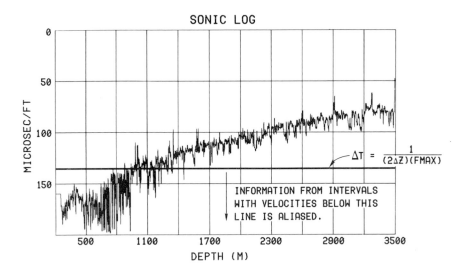

SONIC LOG

$$\Delta T = \frac{1}{(2\Delta Z)(FMAX)}$$

INFORMATION FROM INTERVALS
WITH VELOCITIES BELOW THIS
LINE IS ALIASED.

Figure 4-5   Relationship between the spatial sampling interval, ΔZ, of VSP data and digital aliasing of the data. If the maximum frequency, FMAX, of a seismic wavelet is fixed, the minimum seismic propagation velocity for which no digital aliasing occurs for a particular ΔZ value defines the position of the horizontal cutoff line. See Equation 3. This minimum velocity should be compared to the sonic log recorded in a VSP well to judge the severity of the spatial aliasing problem.

the frequency content of the wavelet cannot or should not be adjusted, then in order to avoid aliasing, the only alternative is to adjust the depth sampling interval, ΔZ, so that the velocity cutoff line occurs at an acceptable level.

VSP data recorded in this well are shown in Figure 4-6. The depth sampling interval was set at 30 meters in this experiment. There is nothing in the character of the data shown in Figure 4-6 which implies that a sampling error or any recording problem exists, except in the interval from the surface down to almost 900 meters. In this shallow interval, though, the change in data character is a borehole noise problem caused by uncemented multiple casing and is not related to data aliasing. There is no apparent explanation why good quality data are recorded inside double casing for almost 200 meters above the measured top of cement at 1150 meters. One can see reflection events originating at several depths (e.g. 1340 and 2150 meters) and also some downgoing multiples; therefore, the data provide valuable insights into the basic behavior of propagating seismic wavelets in the area of this study well regardless of whether or not a correct depth sampling interval was used to record the data.

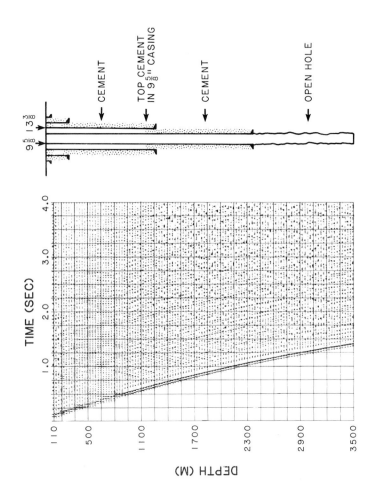

Figure 4-6

These VSP data are recorded at a depth sampling interval which violates the Nyquist sampling theorem; however, the data are still valuable for exploration purposes. Note that data quality deteriorates inside multiple casing strings that are not rigidly cemented to each other and to the formation.

The result of subjecting the sonic log recorded in this well to the test described in Figure 4-5 is shown in Figure 4-7. The maximum frequency, f_{max}, used to position the

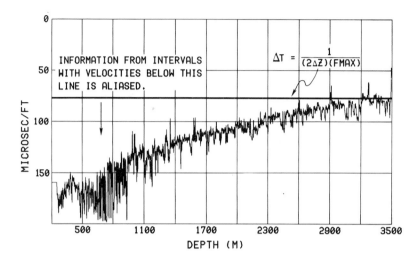

Figure 4-7   Relationship between the spatial sampling of the VSP data in Figure 4-6 and the amount of digital aliasing. ΔZ was set at 30 meters in this experiment, and the effective maximum frequency in the VSP data is 65 Hertz. The resulting velocity cutoff is 12795 ft/sec (78 μsec/ft). The sonic log recorded in the well shows that information from essentially the entire drilled section will be spatially aliased.

velocity cutoff line was set at a value of 65 Hertz after examining power spectra of several downgoing VSP wavelets. It is obvious from Figure 4-7 that some amount of aliasing will occur in the VSP data for most of the drilled section since the majority of the sonic log values lie below the velocity cutoff line. In order to effectively avoid aliasing, the depth sampling interval should be set at about 15 meters, which would move the Δt cutoff line down to 156 microseconds/foot. In this survey, economic considerations and a precarious borehole environment could not permit the data recording to be extended over a long enough period of time to allow this increased data sampling. Frequently, compromises have to be made.

The aliasing problem in these data can be verified by calculating frequency-wavenumber (f-k) Fourier transforms of the data. The transform results are shown in Figure 4-8. In this display, the horizontal coordinate is labeled in wavelength units rather than wavenumber units. Downgoing energy modes plot in the right half plane, and upgoing modes occur in the left half plane. The black areas locate the highest energies in the data, and the dashed region defines where the lowest energies occur. A straight line emanating from the point ($f=0$, $\lambda=\infty$) defines an event propagating with a group velocity given by $V = f\lambda$, where f and λ are values of frequency and wavelength lying on the straight line. Note the aliased events labeled in the figure. Because of spatial aliasing, some downgoing energy, which should be located in the right half plane, is moved to the left half plane and deteriorates upgoing signal quality. The highest frequencies of the upgoing events are aliased into the right half plane and become intertwined with downgoing events.

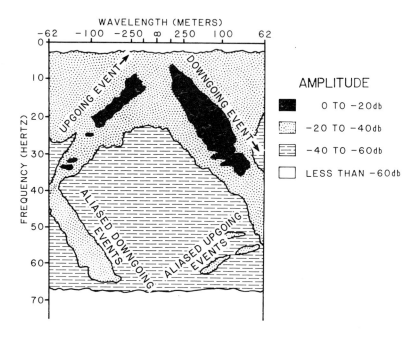

Figure 4-8   The f-k spectrum of the VSP data shown in Figure 4-6. F-K spectra and the locations of upgoing and downgoing events in such displays are discussed in Chapter 5.

A possible velocity filter design which would suppress downgoing events in these data and preserve upgoing events is shown in Figure 4-9. A narrow velocity passband filter like the one shown here creates rather severe spatial mixing of upgoing events. Such wavelet averaging can be an advantage or disadvantage in interpreting the data, depending on what the objectives of the experiment are. The design of velocity filters in f-k space will be further discussed in Chapter 5. This filter is implemented by multiplying the data points in the REJECT area of the frequency-wavenumber space by a small value, usually 0.1, or 0.01, or 0.001, and leaving the data points in the PASS area unchanged.

These altered frequency-wavenumber data can then be inverse Fourier transformed into VSP trace data which are expressed as a function of time and space. In this example, some aliased downgoing data are unavoidably contained in the PASS area specified in Figure 4-9 unless all frequencies greater than 40 Hertz are eliminated. These aliased data contaminate the transform values so that the filtered results are not "pure" upgoing events. One can frequently accept a modest amount of such contamination and still be able to make valuable measurements and interpretations using the filtered upgoing events. Optimum data filtering will result; however, only when spatial aliasing is avoided.

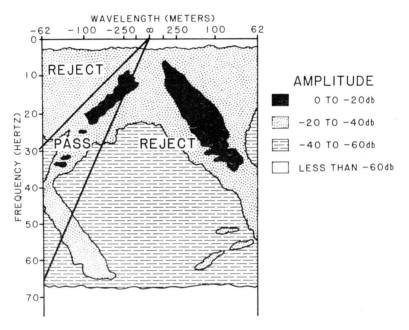

Figure 4-9   A possible velocity filter which can suppress all VSP events except those traveling upward with a velocity between 1800 and 4000 m/sec is shown here.   All data points in the two REJECT regions are attenuated by multiplying them by small numbers. A narrow passband filter like this one creates severe Rieber mixing of the data, as discussed in Chapter 5.

The results of applying the filter specified in Figure 4-9 to the VSP data in Figure 4-6 are shown in Figure 4-10. The filtered data show upgoing events that cannot be defined in the original data in Figure 4-6 and demonstrate why velocity filtering is a valuable procedure for analyzing VSP data. A person confronted with the problem of correlating these upgoing VSP events with a surface seismic line crossing the VSP well site will have to decide whether or not the aliased data contained in this filtered display prohibit reliable interpretations from being made. It appears that the filtered data are

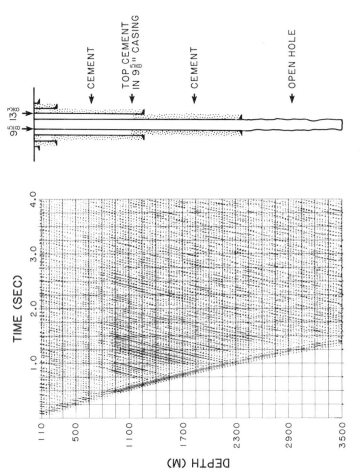

Figure 4-10

These data are the result of applying the f-k velocity filter in Figure 4-9 to the data shown in Figure 4-6 so that all events which are not traveling upward with a velocity between 1800 and 4000 m/sec are attenuated. A numerical AGC function is then applied so that weak amplitudes are amplified.

114

still valuable, even with the attendant aliasing problem, since they give a reasonably good idea of which events are primaries and which are multiples. This example shows that some spatial aliasing can be permitted in a VSP survey without completely sacrificing the quality and value of the data. However, if economics permits VSP data to be gathered at a spatial sampling that will avoid aliasing, then the recordings should be made at these smaller spatial increments. The depth sampling interval defined by Equation 2 should be rigorously enforced as a field procedure whenever possible.

Differential Pressure Sticking

Differential pressure sticking of equipment in fluid-filled, uncased boreholes is a problem that has frustrated drilling engineers for some time (Helmick and Longley, 1957; Outmans, 1958; Annis and Monaghan, 1962; Moore, 1974). In order to avoid well blowouts, the pressure in the fluid column in a wellbore must exceed the fluid pressure in the pore system of the formation being drilled. As long as a piece of equipment inside a wellbore avoids contact with the formation, there is equal pressure on all sides of the device, as shown in part A of Figure 4-11. The force F_A needed to raise the equipment is the difference between its weight and the bouyant force resulting from the volume of fluid that it displaces.

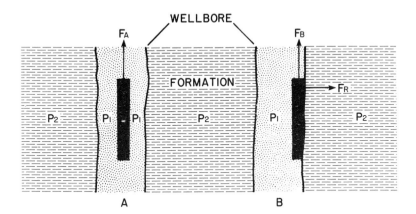

Figure 4-11   An example of differential pressure sticking. Two fluid-filled boreholes having an internal fluid pressure P_1 are shown penetrating a formation in which the pressure in the pore system is P_2. The piece of equipment shown in A can be raised with a modest force F_A. In B, the device has broken through the protective mud cake and is exposed to the formation pressure P_2. The pressure difference (P_1-P_2) pushes the equipment against the borehole with a force F_R so that an increased vertical force F_B is required to overcome sliding friction and raise the device.

Anytime a piece of downhole equipment touches a formation, it is subjected to a pressure imbalance as shown in part B of Figure 4-11. Since the borehole fluid pressure, P_1, is greater than the formation fluid pressure, P_2, the device is pushed against the borehole wall with a force

$$F_R = (P_1-P_2) \cdot A \tag{4}$$

where A is the contact area between the device and the formation. A larger force, F_B, is now required to raise the equipment in order to overcome the sliding friction generated by the radially directed force F_R. This phenomenon is called "differential pressure sticking." In some instances, the radial force F_R may be so large that the vertical force F_B needed to raise the equipment either cannot be generated by the surface hoisting system or breaks the cable connected to the device.

In vertical seismic profiling, there are two pieces of equipment that can be subjected to differential pressure sticking. One of these equipment items is the downhole geophone package; the second is the logging cable that connects the geophone to the surface. Even though a geophone is pressed firmly against the borehole wall during data recording, the geophone package is not considered to be an important factor in differential pressure sticking because the device is short, and its high rigidity means that only a few small contact areas exist between it and a rough borehole wall. If the borehole wall is extremely smooth, or if the formation is so soft that it deforms around the geophone barrel, then a large contact area may exist, and differential pressure sticking of the geophone package becomes an operational reality.

The second downhole equipment item, the logging cable, is the more common source of differential pressure sticking for the following reasons:

1. Even though logging cable has a small diameter, it can have a large contact area with the formation because of its great length. Several hundred meters of cable may pass through an uncased interval of a VSP well.

2. Logging cable is flexible, particularly when it is slacked while recording VSP data, thus it can fit into the contours of a rugous borehole. This fact means that cable slacking, which is essential in VSP data recording, increases the contact area between the cable and the formation.

3. Because boreholes are not perfectly vertical, a logging cable seldom stays in the center of a borehole. Instead, it tends to rub against the borehole walls and break through the protective mud cake, cutting grooves in the formation, as shown in Figure 4-12.

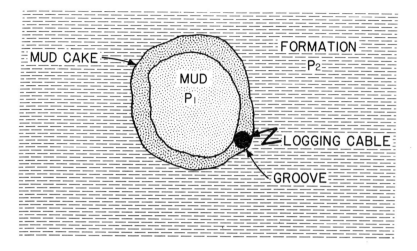

Figure 4-12   This is a view looking down a vertical borehole which shows that if logging cable is left immoble in an uncased borehole too long, it may be permanently lost due to differential pressure sticking. The lateral force pressing the cable to the formation is $(P_1-P_2)A$, where A is the contact area between the cable and the formation. The contact area, A, tends to increase with time if the logging cable is not moved to break the contact.

When the cable cuts through the mud cake, it is exposed to the reduced formation pressure, P_2, and the groove it cuts increases the contact area between the cable and the formation.

Mud filter cake is capable of withstanding large pressure differentials, especially if the force created by the pressure difference is in the direction from the wellbore to the formation. For example, mud filter cake has been known to withstand up to 2000 psi (13790 kPa) without being discharged from a perforation hole (Allen and Atterbury, 1954). Consequently, breaking through mud cake can sometimes expose logging cable to a surprisingly large pressure difference. To illustrate the large force that can be exerted on a cable, assume that a 3/8 inch (0.95 cm) diameter cable has 25 percent of its surface area in contact with a formation over an interval of 1000 feet (304.8 m). If the difference between the mud column pressure and the formation pore pressure is 100 psi (6689 kPa), then the radial force pressing the cable to the formation is approximately 350,000 pounds (1.56×10^6 N).

A radial profile of fluid pressure in a mud-filled, uncased borehole is shown in Figure 4-13. The pressure decreases linearly across the mud cake layer from a high value P_1 in

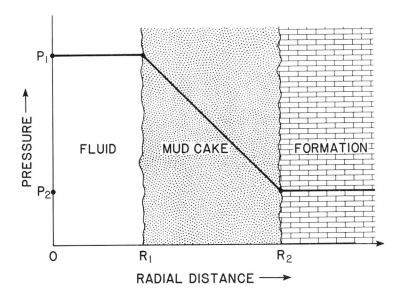

Figure 4-13   Fluid pressure decreases linearly across a mud cake layer as shown by the heavy line. When mud circulation is stopped during VSP experiments, mud cake thickness continually increases. Stationary equipment is susceptible to differential sticking because mud cake builds up around it, and thereby introduces it into the pressure imbalance regime between R_1 and R_2.

the borehole to a lower value P_2 in the formation (Outmans, 1958). Thus, if some downhole device penetrates a distance, X, into the mud cake, it is subjected to a radially directed pressure imbalance of

$$\Delta P_x = X \ \frac{P_1 - P_2}{R_2 - R_1}. \tag{5}$$

Downhole equipment, therefore, does not have to make actual physical contact with a formation in order to be subjected to potentially serious sticking problems; it simply has to break into the protective mud cake. Differential pressure sticking is most likely to occur when a logging cable is stationary. During a VSP experiment, a logging cable is stationary while data are being recorded; thus, the potential for differential pressure sticking of the cable occurs often. It is important to note that the likelihood of differential pressure sticking increases the longer that a cable remains stationary because the cable continues to conform to the contours of the borehole and because mud cake continually builds up around it. Laboratory studies show that the force needed to pull any equipment or drill pipe out of a mud-filled borehole depends strongly on how long the equipment is allowed to remain stationary in the hole (Annis and Monaghan, 1962; Helmick and Longley, 1957). The behavior discovered in their tests is illustrated in Figure 4-14.

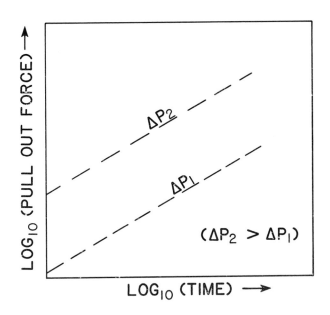

Figure 4-14 The force needed to overcome differential pressure sticking increases with time because of mud cake buildup. It also increases as the difference in borehole and formation pressure, ΔP, increases. Specific time and force values are given by Annis and Monaghan, 1962, and by Helmick and Longley, 1957.

Note also that the pullout force depends on the magnitude of the pressure difference between the borehole and formation fluids. These studies conclude that the fundamental reason causing pullout force to increase with time is that mud cake builds up around the downhole equipment, and this increased mud thickness, X, introduces the pressure imbalance specified by Equation 5. Because of the danger of differential pressure sticking, some well logging service companies refuse to let a logging cable remain stationary in an uncased hole for more than 15 minutes. This precaution is a sound field practice and should be followed in VSP work when possible.

Depth Control

The cable supporting the downhole geophone device used in a VSP survey is moved up and down many times during data collection. The data gathering procedure usually consists of the sequence:

1. Raise the geophone to depth level N and clamp it to the borehole wall.
2. Slack the cable and record data.
3. Take up the cable slack.
4. Unclamp the geophone and raise it to depth level (N+1).
5. Clamp the tool to the borehole wall and slack the cable for the next data recording.

This procedure is repeated over and over, and the cable is moved up and down in increments and decrements of a few meters many times during the course of a VSP survey. If the wireline depth measuring system consists only of a mechanical odometer, then considerable depth error can build up by this repetitive up and down cable motion. Errors of 8 to 12 meters are often observed in surveys covering 2500 or 3000 meters of borehole. Errors of this magnitude should be avoided since they make it difficult to correlate VSP data with other borehole information such as core analyses, well logs, and cutting samples. Even larger depth errors can occur in situations where extremely adverse field conditions exist. Inclement weather and operator fatigue are two factors that can combine to create severe depth measurement inaccuracies. For example, in one wintertime VSP, icing conditions froze the measuring wheel in the cable odometer after data were recorded at a few depth levels in the bottom portion of a deep well. As the experiment progressed to shallower depth levels, several meters of cable motion occurred before the fatigued operator was aware that the motion was not being properly measured by the sluggish odometer wheel. Obviously, some type of a backup procedure for confirming geophone depths, other than complete reliance on a mechanical odometer, is advisable.

The repetitive up and down cable movement that occurs during vertical seismic profiling is quite different from the smooth, continuous upward cable movement used when recording sonic, SP, resistivity, and other standard well log data. In order to insure that the wireline depths measured for VSP data recordings are identical to the depths measured for well log data, it is advisable to use the same wireline system for the VSP work which was used when recording well log data. Most logging cables have magnetic markers at specified depth intervals which serve to check the accuracy of the odometer measurements. However, while recording well log data, a physical mark such as a wrap of strong tape or a painted stripe should also be made on the logging cable every 300 or 400 meters, and the VSP tool run downhole immediately after the logging runs on this same cable. When the wireline odometer indicates that the geophone, after numerous up and down movements, is at a depth equal to one of the premarked depths established during the logging runs, then an operator should visually confirm that the physical mark made on the wireline aligns with a preselected fixed reference mark located on the wellhead or the derrick. (See Figure 4-15.) If the marks do not align, the wireline operator should move the geophone up or down so that the marks are at the same elevation. Readjusting VSP depth registration in this manner every 300 to 400 meters insures optimum depth correlation between the VSP data and log data. This technique usually works satisfactorily in a vertical cased borehole, but physical cable marks are easily destroyed or altered when data are recorded in uncased holes or in highly deviated cased wells because of the friction created by the cable cutting grooves in an uncased formation or rubbing against the non-vertical walls of a deviated hole.

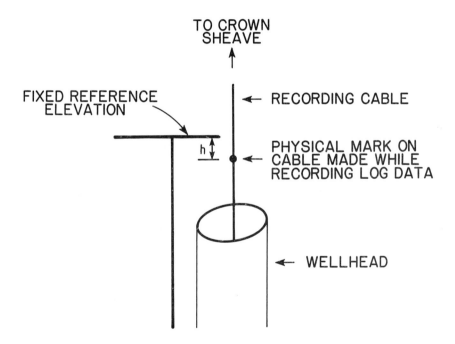

Figure 4-15   Depth calibration of VSP recording levels. If a VSP geophone is at the same depth where a logging tool was when a physical mark was made on the cable at the wellhead, then this cable mark should return to the same elevation as the selected reference point. In this example, the wireline operator needs to move the geophone up a distance h in order for the VSP depths to be in register with logging depths.

A common VSP recording situation is that one wireline system is used to record well log data, and a few days after these logging runs, VSP data are recorded with a different wireline system. In such a case, it is advisable to run a gamma-ray tool in the well immediately before the VSP survey using the same wireline system that will support the VSP geophone tool. Thirty or forty meters of gamma-ray log should be recorded at intervals of 300 or 400 meters starting at TD and extending to the shallowest depth where VSP data will be recorded. This gamma-ray curve can then be depth-correlated with the original gamma-ray log made in the well, and a physical mark made on the wireline at each of these 300 or 400 meter intervals to establish depth registration between the original wireline system used to record log data and the wireline used for the VSP survey. The wireline odometer reading can then be readjusted during the VSP survey to stay in register with these marks in the same manner as described above.

Combating Tube Waves

Tube waves are probably the strongest coherent noise encountered in vertical seismic profiling, so it is important to know which field procedures eliminate or reduce this type of borehole wave. The physics of tube wave behavior is discussed in detail in Chapter 3. Only a brief background will be given here in order to establish the principles of some field procedures that are effective in combating these fluid-borne waves.

Tube waves are created when any part of the fluid column in a wellbore is disturbed so that the fluid particles oscillate at frequencies within the seismic frequency band. The term "fluid" in this context means liquid or mud; air-filled boreholes cannot support tube waves. Tube waves are guided waves that propagate along the axis of a borehole and do not radiate outward away from the wellbore as body waves. Because tube wave energy propagates in only one direction; i.e., along the wellbore axis, tube wave amplitudes remain robust over long travel distances. Essentially all of the particle displacements created by tube waves occur in the mud column and in a small annulus of earth around the borehole. Because tube waves are confined to a small region around a borehole, they provide a smaller amount of information to explorationists than do compressional and shear body waves, which sample stratigraphic and structural conditions far away from a borehole and carry that information to a borehole geophone by means of transmitted and reflected signals.

In order to establish field procedures that combat tube waves, the mechanisms that create these waves must be understood. These mechanisms are illustrated and discussed in Chapter 3. The key points established there which pertain to preventative tube wave field procedures are:

1. Tube waves can be created by compressional body waves when they interact with a strong impedance contrast within a borehole, such as a casing point or a dramatic change in borehole diameter.

2. The dominant source of tube waves is the surface Rayleigh wave that moves across a wellhead and vertically vibrates the top of the mud column in a fluid-filled borehole.

There is really no defensive field procedure that can reliably prevent propagating body waves from creating tube waves at large impedance contrasts in boreholes. The casing and cementing techniques used to complete oil and gas wells will always result in significant borehole impedance changes at various depths. Consequently, any compressional body wave intersecting these wellbores can, under proper circumstances, generate a tube wave at one or more of these impedance anomalies.

Since the major cause of tube waves is the ground roll that sweeps across a wellhead and creates a vertical motion in the mud column in a VSP well, then defensive field procedures should focus on either reducing ground roll at the wellhead, or on preventing the transfer of energy from the ground roll to the mud column. Several procedures have been used with some success to reduce the amount of ground roll energy reaching a wellhead. One of these techniques is simply to increase the distance between the energy source and the wellhead. Ground roll amplitude decreases with propagation distance; thus, increasing the source-to-wellhead distance means that ground roll amplitudes will be diminished when a surface wave reaches a VSP study well. An example of the effect of source-to-wellhead distance on tube wave amplitude is shown in Figure 4-16. These data were recorded at two geophone depths. In each case, two compressional Vibroseis® sources were offset at distances of 200, 400, and 685 feet from the wellhead. The vibrators were positioned side-by-side except at an offset distance of 400 feet. Here they were also placed in-line with the wellhead and separated a distance of $\lambda/2$, where λ was the measured dominant wavelength of ground roll at this well site. The VSP response measured with this in-line source geometry is labeled "400 FT. SOURCE ARRAY OFFSET". The data are plotted so that the compressional first arrival amplitudes are identical; thus, tube wave amplitudes for each source offset distance can be directly com-

---

®Registered Trademark of Conoco, Inc.

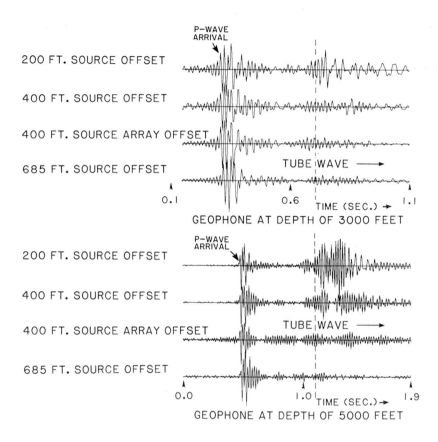

200 FT. SOURCE OFFSET

400 FT. SOURCE OFFSET

400 FT. SOURCE ARRAY OFFSET

685 FT. SOURCE OFFSET

P-WAVE ARRIVAL

TUBE WAVE ⟶

0.1 0.6 TIME (SEC.) ➔ 1.1

GEOPHONE AT DEPTH OF 3000 FEET

200 FT. SOURCE OFFSET

400 FT. SOURCE OFFSET

400 FT. SOURCE ARRAY OFFSET

685 FT. SOURCE OFFSET

P-WAVE ARRIVAL

TUBE WAVE ⟶

0.0 1.0 TIME (SEC.) ➔ 1.9

GEOPHONE AT DEPTH OF 5000 FEET

Figure 4-16 Tube waves created by Rayleigh waves propagating across a wellhead and vertically shaking the top of the fluid column in a wellbore can be reduced by moving the source farther from the wellhead. The approximate onset of the tube wave event in this well is marked by the vertical dashed line. The energy source in this test was two compressional vibrators. Rather severe correlation noise exists before the compressional arrivals at a depth of 3000 feet. (An example of correlation noise is shown in Figure 3-2.) These data were recorded with a vertically oriented geophone.

pared. The experiment illustrates that increasing source offset distance so that ground roll amplitude is reduced at a wellhead will also reduce tube wave amplitudes recorded in the borehole. The simple in-line source array is not a great improvement in canceling ground roll in this particular instance.

The data in Figure 4-17 further document the reduction in tube wave amplitude that results when a VSP energy source is moved farther away from the wellhead of a VSP study well. In this experiment, VSP data were recorded by positioning a single compressional vibrator at distances of 200, 500, and 1500 feet from a vertical, fluid-filled borehole. The resulting vertical seismic profiles are shown in Figure 4-17 after each trace is delayed by its first break time, and after all traces are velocity filtered to attenuate downgoing events. When the vibrator is only 200 feet from the wellhead, the downgoing tube wave is so strong that it cannot be completely removed by the velocity filter. One remnant of the downgoing tube wave is the event in panel C extending from 2.1 seconds at a depth of 4000 feet to 3.0 seconds at a depth of 7400 feet. Because the tube wave is reduced in amplitude as the source moves farther away from the well, the velocity filter successfully cancels the downgoing tube wave mode in the top two data panels. The upgoing tube wave mode is preserved by the velocity filtering. This upgoing mode is a high amplitude event when the source is 200 feet from the well, a weak amplitude event starting at about 3.4 seconds at a depth of 7400 feet when the source is 500 feet away, and essentially non-existent when the source is 1500 feet from the wellhead.

A second field technique that has been used to reduce the ground roll arriving at a wellhead involves placing a physical barrier between the source and the wellhead which will block surface waves traveling toward the well. One possible barrier is a ditch positioned a distance $\lambda/2$ from the source, where λ is the dominant spatial wavelength contained in the Rayleigh wave. The concept is shown in Figure 4-18. The ditch can usually be as narrow as one or two feet, and as shallow as five or six feet, and still block a reasonable amount of the ground roll. The length of the ditch must be long enough so that the Rayleigh wavefront will not heal on the backside of the ditch (relative to the source) and continue on toward the wellhead. In some instances, topographic features such as ravines, creek beds, farm ponds, hills, or natural depressions can be used as a ground roll barrier so that a ditch does not have to be constructed.

A third way to reduce ground roll is to create a source array which attenuates the dominant wavelength of the ground roll. Several source elements are needed in order to construct an effective array. Commonly, only one vibrator, one weight dropper, or one air gun is used in vertical seismic profiling. Occasionally, two source elements may be used. Rarely are three or more sources used. It is impossible to construct a source array with one source element, and only a modestly effective array can be made with two sources. A comparison of the response created by a two-element source array and that measured for a single point source is shown in Figure 4-16. If ground roll is anticipated in a VSP experiment, as many sources as possible should be on-site in order to create an optimum array for canceling it out.

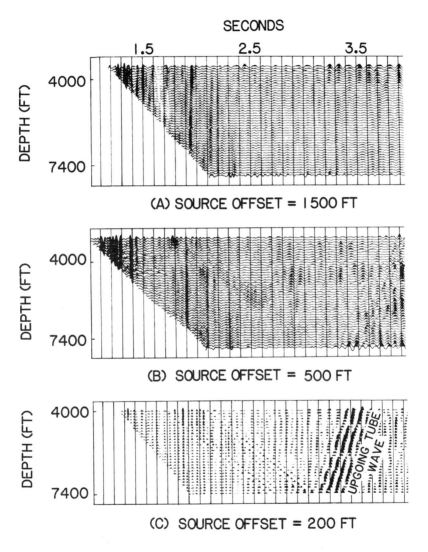

Figure 4-17   Effect of source offset distance on tube wave generation. These data were recorded with a vertically oriented geophone. The source was a compressional vibrator. (After Poster, 1982)

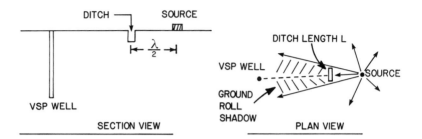

Figure 4-18   Some ground roll reaching a VSP well can be blocked by a narrow ditch 5 or 6 feet deep located between the source and the well at a distance λ/2 from the source, where λ is the dominant wavelength in the ground roll. The length, L, of the ditch should be long enough so that the ground roll wavefront does not heal itself before reaching the well.

One effective way to reduce tube waves is to concentrate on preventing Rayleigh wave energy from being transferred to the mud column in a wellbore rather than trying to attenuate the Rayleigh waves before they reach the mud column. Rayleigh waves are interfacial waves which propagate along the earth-air boundary in all directions away from a surface source. The amplitude of a Rayleigh wave decreases exponentially in a direction normal to the earth-air interface in a manner that can be expressed mathematically as

$$e^{-kz} = \exp(-2\pi z/\lambda) \tag{6}$$

where λ is the wavelength of the Rayleigh wave, and z is depth below the earth's surface. At a depth of one-half of a Rayleigh wavelength, the amplitude of the particle displacement created by the Rayleigh ground roll wave is decreased by an amount

$$e^{-\pi} = 0.043 \tag{7}$$

relative to the amplitude at the surface. If the top of the mud column is at least a distance λ/2 below the surface, then very little energy can transfer from the Rayleigh wave to the mud column. This concept is illustrated in Figure 4-19. Thus, lowering the top of the mud column in a well can effectively prevent tube waves from being generated in the wellbore even if strong ground roll waves arrive at the wellhead. Sometimes safe drilling procedures will not allow any decrease in mud column height if part of the

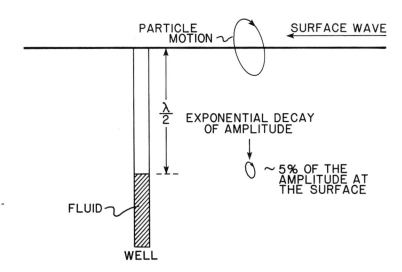

Figure 4-19   The transfer of energy from a surface Rayleigh wave to the fluid column in
a well can be reduced by lowering the fluid level since the amplitude of the
particle displacement created by the Rayleigh wave decays exponentially
with depth. At a depth equal to half the dominant vertical wavelength of
the Rayleigh wave, the particle amplitude is less than 5 percent of the
amplitude at the surface.

borehole is uncased. In cased holes, lowering the top of the liquid column usually presents
no safety problem. In fact, in some studies conducted in shallow cased holes, researchers
have removed all fluid from a well in order to prevent tube waves. A recommended field
procedure to combat tube waves is to lower the liquid column in a VSP well as much as
safety and practicality will allow.

VSP Data Recording in Deviated Holes

Many oil and gas wells are deliberately deviated away from a true vertical line.
Deviated wells are particularly common in offshore drilling, so the field procedures
described here concentrate on marine vertical seismic profiling. The methodology can be
transferred to onshore VSP work with little difficulty.

128

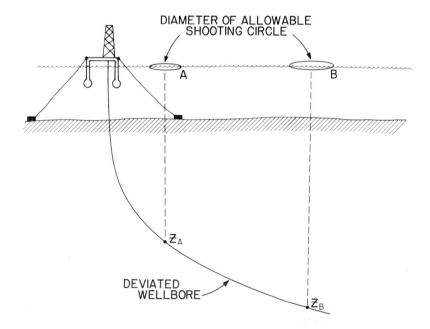

DIAMETER OF ALLOWABLE
SHOOTING CIRCLE

A   B

Z_A

DEVIATED
WELLBORE

Z_B

Figure 4-20   When recording VSP data in a deviated borehole, the energy source should be
positioned vertically above the geophone. In practice, the energy source is
allowed to be anywhere within a circle centered directly above the geophone
and having a diameter that is 2 or 3 percent of the vertical depth to the
geophone. These shooting circles thus decrease in size as the geophone
proceeds up the well toward the surface.

The interpretation of VSP data is simplified when the travel path from source to
geophone is as near to vertical as possible. This requirement can usually be achieved with
a fixed surface source position if the borehole is vertical. If the borehole is deviated,
then the source must be moved continuously throughout the data collection in order to
maintain vertical travel paths.

The concept of a "shooting circle" is used to describe the position that a surface
energy source must assume while recording VSP data in a deviated hole. A cross-sectional
view of a typical offshore deviated well is shown in Figure 4-20. Two shooting circles are
indicated centered directly above recording depths Z_A and Z_B. When the VSP geophone is
at depth Z_B, the position of the energy source should be confined to the interior of circle
B. In order to insure a vertical travel path, the diameter of a shooting circle should be no
larger than 2 or 3 percent of the geophone depth. This restriction requires precise
navigational control between the drilling platform and the shooting boat, as well as
skillful boat handling. Note that the shooting circle decreases in size as the geophone
nears the surface.

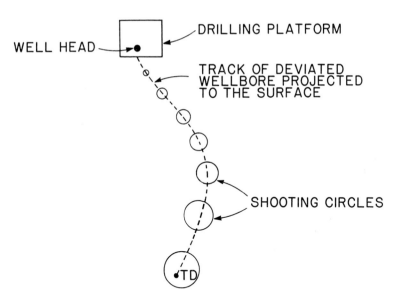

Figure 4-21   This plan view of an offshore deviated hole VSP shows how the allowable shooting circles must follow the surface track of a deviated wellbore and become smaller as the VSP geophone nears the surface. Precise navigation control between the drilling platform and the shooting boat is essential in order to keep shotpoints within the circles.

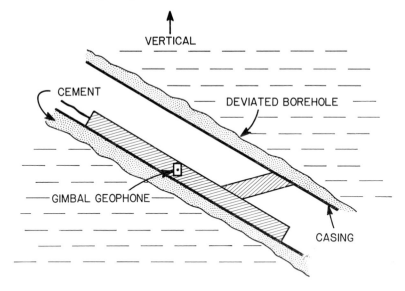

Figure 4-22   Gimbal mounted geophone elements are commonly used when recording VSP data in a deviated borehole. This mounting concept allows the geophone to orient itself vertically regardless of the repose angle of the geophone package. Note that highly deviated wells are almost always cased before any logging tool is allowed in the borehole.

A plan view of this same marine deviated well is shown in Figure 4-21. Good borehole directional survey data must be available in order to calculate the surface X-Y coordinates of the center of the shooting circle corresponding to each recording depth. Only a few shooting circles are indicated in Figure 4-21. In actual practice, shooting circles corresponding to successive recording depths will greatly overlap.

Well logs are rarely recorded in a highly deviated uncased borehole because there is a possibility that a logging tool will become wedged against small ledges or other borehole irregularities as it is being lowered downhole. Consequently, highly deviated wells are almost always cased before any logging instrument enters the borehole. The presence of casing restricts the type of log data that can be recorded since most sonic and electrical logging tools cannot function correctly inside casing. This fact enhances the value of any borehole measurement, such as vertical seismic profiling, that can record useful stratigraphic and formation evaluation data inside cased holes. Downhole VSP geophone packages used in deviated holes commonly employ gimbal mounted geophones as shown in Figure 4-22. These geophones orient themselves vertically regardless of the repose angle of the downhole tool.

Testing VSP Geophone-to-Formation Coupling

One of the more valuable VSP field procedures that can be used is one which determines the quality of the geophone-to-formation coupling in situ at each depth level before VSP data are recorded. Wuenschel (1976) describes a three-component downhole geophone system which uses two internal electrically driven oscillating masses to vibrate a geophone tool either vertically or horizontally while it is in a locked position in a wellbore. By vibrating these internal inertial elements and observing the resonance peaks of each geophone output, one can determine how the geophone-to-formation coupling varies from recording level to recording level. Estimates of the frequency changes and phase shifts created by variations in geophone-to-formation contacts are essential in order to numerically construct precise VSP wavelet character during data processing. No in situ coupling measurement capability presently exists in geophone tools provided by VSP data recording contractors, but such capability is being considered. Until geophone systems are available that allow one to measure geophone-to-formation coupling in situ, an attractive alternative is to conduct rigorous laboratory measurements to determine what mechanical features and which operating principles of a VSP geophone device affect the coupling of a geophone to a formation. This knowledge can then be used to establish appropriate field procedures which minimize variations in geophone-to-formation coupling.

The mechanical and electrical characteristics of a downhole VSP geophone package are fundamental properties that determine the quality of the data that are input to a recording cable. Some mechanical designs of tools allow good quality three-component data to be recorded; whereas, other designs result in usable data being recorded only by vertically oriented geophones. Most VSP geophone designers think that a lateral force of two to five times the tool's weight is needed in order to properly couple non-vertical geophones to a borehole wall. However, only sparse field data exist that document how much geophone locking force should be used. The effect of lateral locking force is only one example of a mechanical tool characteristic that is an important factor affecting VSP data quality. The effects of numerous other tool characteristics on data quality should also be investigated. For example,

1. Where should geophone elements be positioned in a downhole geophone package?

2. Which geophone geometry is better, a Cartesian XYZ arrangement or a triaxial 54 degree type?

3. What is the optimum tool weight?

4. Is there a preferred length-to-weight ratio for a geophone tool?

5. What type of tool locking mechanism should be used?

All of these questions, plus others that could be added, are various ways of asking, "What mechanical and physical properties of a VSP geophone tool optimize geophone-to-formation coupling?" A convenient and definitive way to answer these questions for a specific VSP geophone tool is to conduct geophone impulse tests with the tool locked in a shallow test hole. These test holes should represent realistic downhole borehole conditions as much as possible. Tests should be made in uncased holes, in holes that are cased but uncemented, and in properly cemented casing as shown in Figure 4-23. The holes should be prepared in a wide variety of lithological outcrops, if possible, and be deep enough to accept at least one joint of casing.

Impulsive testing of standard land geophones has been documented by Washburn and Wiley (1941), and Hoover and O'Brien (1980). These tests were executed by dropping a small weight on the top of a geophone case when the geophone was planted in a representative soil type, and then recording the geophone response to this impulse. The quality of the geophone-to-earth coupling can be determined by calculating the

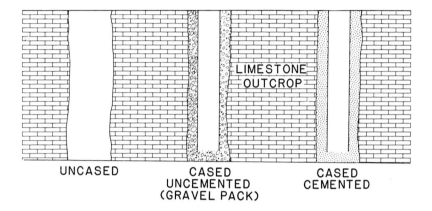

Figure 4-23   Shallow holes for impulse testing of VSP geophone tools.

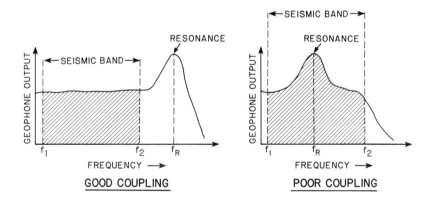

Figure 4-24   Determining a geophone's coupling to the earth by measuring its frequency response in a "planted" condition to a mechanical impulse applied to the geophone case.

frequency spectrum of the geophone output signal. Idealized frequency spectra of two hypothetical impulse tests are shown in Figure 4-24. A geophone-to-earth coupling will normally create some type of a spectral resonance peak. The quality of the geophone coupling is determined by where this resonance frequency, f_R, occurs relative to the desired seismic bandwidth, which is defined as extending from f_1 to f_2. If the resonance occurs outside the seismic band, then all frequencies between f_1 and f_2 couple from the earth to the geophone case with minimal distortion. If the resonance occurs within the seismic band, then frequency components near f_R will be excessively amplified and have distorted phases. This latter type of geophone output is not representative of the seismic signal that traveled through the earth to the geophone. These testing procedures have helped investigators select the optimum weight, spike length, contact area, and other design characteristics of velocity-sensitive land geophones used to record surface reflection data.

This same testing concept can be extended to VSP geophone systems. Mack (1966) documents one study of the coupling behavior of a borehole geophone in a cased well. Any testing procedure of a VSP geophone is complicated by the fact that the device is large and bulky and is rather inaccessible when it is "planted" in a borehole. One possible impulse testing procedure for a three-component VSP geophone is shown in Figure 4-25. This particular concept, involving small weights falling down tubes to the geophone assembly, may not be practical, but does illustrate the concept. More feasible impulsing schemes could use water-tight, electrically activated solenoid plungers attached to the outside of the geophone case which lightly ping the tool at selected angles of impact, or could use internal oscillating masses as in Wuenschel's (1976) design.

Idealized test results are sketched in Figure 4-26. Each spectrum repesents the signal output of a VSP geophone as only one physical parameter of the tool is altered. For example, spectrum #1 could be the geophone response of a horizontal geophone when the lateral locking force is F, spectrum #2 the response of this same geophone when the force is increased to 2F, and spectrum #3 the response when the locking force is 3F. In this hypothetical example, the locking force should be at least 3F in order to move the resonance peak outside the seismic band. Tests of this nature can help answer questions about proper VSP geophone design and specify preferred field operating procedures for particular types of VSP geophones. Some actual test measurements which illustrate the effects of tool mass and locking force on VSP data quality are shown in Chapter 9, Figures 9-6, 9-7, and 9-8.

134

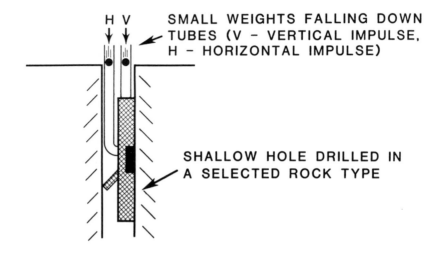

H V   SMALL WEIGHTS FALLING DOWN
TUBES (V - VERTICAL IMPULSE,
H - HORIZONTAL IMPULSE)

SHALLOW HOLE DRILLED IN
A SELECTED ROCK TYPE

Figure 4-25   Impulse testing of a planted VSP geophone to determine the quality of the geophone-to-formation coupling.

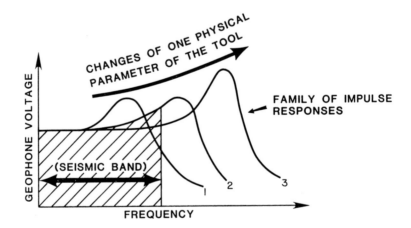

CHANGES OF ONE PHYSICAL PARAMETER OF THE TOOL

FAMILY OF IMPULSE
RESPONSES

GEOPHONE VOLTAGE

(SEISMIC BAND)

FREQUENCY

Figure 4-26   An example showing how the spectra of geophone impulse tests can determine how a specific mechanical or physical property of a VSP geophone affects the geophone-to-formation coupling.

Recording Near-Field Wavelets

   A near-field geophone serves as a source monitor to document the consistency of the seismic shot wavelets that are created during the course of a VSP survey. It is essential to know if a wavelet change recorded by a borehole geophone is a manifestation of an in situ rock or fluid property, an effect of local stratigraphy and structure near the recording depth, a variation in geophone-to-formation coupling, or simply due to a different shot wavelet being created by the source. A properly positioned, near-field geophone allows the last of these possibilities to be confirmed or rejected as the reason for a wavelet variation being recorded downhole. In addition, recording near-field source wavelets allows numerical wavelet shaping operators to be designed and applied to all data recorded downhole so that numerically identical source input wavelets can be synthesized at each recording depth. This concept of wavelet processing of VSP data will be discussed in Chapter 5.

   Several of the figures in Chapter 2 which illustrate VSP energy sources show a hydrophone or geophone placed near the energy source, and the term "near field" is used to describe these transducers. The near field of a seismic energy source can be defined in several ways, and geophysicists have varying opinions as to what is meant by the term. One possible definition of a near field is, "that region outside of an airgun bubble or plastic deformed zone of the earth in which the measured elastic seismic waveshape changes as a function of distance from the source." Although a seismic wavelet always diminishes in amplitude as it propagates away from its point of origin, there will be some travel distance beyond which the shape of a wavelet is invariant, save for non-source related effects such as short period multiples which are generated throughout a stratigraphic section. The region in which this invariant waveshape exists is, by definition, the "far field" of that particular seismic source.

   A near field may extend no farther than 1 or 2 meters beyond the boundaries of a single airgun bubble or a single plastically deformed zone. However, if an array of several sources is used, or if ghost events are created near a source, then the near field may extend for some distance. Obviously the concept of a near field is relatively easy to define in words, but can be rather difficult to determine in a real earth experiment.

   In this text a near field will be arbitrarily defined as that region surrounding a VSP energy source which is less than a distance, λ, from the source, where λ is the dominant wavelength of the source wavelet in the earth volume immediately around the source position. A geophone placed inside this region is defined as a near-field geophone, as illustrated in Figure 4-27.

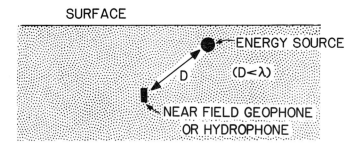

Figure 4-27 A geophone will be arbitrarily defined as being in the near field of a VSP energy source if it is located a distance, D, from the source such that D is less than the dominant wavelength of the source wavelet.

A major emphasis of any field procedure whose objective is the capture of near-field source wavelets should be to create a fixed, invariant environment encompassing both the near-field transducer and the VSP energy source. This environment is best achieved by (a) burying the transducer rather than placing it on the surface, (b) maintaining a constant distance between the source and the transducer, (c) maintaining a constant distance between the earth (or water) surface and the transducer, and (d) insuring that a good coupling exists between the transducer and the earth. The difficulty in achieving these conditions depends on whether a VSP well is onshore or offshore and whether or not one or several source positions are employed while recording data.

For an onshore VSP experiment, a near-field geophone should be buried 10 to 50 meters deep in a hole placed as close as possible to the source position. At least two geophones should be buried, and each geophone response recorded on a separate data channel. It is essential that land geophones be correctly oriented in their buried positions. For instance, vertical geophones must remain vertical after being buried. Usually proper geophone orientation is assured by hanging a heavy weight below the geophone, as shown in Figure 4-28, and keeping the supporting cable taut while carefully filling the hole. The backfill must be some type of material that provides a good acoustic coupling between the geophone and the earth. Cement is probably the best backfill, but fine sand can be used. Horizontally oriented geophones can be cemented inside a short plastic pipe, and this geophone package can then be lowered with a weight and backfilled as shown in Figure 4-28. If several source positions are used in a VSP experiment, then a near-field geophone must be buried at each source location. Obviously, considerable planning and site preparation must precede an onshore VSP experiment if near-field wavelets are to be successfully recorded.

TO SURFACE

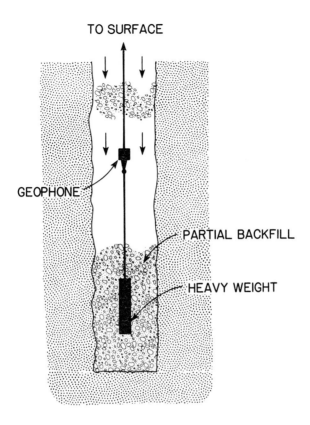

GEOPHONE

PARTIAL BACKFILL

HEAVY WEIGHT

Figure 4-28   A near-field land geophone must be planted below the surface so that it is properly oriented. A vertical geophone is shown being backfilled to create a coupling between it and the formation.

In marine VSP work when an airgun is suspended from a drilling platform, near-field wavelets can be recorded by one or more suspended hydrophones such as shown in Figure 2-13. However, if variable source offset distances are needed and a shooting boat is used, then the situation becomes slightly more complicated. Usually a near-field hydrophone can be suspended from a shooting boat in some manner, but if the boat has to maneuver forward and then in reverse in order to stay within its assigned shooting circle (Figures 4-20 and 4-21), it is likely that a constant source-to-hydrophone distance cannot be maintained due to the induced to-and-fro swinging motion of the airguns and the hydrophone. Consequently, variable near-field wavelets are recorded because different surface ghost times are created by the change in hydrophone and airgun positions relative to the water surface and to each other. It is difficult to separate this effect from a true waveform change in the airgun pulse. A rigid suspension system that keeps the airguns and hydrophones fixed relative to each other is desirable in this type of shooting.

Instrument Tests

Detailed numerical analyses of the phase, frequency, and amplitude characteristics of VSP data are required in order to determine how rock and pore fluid properties affect propagating seismic wavelets. Thus, any frequency, phase, or amplitude changes of recorded data that are not created by rock or pore fluid conditions in the earth must be recognized, and their effects not falsely assumed to be geological information. For example, the electrical properties of the logging cable that transmits VSP data to the surface can alter the shape of wavelets that were fed into the cable by the downhole geophone. Likewise, the amplifiers, filters, analog-to-digital converter and other components of the surface recorder can alter signal waveshapes. For these reasons, the electrical characteristics of the complete VSP recording system, including the logging cable, should be measured and documented each time VSP data are recorded. Ideally, these tests should be made immediately before and after a VSP survey. As a minimum, the tests should be performed once before leaving a VSP well site.

A simplified illustration of the testing procedure is shown in Figure 4-29, and brief descriptions of critical measurements are given in Table 4-1. Basically, the tests consist of feeding accurately calibrated square wave or sine wave signals into the logging cable at its downhole end (connection A in Figure 4-29), or into the input terminal of the surface recording system (connection B in Figure 4-29). As many system responses as possible should be documented in the field by camera playouts or oscilloscope displays. All system responses should be recorded on digital field tapes so that a permanent record of the equipment behavior exists, and so that precise numerical analyses of the test results can be performed in a data processing center.

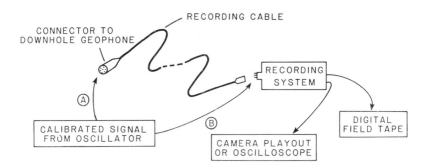

Figure 4-29   A simplified representation of instrument tests that should be performed in conjunction with a VSP survey. Specific measurements that should be made are described in Table 4-1.

TABLE 4-1

SUMMARY OF TYPICAL INSTRUMENT TESTS THAT SHOULD BE MADE
IN CONJUNCTION WITH A VERTICAL SEISMIC PROFILE

TEST	INPUT	OUTPUT	INTERPRETIVE ACTION
Gain Accuracy	Connect a 25 Hz sine wave signal to each amplifier input and make a series of records with the amplifier in fixed gain setting.	Record on magnetic tape one high amplitude output for each fixed gain setting.	Field analysis is limited to comparing deflection of all galvos at a high galvo gain to assure relative gain levels. A computer analysis should compare the numerical gain values between channels before and after the survey.
Dynamic Range (DRD)	Connect a 25 Hz sine wave signal to each amplifier input at a level 3 db below converter full scale.	Make one fixed gain record at 3 db below converter full scale. Decrease oscillator signal in 6 db steps and make fixed gain recordings covering the full range of the converter. Last record should have zero signal input. Playback all recordings with appropriate fixed gain settings.	Field analysis is: $DRD = 3+G + 20 \log (S/N)$ where DRD = Converter dynamic range in DB. G = Difference in system gain when playing back the first record with maximum converter signal and the last record with zero converter signal. S = Peak-to-peak galvo deflection of the maximum converter signal. N = Peak-to-peak galvo deflection of the zero converter signal.
Cable Attenuation	Single DC pulse of 1/2 sample period duration input to pin (N) of downhole cable connector. Repeat for all pins.	Record on magnetic tape.	No field analysis is made. Calculate Fourier transform of output to determine phase and frequency changes introduced into data by transmission through cable.

TABLE 4-1 - Continued

TEST	INPUT	OUTPUT	INTERPRETIVE ACTION
Cable Impedance	None	None	Use ohm meter to determine that impedance between any two cable pins is at least 1 megohm.
Equivalent Input Noise (EIN)	(A) Terminate amplifier inputs with resistance value near that of the amplifier source impedance. No oscillator input. (B) 25 Hz sine wave having one microvolt RMS level.	(A) At maximum fixed gain make a paper record and also save data on magnetic tape. (B) Make a second paper record and save data on magnetic tape at the same maximum fixed gain.	Field analysis can approximate system input noise by: $EIN = (Vc)(Vu/Vo)$ where EIN = Input noise in volts RMS. Vo = Peak-to-peak measurement of oscillator signal on paper copy (step B). Vu = Peak-to-peak measurement of noise on paper copy (step A). Vc = Oscillator voltage (RMS).
Crossfeed	Input maximum amplitude oscillator signal allowed by system to channel N. Terminate channel M with resistor equivalent to input impedance.	Record #1 = Adjust gain until output of channel N is maximum but undistorted. Write output on paper playout and magnetic tape. Record #2 - Increase gain on channel N until deflection is observed on channel M. Repeat until all channels are checked.	Crossfeed isolation is: $C = G + 20 \log (Vo/Vu)$ where C = Crossfeed of tested channel in DB (Desire at least 60 db isolation). G = Change in gain of channel N between record #1 and record #2. Vu = Peak-to-peak measurement of the test channel (M) on record #2. Vo = Peak-to-peak measurement of the oscillator signal on record #1.

TABLE 4-1 - Continued

TEST	INPUT	OUTPUT	INTERPRETIVE ACTION
Filter Response	Input a DC pulse of 1/2 sample period duration at cable connection of surface recorder.	System response to impulse written on magnetic tape. Make paper record also.	Field analysis is limited to phase agreement of waveforms between channels. A frequency spectral analysis of each channel can be made in a computing shop to check response characteristics.
Geophone Tap Test	See first section of Chapter 4.		
Harmonic Distortion	Input a low distortion sine wave signal to each recording channel and make a broadband recording using the maximum allowable oscillator signal value for the selected pre-amp gain.	Make several recordings with oscillator frequency changed from a low value to a high value.	Numerical frequency analysis should measure the value of all harmonics of the fundamental frequency. Total harmonic distortion in percent is: $$THD = 100 \ (V_{h1} + V_{h2} + V_{h3} + \ldots) \ /V_f$$ where V_f is the height of the spectral peak of the fundamental frequency and V_{hn} is the height of the spectral peak of harmonic n.
Tape Speed	None	None	Measure tape transport speed with a tachometer and compare to manufacturers specifications.

TABLE 4-1 – Continued

TEST	INPUT	OUTPUT	INTERPRETIVE ACTION
Source Synchronization			
Dynamite	Dummy shot using a cap.	Make paper record of dummy shot	Compare the time break generated from the dummy shot with the start of data.
*Vibrator	Baseplate accelerometer signal	Make autocorrelation of the reference sweep.	Determine that the peak of the correlated pulse agrees with the start of data.
*Airgun	Test pops	Make paper record of test pops of a single airgun.	Compare the recorded time break from the test pops with the start of data.
Vibrator Summing	Sum 8 sweeps from baseplate accelerometer. Invert input and sum 8 more sweeps.	Make playout of summed sweeps.	Output should sum to zero.

*Multiple sources should have phase agreement within 15 degrees.
For vibrators – Correlate baseplate accelerometer outputs with the reference sweep in order to inspect phase alignments.

Geophone Calibration

1. Verify that the geophone damping is 70 percent of critical.
2. Determine geophone amplitude and phase response curves over the temperature range expected in the borehole.

Comparative Field Tests of VSP Energy Sources

In any exploration region, it is important to know which types of energy sources are most effective for vertical seismic profiling. The most reliable way to acquire this information is to assemble a variety of energy sources at a VSP well site and record downhole data from each one of them. Although this approach is rigorous, it is so costly that it is rarely done. Consequently, any comparative field study of VSP energy sources is extremely valuable, and especially so if the comparison is made in the same borehole with the same geophone recording system. Some energy source characteristics that should be investigated in such field tests are:

1.   What frequency bandwidth does the source create?

2.   Is the energy output sufficient to image deep geological anomalies?

3.   Are source wavelets consistent from shot-to-shot?

4.   Are extraneous VSP noise modes associated with the source?

5.   How many airgun pops, vibrator sweeps, weight drops, or kilograms of dynamite are required to create a specified amount of body wave energy?

An example of a field test comparing a Litton Model 309 compressional vibrator and a 200 in^3 Bolt airgun operating in a water-filled pit is shown in Figure 4-30 (Poster, 1982). Both sources were positioned 200 feet from the wellhead in this test. The airgun pit was 5 feet wide, 10 feet deep, and lined with a steel culvert. Every vibrator trace represents the sum of three upsweeps, each 12 seconds long, extending from 8 to 82 Hz. Each airgun trace represents either three or four pops with the chamber operating at 2000 psi (13790 kPa). Both data sets are bandpass filtered to 8-40 Hz and then plotted with a fixed gain so that the maximum amplitude in each trace has the same height in the display.

In this test, these two sources compare favorably. The energy outputs are approximately the same, and source wavelets exhibit essentially the same degree of consistency from trace to trace. The vibrator, however, creates more tube wave noise because it generates a larger Rayleigh ground roll than does the airgun in the water pit.

There is a measurable difference in the frequency behavior of the vibrator and airgun wavelets which could be important if the objective of the vertical seismic profile is

144

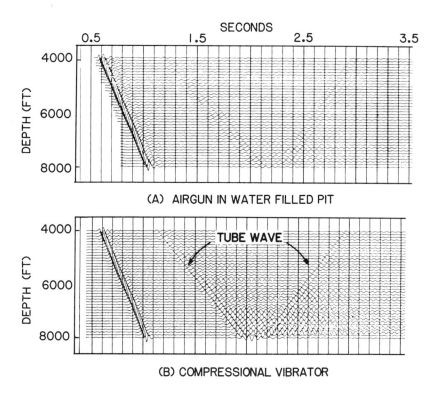

Figure 4-30   Field comparison of VSP energy sources (After Poster, 1982).

to image small anomalies at great depth at this well site. Shown in Figure 4-31 are
frequency spectra calculated from every recorded trace in a time window encompassing
only the downgoing first arrival wavelet. At each recording depth, data from both sources
are recorded before the borehole geophone is relocated, thus frequency differences seen
in the data at any one depth cannot be caused by differences in geophone-to-formation
coupling. These data have not been bandpass filtered as are the data in Figure 4-30. Near
a depth of 4000 feet, the airgun and vibrator wavelets have comparable bandwidths if the
positions of -18 db points are used as a measure of bandwidth. However, the vibrator
wavelets have whiter spectra (i.e., the 0 to -6 db amplitude range extends over more
frequencies) than do the airgun wavelets. As the wavelets propagate to deeper depths,
the airgun wavelet loses some of its higher frequency components. In the deeper parts of
this particular stratigraphic section, the vibrator wavelet would provide better resolution.
Thus, in this case, one must decide whether to use a broader band vibrator

wavelet with its attendant tube wave, or to use a lower frequency airgun wavelet and avoid tube waves. The choice would, of course, depend upon what the objectives of the VSP experiment were.

This source comparison is only one specific test in a single geological province and in a particular type of wellbore environment, and it should not be considered to be a universally applicable characterization of these two particular VSP energy sources. The intent is simply to demonstrate the value of comparative source testing and illustrate some of the features of the test data that should be considered. The energy transfer between a vibrator and the earth depends upon specific soil conditions at the vibrator location; thus, at a different well site, or at this same well site when the soil is wetter or drier, this vibrator wavelet could be different. Likewise, the characteristics of an airgun wavelet can be altered by changing the physical dimensions of the airgun shooting pit. The point to be emphasized is that conducting these types of tests in a variety of geological environments will result in optimum selections of VSP energy sources for a wide range of vertical seismic profiling conditions.

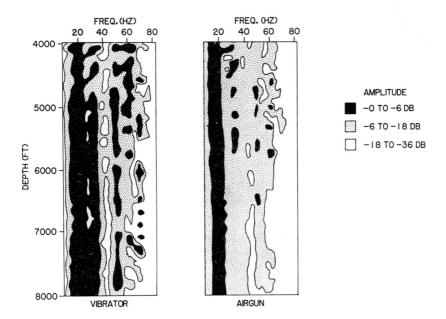

Figure 4-31   Comparison of frequency content of vibrator and airgun first arrival wavelets (After Poster, 1982).

The borehole in which these data were recorded was cased to a depth of 8000 feet, but no cementing was done below a depth of 2000 feet. However, the well was 27 years old at the time this VSP experiment was performed. Note that the first arrival wavelets in Figure 4-30 have a consistent waveshape throughout this uncemented interval, and that no high amplitude resonances exist in the tube wave, as they did for the data shown in Figure 3-6. The data in Figure 4-30 are an example of some of the field evidence which makes VSP researchers conclude that the casing-to-formation bonding of uncemented casing increases with time. In contrast, the data in Figure 3-6 were recorded in a well that was cased only a few days before that VSP experiment was performed. Consequently, there was not enough time for the uncemented casing in that well to become firmly bonded to the formation.

CHAPTER 5

VSP DATA PROCESSING

Several problems become apparent when an interpreter is confronted with a set of unprocessed VSP field data that needs to be analyzed. There is first the realization that the downgoing wavefield is so dominant that any interpretation involving upgoing primary reflections is difficult and often impossible to make. In addition, both random and coherent noises contaminate the data, source wavelets are longer than desired, the shape of the source wavelet can vary from trace to trace, and numerous surface and intrabed multiples exist. These factors often combine to prevent an interpreter from achieving a desired interpretation objective if only raw, unprocessed VSP data are used. Extensive data processing is thus required to achieve the maximum utilization of VSP field data, particularly if the application requires an analysis of upgoing primary reflection events.

The following sections illustrate several general data processing procedures that need to be applied to VSP data recorded with a single-component, vertically oriented geophone so that the data can be used to study rock properties, infer stratigraphic relationships, depict structural geometries, and support the interpretation of surface-recorded seismic data. Only brief references to processing techniques required for analyzing data recorded with orthogonal, three-component geophone elements will be made in this chapter. Some three-component data analyses will be discussed in Chapters 7 and 9.

Several numerical techniques can be used to emphasize upgoing VSP reflections. Most of these procedures are common seismic data processing concepts that are widely used by explorationists. One or more of the following processes can be involved in separating upgoing VSP events from downgoing events:

> vertical summation
>
> restricted vertical summation
>
> velocity filtering
>
> deconvolution.

Each process will be described and illustrated with data examples in the following sections. An overview of VSP processing techniques, with special emphasis on onshore VSP data, has been published by Lee and Balch (1983).

Stacking at Constant Recording Depth

In offshore vertical seismic profiling, one or two small airguns are usually used as energy sources. These airguns are weak compared to the larger airgun arrays used in standard marine reflection profiling. Thus, several airgun pops usually need to be individually recorded and then summed at each geophone depth in order to simulate a stronger energy source. Onshore, surface energy sources such as vibrators, airguns, and weight droppers are widely used in vertical seismic profiling. Environmental constraints in highly populated areas, and in many public lands, dictate that surface sources be used rather than drilling holes for dynamite shots. The body wave energy output of surface sources is often weak compared to that created by buried explosives. In order to get a sufficient amount of energy into the ground for a VSP experiment, surface sources are activated several times, and each of these surface shots is recorded individually. These individual shots can then be summed, as in the marine airgun situation, to create the equivalent of a single, high energy seismic shot.

Examples of individual surface shots recorded at three consecutive geophone depths of a vertical seismic profile are plotted in Figure 5-1. Often one or more traces recorded at a fixed geophone position contain excessive noise or spurious events. Examples of such traces are flagged with a check mark in Figure 5-1. These traces must be omitted from the stacking process that creates the final composited trace representing the geophone response at each depth. If N traces are summed, and there is no coherent noise in the data, the stacked trace has a signal-to-noise improvement of $\sqrt{N}$ relative to the signal-to-noise character of any individual trace.

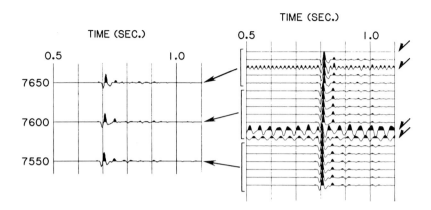

Figure 5-1   Editing and summing VSP field records.

The stacking process is effective if the wavelet character of each trace is identical. If coherent noise modes exist in the data, this noise will be amplified by the stacking process just as are coherent signal components.   If the energy source creates variable source wavelets from shot to shot, stacking will not create a composited trace with significantly improved signal content.

It may be necessary to apply static time corrections to the individual traces so that they are optimally aligned in time before they are stacked.   The need to time shift traces is more common when using buried energy sources.   Surface energy sources usually exhibit a high degree of time coherency from shot to shot.

Static Time Shifting

Some of the raypaths by which seismic energy can propagate from a surface energy source to a geophone in a well are shown in Figure 5-2.   Both primary and multiple reflections are shown, and all events are upgoing when they reach the geophone position. Note that the reflectors involved in this analysis are assumed to be flat and horizontal. The source is assumed to be near the well so that the rays travel vertically,

150

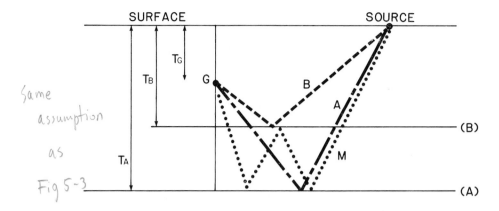

Same assumption as Fig 5-3

Figure 5-2   Raypaths corresponding to upgoing primary and multiple reflections arriving at VSP geophone position G. Reflectors A and B are flat and horizontal.

but it is shown at a large offset in order to separate the rays for visual clarity. If T_A, T_B, and T_G are the one-way vertical travel times to reflector A, reflector B, and the geophone depth, then the arrival times of reflection A, reflection B, and multiple M at the geophone are:

$$t_A = T_A+(T_A-T_G) = 2T_A-T_G$$
$$t_B = T_B+(T_B-T_G) = 2T_B-T_G \qquad\qquad (1)$$
$$t_M = T_A+3(T_A-T_B)+(T_B-T_G) = 2T_A+2(T_A-T_B)-T_G$$

Note that all of these expressions have the form:

Time of upgoing event 'N' at geophone depth	=	Two-way arrival time of event 'N' at surface	−	One-way time to geophone depth

Raypaths describing the propagation of downgoing surface and intrabed multiples that arrive at this same VSP geophone position are shown in Figure 5-3. The notations used for travel times in this situation, and the assumption of flat, horizontal reflectors and small source offset distance, are the same as those used in Figure 5-2. The arrival times of these downgoing events at the geophone are given by:

$$t'_A = 2T_A+T_G$$
$$t'_B = 2T_B+T_G \qquad\qquad (2)$$
$$t'_M = T_A+(T_A-T_B)+(T_G-T_B) = 2(T_A-T_B)+T_G$$

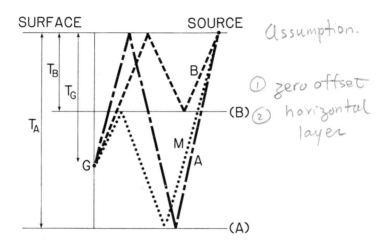

Figure 5-3   Raypaths describing the propagation of downgoing surface and intrabed multiples that arrive at VSP geophone position G. Reflectors A and B are flat and horizontal.

These expressions all have the form:

Time of downgoing event 'K' at geophone depth	=	Two-way time when event 'K' was reflected downward	+	One-way time to geophone depth

Equations 1 and 2 show that a static time shift of either $+T_G$ or $-T_G$ will position selected VSP events to the same times at which they would be recorded by surface geophones. By definition, T_G is the first break time for the VSP trace recorded at geophone position G. Thus, delaying each VSP trace by an amount equal to its first break time will position all upgoing events in that trace to the same time at which they would arrive at the surface (i.e. add $+T_G$ to both sides of Equation 1). Advancing a VSP trace by an amount equal to its first break time will position all downgoing events to the two-way times at which these events leave the interface where they are last reflected in a downward direction (i.e. add $-T_G$ to both sides of Equation 2). This interface may or may not be the earth's surface. These two static time shifts are illustrated in Figures 5-4 and 5-31.

152

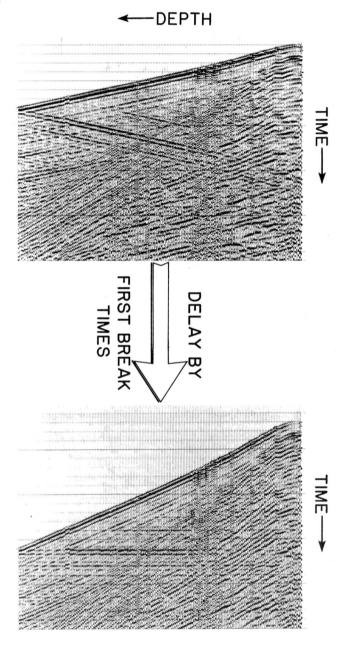

←—DEPTH

TIME—→

DELAY BY

FIRST BREAK
TIMES

TIME—→

Figure 5-4   For small source offset distances, all upgoing events from flat, horizontal reflectors
can be positioned to the two-way times at which they arrive at the earth's surface,
irrespective of the depth at which the events are recorded, if each VSP trace is
delayed in time by an amount equal to its first break time. A few of the traces in the
left panel have slight static time shift errors that should be corrected.

Vertical Summation

One numerical procedure that emphasizes upgoing VSP events and attenuates downgoing events is a simple vertical summation of VSP traces which have upgoing reflections from horizontal strata aligned in phase along equal time lines, and downgoing events misaligned. This alignment can be accomplished by delaying each trace by its first break time as illustrated in Figure 5-4. Vertically summing the set of time-shifted traces shown at the right side of Figure 5-4 will yield a single trace containing all upgoing events and a small amount of contamination due to the out-of-phase summation of the downgoing wavefields. Anstey (1980) refers to this process as upstacking. Three points should be emphasized about this summation:

1. The process assumes that reflectors are flat and horizontal. Reflections from dipping interfaces will not be vertically aligned along equal time lines by the static time shifts described by Equation 1. The time stepout behavior of a reflection created by a dipping interface will be illustrated by model calculations in Chapter 6.

2. The time shifting aligns upgoing multiples as well as upgoing primary reflections; consequently, the vertical sum of the traces contains both primary and multiple events unless multiples are first removed by deconvolution procedures.

3. All upgoing events, regardless of whether they are primary or multiple reflections, are positioned at their proper two-way surface arrival times.

Restricted Vertical Summation

In order to create a vertically summed VSP trace that contains only primary reflections, one should note which part of a vertical seismic profile is most contaminated by upgoing multiples, and which part is least affected by multiples. These two VSP data regimes are illustrated in Figure 5-5. The position of the boundary AB between these two primary-multiple regimes is arbitrary; a visual inspection of a VSP data set will usually suffice to indicate how the data should be divided.

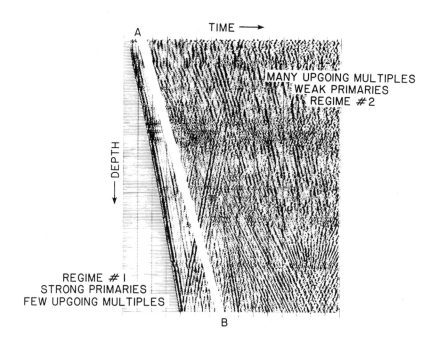

Figure 5-5   A VSP data set can be divided into two regimes by the line AB. Regime #1,
which is below and to the left of line AB, contains strong primary
reflections and a few multiple reflections. Regime #2, which is above and
to the right of line AB, contains weak primary reflections and numerous
multiple reflections. Many analyses involving primary reflections are easier
to perform if they are limited to Regime #1. The position of line AB is
arbitrary and varies from one VSP data set to another.

A static time shift of VSP data that vertically aligns upgoing events, such as shown
in Figure 5-4, followed by a vertical summation between first breaks and the shifted
position of boundary line AB will create a composite trace containing predominately
primary reflections.   Such a summation will be referred to as "restricted vertical
summation".   A small amount of noise caused by early downgoing and upgoing multiples
will exist in the composited trace depending on the nature of any given VSP data set. This
restricted vertical summation process is illustrated in Figure 5-6.   Data regime #1 in
Figures 5-5 and 5-6 is called the "front corridor" by some processors. Thus, this restricted
vertical summation can also be referred to as "summing the front corridor".

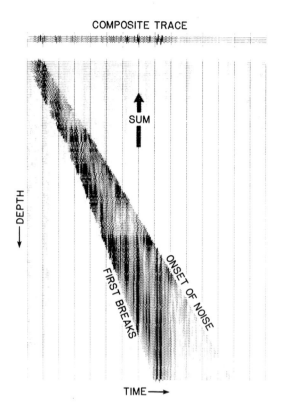

COMPOSITE TRACE

SUM

DEPTH

FIRST BREAKS

ONSET OF NOISE

TIME →

Figure 5-6   The concept of restricted vertical summation is illustrated by this example. All data outside of Regime #1 in Figure 5-5 are muted to zero. The remaining data in Regime #1 are filtered to remove downgoing events, shifted to vertically align upgoing events, and vertically summed. The resulting composite trace is an accurate estimate of all upgoing primary reflections and contains little contamination from multiple reflections. Some of the traces shown here should be statically adjusted in time before summing.

Amplitude Processing

VSP data can be used in some valuable exploration applications if the amplitudes of the recorded subsurface wavelets are properly reconstructed. Surface-recorded seismic data are often numerically processed so that reflection amplitudes can be used as an interpretive tool for explorationists. The seismic bright spot technology developed in recent years is one example of this concept being used to estimate the types of rocks generating a seismic reflection, and the types of fluid contained in the pore system of these rocks. Proper amplitude reconstruction of the subsurface seismic response measured in vertical seismic profiling should lead to an improved calibration of surface-recorded seismic reflection amplitudes in the vicinity of a VSP well, and thereby improve the interpretive value of these surface data.

Seismic wavelet amplitudes also play an important role in synthetic seismic modeling of the subsurface. VSP data serve as an intermediate type of data between synthetically constructed seismic responses of subsurface reflectors, and actual surface measurements of these reflectors. Since a vertical seismic profile, in concept, allows one to follow primary and multiple reflections through a stratigraphic section, VSP data should explain why synthetic calculations and actual surface seismic reflection measurements do or do not agree. Again, this understanding is best accomplished if the amplitudes of the VSP data are correctly reconstructed during data processing.

Several physical processes affect the amplitude of a propagating seismic wavelet, and the amplitude effects created by these processes must be compensated for by the numerical process used to adjust VSP wavelet amplitudes. A brief description of the major physical processes that influence seismic wavelet amplitudes follows.

Spherical Divergence - Geometrical spreading of seismic energy away from a seismic energy source is the dominant physical process that creates reduced wavelet amplitudes at subsurface receiver points. Most numerical algorithms that adjust surface-recorded seismic reflection data for spherical divergence of the propagating wavefield are designed to adjust only the amplitudes of the propagating compressional wave mode. VSP experiments commonly record the compressional wavefield as well as other wave modes, particularly if the downhole instrumentation contains triaxial geophones. If a VSP measurement does record compressional, shear, and tube wave events, then one must account for at least three different types of geometrical spreading when processing VSP data. A numerical reconstruction process designed to adjust data amplitudes for the expansion of a compressional wavefront may not properly reconstruct shear wave amplitudes, since the shear mode can be expanding in a different way than the compressional wavefront. Each distinct body wave, and each interfacial wave, recorded by a vertical seismic profile can have a unique rate at which its amplitude diminishes due to geometrical spreading, as shown in Figure 5-7. A tube wave exhibits no spherical divergence, so its ray paths are shown as parallel, non-divergent lines. The P-wave and S-wave divergences shown in this illustration will be discussed later in this chapter.

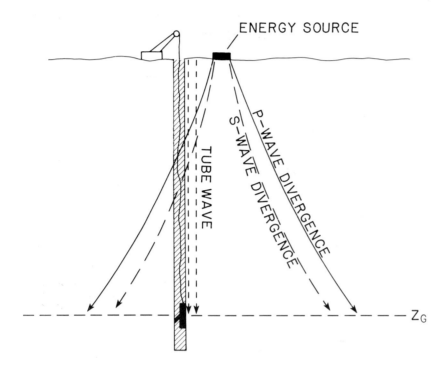

Figure 5-7   Compressional waves and shear waves may experience different amounts of amplitude loss due to geometrical spreading by the time they reach a VSP geophone at depth Z_G. Because of refraction, both compressional and shear wavefronts expand according to a V^2T relationship. Tube waves do not experience geometrical spreading.

Transmission Losses - As seismic energy propagates through the earth, some fraction of the energy is reflected at each impedance contrast. Each reflection, depending on the algebraic sign of the reflection coefficient, causes the amplitude of the transmitted wave to either increase or decrease. In general, energy is extracted from an expanding wavefront, and the amplitude of a propagating wavelet experiences a decay in addition to that caused by spherical divergence. These transmission losses are difficult to account for unless one has independent subsurface data, such as well logs, that define the reflectivity function of the earth. These log data are usually available in a well where a VSP experiment is conducted.

158

An example of a sonic log recorded in an experimental VSP well, and a calculation of the transmission losses that a downgoing particle velocity impulse propagating through this stratigraphic section at compressional velocity could experience, are shown in Figure 5-8. Note that the amplitude of the transmitted wave decays slowly except in a few low velocity intervals, where its amplitude momentarily increases. An example of an amplitude increase would be the behavior in the depth interval between 9100 and 9200 feet.

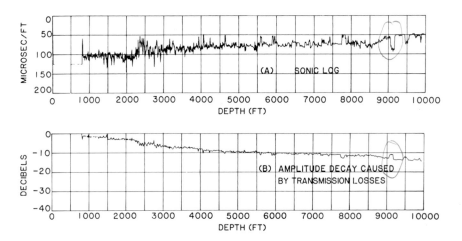

Figure 5-8   The amplitude decay (B) of a plane particle velocity wave resulting from transmission losses calculated from a sonic log (A) at each millisecond of one-way travel time.

The following assumptions are involved in this calculation:

1.   Only particle velocity disturbances propagating with P-wave velocity are analyzed since this sonic log measures only compressional velocities.

2.   The propagating wavefront can be approximated by a plane wave.

3.   Reflecting interfaces are flat and horizontal, and the wavefront propagates normally to these interfaces. This assumption, together with assumption 2, allows Equations 1 and 2 of Chapter 7 to be used for calculating reflection coefficients.

4.   Reflecting interfaces occur at intervals of one millisecond of one-way travel time.

5.   Reflection coefficients can be calculated from just the compressional velocities provided by the sonic log.

6.   No downgoing short period (pegleg) multiples are allowed to follow the downgoing first arrival and rebuild the amplitude decay caused by the repeated reflections.

Under these assumptions, the amplitude decrease resulting from transmission losses can be quite large. Assumption 6 is probably the most unrealistic condition imposed on the calculation. In real earth seismic wave propagation, short period multiples continually redirect upward traveling reflected energy back downwards. Thus, many weaker events follow a VSP first arrival at varying short time delays. The result is that the energy distribution in a downward propagating wavelet shifts later in time, high frequency components of the wavelet are suppressed, and the overall amplitude of the wavelet remains higher than otherwise expected. Consequently, the amplitude loss shown in Figure 5-8 is an overestimate, but nonetheless, the calculation provides an important insight into the possible magnitude of amplitude decay that can be caused by transmission losses. If short period multiples were included in the calculation, the amplitude loss at the deepest calculated value would be considerably less than that shown.

Scattering - The transmission losses discussed above occur at impedance contrasts which are laterally extensive. Often there are small earth volumes of limited areal extent, i.e. smaller than a first order Fresnel zone, within a stratigraphic section which have impedances that are different from their surroundings. These small subsurface anomalies cause energy to scatter out of a seismic wavefront as it passes by them. This scattering tends to remove the higher frequency components from a propagating wavelet and creates an overall decrease in seismic wavelet amplitude. Scattering behavior, like the transmission loss process, is difficult to account for numerically unless one knows the physical properties of the subsurface in detail. Generally, the effects of scattering are ignored in seismic amplitude reconstruction.

Absorption - Absorption will be defined as that process which removes energy from a propagating seismic wavefront by means of (a) frictional losses between oscillating rock particles, or (b) pore fluid movement inside the vibrating rock pore system. Absorption is similar to scattering in that it removes more high frequency components from the seismic wavelet than it does low frequency components. One important factor that has to be accounted for in absorption measurements is the effect of pegleg multiples on the propagating wavelet. These short period multiples create a pseudo absorption effect by lengthening the seismic wavelet and reducing the higher frequency components of the wavelet signal. This topic is discussed by Schoenberger and Levin (1974) and Spencer, et al. (1982). .

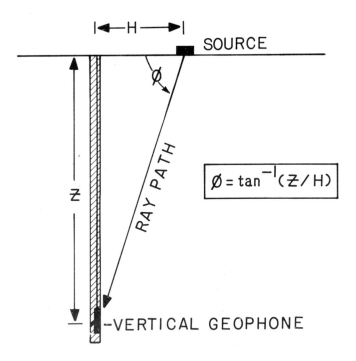

Figure 5-9   In a homogeneous medium where a straight ray path can be assumed, the vertical component of compressional particle motion created by a downgoing first arrival at a borehole geophone is proportional to sin (∅). This amplitude effect should not be associated with subsurface geological conditions.

Source-Receiver Directivity - If a VSP energy source directs more energy in some directions than it does in others, then a measured downhole response will contain amplitude effects that are purely a result of this source directivity and are not an indication of a physical earth property. A downhole geophone also has a directivity behavior. For a seismic wavefront spreading with a uniform amplitude in all directions in a homogeneous medium, the response of a vertically oriented, moving-coil type borehole geophone is proportional to $\sin(\emptyset)$, where $\emptyset = \tan^{-1}(Z/H)$, H is the offset distance from the wellhead to the seismic source, and Z is the depth of the geophone. These receiver directivity parameters are diagrammed in Figure 5-9.

Of all the physical processes described in the preceeding paragraphs, spherical divergence of a seismic wavefront creates an amplitude decay that is greater than the effects of all of the other processes combined. Consequently, it is particularly important that any numerical algorithm which creates "true" amplitude VSP data properly adjusts the data for amplitude losses caused by spherical divergence. A major step in this amplitude restoration is the determination of a mathematical function that represents the loss caused by spherical divergence. Once this function is known, its inverse can be applied to VSP data to amplify each data point of each trace so that wavelet amplitudes are equivalent to those that would be created by plane waves propagating through the same stratigraphic section.

When a seismic wavefront expands as a uniform sphere in a homogeneous medium, the amplitude of the first arrival diminishes as $1/R$, or $1/(VT)$, where R is propagation distance, V is average propagation velocity, and T is travel time. This amplitude decay is strictly a result of the geometrical effect of distributing a fixed amount of energy uniformly over a continually enlarging spherical surface. However, extensive sections of homogeneous material are rarely encountered in the real earth. Instead, most stratigraphic sections consist of stacked rock layers having different seismic velocities and densities. In such a stratified impedance sequence, refraction causes an increased divergence of seismic ray paths, and the amplitude decay, D, caused by spherical divergence is given by

$$D = V_o/\left(\left(V_{rms}(T)\right)^2 * T\right) \tag{3}$$

where $V_{rms}(T)$ is the root-mean-square value of the layered velocities traversed by the wavefront during travel time T (Newman, 1973). The symbol * means multiplication, not convolution. V_o is a constant which has units of velocity, thus D has units of $(length)^{-1}$.

If the initial maximum amplitude of a spherically spreading seismic wavefront at its point of origin is A_0, then, ignoring all other effects except spherical spreading, its maximum amplitude, $A(Z)$, at depth Z is

$$A(Z) = A_0 \, D(Z). \tag{4}$$

At a given VSP recording depth, Z_G, such as shown in Figure 5-7, the amplitude decay of a compressional wavefront in a stratified medium due to spherical divergence is given by

$$D_c(Z_G) = V_{oc}/\left((V_{rms}(Z_G))_c^2 * T_c(Z_G) \right) \tag{5}$$

where subscript c implies compressional wave velocities and travel times. The amplitude decay of a shear wavefront propagating through this same stratigraphic section to depth Z_G is given by

$$D_s(Z_G) = V_{os}/\left((V_{rms}(Z_G))_s^2 * T_s(Z_G) \right) \tag{6}$$

where subscript s implies that shear wave velocities and travel times must be used. In well consolidated rocks, the velocity and travel time relationships,

$$(V_{rms}(Z_G))_c = 2(V_{rms}(Z_G))_s, \text{ and} \tag{7}$$

$$T_c(Z_G) = 0.5 \, T_s(Z_G),$$

are approximately correct. Using these values in Equations 5 and 6 yields the following relationship between compressional and shear wave amplitude decay due to spherical divergence

$$D_c(Z_G) = \left(V_{oc}/(2V_{os}) \right) D_s(Z_G). \tag{8}$$

The constants, V_{oc} and V_{os}, are the compressional and shear velocities in the topmost layer of the sequence of refracting layers through which the wavefronts travel (Newman, 1973). There is considerable arbitration in deciding exactly where the topmost layer of a stratigraphic sequence is located, and consequently, exactly what the appropriate values for V_{oc} and V_{os} should be. Since these constants are arbitrary, V_{oc} is usually set to unity when numerically reconstructing "relative amplitude" reflections of surface recorded compressional seismic data.

However, when one has recorded both compressional and shear wave events in the same data set, as is done in three-component VSP data, and wants to accurately

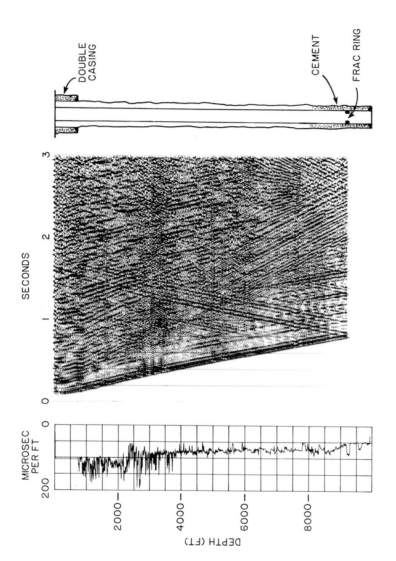

Figure 5-10   Casing and cementing conditions and sonic velocity profile existing in a study well at the time VSP data were recorded. A dynamic automatic gain control function is applied to the VSP data to make all events have the same display amplitude. The borehole diameter is drawn too large; the VSP data quality implies that the casing and formation frequently touch each other in the uncemented interval. These same data are shown in true amplitude form in Figure 5-16.

reconstruct the amplitudes of both wave modes simultaneously, then it is critical to properly define the constants V_{OC} and V_{OS}. If both V_{OC} and V_{OS} are set to unity then Equation 8 shows that shear wave amplitudes decay twice as much as do compressional wave amplitudes if the conditions in Equation 7 are true. In shallow unconsolidated sediment, V_{OC} may be three, four, or five times larger than V_{OS}, and in such a case, spherically spreading compressional waves lose amplitude more rapidly than do spherically spreading shear waves. This condition is assumed in Figure 5-7 so that the shear wavefront is shown spreading less than is the compressional wavefront at depth Z_G. If $V_{OC} = 2V_{OS}$, then both compressional and shear wave events decay at the same rate. Whenever wavelet amplitude behavior in the shallow, unconsolidated part of a stratigraphic section can be ignored in a data analysis, then amplitude processing of VSP data can usually be commenced at a depth where rocks are sufficiently consolidated so that the assumption, $V_{OC} = 2V_{OS}$, is valid.

In order to account for the amplitude decay caused by the spherical spreading of body waves recorded in vertical seismic profiles, the root-mean-square velocity and travel time of each wavefront must be known as a function of depth. There are two convenient ways that these parameters can be determined from data recorded in VSP study wells:

1. The RMS velocity and one-way propagation time at any borehole depth can be determined from sonic log data which have been adjusted to agree with seismic checkshots, or

2. First break times provided by VSP data can be used to determine the RMS velocity and travel time at any borehole depth.

These two methods of determining VSP amplitude processing parameters will be evaluated by means of the VSP data and sonic log data recorded in the study well shown in Figure 5-10. The casing and cementing conditions that existed in this well at the time the VSP data were recorded are shown on the right. A sonic log recorded in the well before it was cased is shown on the left. These data were recorded in a vertical hole with a vertically oriented geophone. The VSP data are displayed via a numerical automatic gain control (AGC) function that forces all events to have the same height. Consequently, relative amplitude changes between two depths, or between two different events at the same depth, cannot be made using this display. The data are shown in this AGC form so that all downgoing and upgoing modes, regardless of their amplitude strengths, can be seen. Any amplitude adjustment procedures must, of course, be applied to raw VSP field traces which have recording gains removed and no variable numerical gains applied, rather than to these AGC'ed data. A relative amplitude version of these same data, in which spherical divergence effects are removed, is shown in Figure 5-16, but first, the gain function that is used to adjust the data for this spherical divergence needs to be developed.

Compressional amplitude decay functions of the form, $V^2_{rms}T$, calculated from both the sonic log and the VSP compressional first arrivals, are shown in Figure 5-11. These functions are plotted as parameters of one-way travel time rather than depth because the amplitude calculations will be performed in the time domain. A comparison of the two functions shows that there is essentially no difference in the $V^2_{rms}T$ behavior determined from the sonic log and that calculated from the VSP first break times. Either function can be used to account for the spherical divergence of a propagating compressional wavefront at this well site.

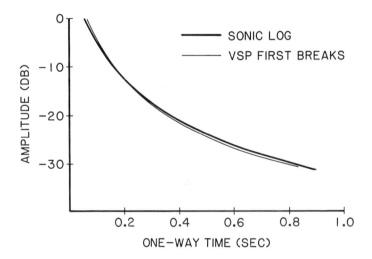

Figure 5-11   Spherical divergence, $(V^2T)^{-1}$, calculated from sonic log data and VSP first break times recorded in the study well illustrated in Figure 5-10. The VSP derived function terminates before the sonic log function does because VSP data were not recorded in the bottom 750 feet of the well.

A verification of the reliability of these $(V^2T)^{-1}$ estimates of spherical divergence is shown in Figure 5-12. The amplitude of the downgoing compressional first arrival is determined as a function of propagation distance by measuring the magnitude of the first trough of the compressional arrival at each recording level. This method of determining wavelet amplitude decay through a stratigraphic section is arbitrary. Other measures of the magnitude of the compressional arrival, such as the RMS amplitude, or the extremum

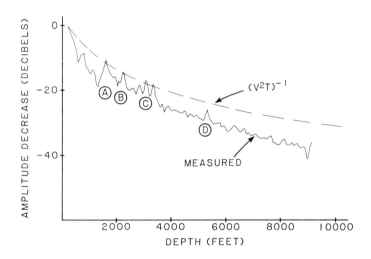

Figure 5-12 Comparison between the measured amplitude decrease of the VSP first arrivals shown in Figure 5-10 and the decrease predicted by a $(V^2T)^{-1}$ spreading function determined from a sonic log recorded in the well. The amplitude increases labeled A, B, C, D are caused by unbonded casing.

of the wavelet envelope function determined via Hilbert transforms, are equally valid schemes by which amplitude behavior can be determined. The measured first arrival amplitudes exhibit anomalously high values in some depth intervals, labeled A, B, C, D, where the casing is not adequately coupled to the formation. This amplitude behavior is camouflaged in Figure 5-10 by the time varying gain function applied to the data in that display. In general, the measured amplitude decay shows that the single casing between 750 and 9700 feet is reasonably coupled to the formation even though it is not cemented.

As shown in Figure 5-12, the amplitude decay, $(V^2T)^{-1}$, caused by spherical divergence accounts for almost all of the amplitude decrease of the compressional wavefront observed in this VSP well. The difference between the $(V^2T)^{-1}$ curve and the measured amplitude decrease is due to transmission losses, absorption, scattering, and any seismic energy loss mechanisms other than spherical divergence.

The sonic log function in Figure 5-11 is arbitrarily chosen as the spherical divergence correction to use to adjust the amplitudes of the VSP data shown in Figure 5-10. The gain function (i.e., the inverse of this $(V^2T)^{-1}$ curve) that restores the subsurface VSP amplitudes is shown in Figure 5-13. The one-way time to the total depth of the well is 892 ms. For one-way times greater than 892 ms, the linear extrapolation of the function shown by the dashed line is used.

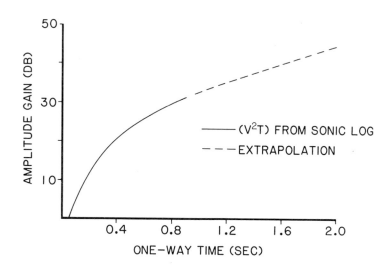

Figure 5-13   Gain function used to remove the effect of spherical divergence on the VSP amplitudes shown in Figure 5-10.

Once a gain function is known in the form shown in Figure 5-13, it is often applied to VSP data differently than is an amplitude restoration function used to adjust the amplitudes of surface-recorded reflection data.   A principal difference between VSP data and surface-recorded reflection data is that the recording times at which many events are measured in a vertical seismic profile are neither one-way nor two-way travel times.   In contrast, the measured times for surface data are always two-way travel times.   The time at which a first arrival is measured in a borehole at some depth, Z, is one-way time, but the time at which an upgoing primary reflection, or an intrabed multiple, is measured at that same depth is a value greater than one-way time but less than two-way time.   This difference in the travel times observed in VSP data and surface reflection data allows one to use a different kind of gain restoration procedure than is used in standard surface seismic data processing.   An approach uniquely applicable to VSP data will be presented first (Equations 9-11), and then an alternate approach for adjusting VSP amplitudes, similar to what is used when processing surface-recorded reflection data, will be described (Equations 12-14).

168

The problem of adjusting amplitudes of VSP data can be illustrated by the diagram in Figure 5-14. In this figure, a single VSP trace at a depth Z_0 is shown. The compressional first arrival, DA, at depth Z_0 occurs at a one-way time of T_0. The function shown in Figure 5-13 can thus be used to describe the amplitude decay for this compressional first arrival since this gain function is defined strictly in terms of one-way travel time. However, this function cannot be used, without modification, to describe the effect of spherical divergence on the primary reflection event, R, generated at depth Z_R but recorded at depth Z_0. The modification is demanded because the time, T_1, at which the reflection is measured at depth Z_0 is not a one-way time, which is the only parameter used to define the gain value in Figure 5-13.

A procedure that allows the spherical divergence of all events recorded at depth Z_0 to be properly accounted for, irrespective of whether or not those events are downgoing direct arrivals or upgoing primary reflections from flat, horizontal reflectors, has been outlined by Morris (1979). Basically, the procedure looks at the divergence that occurs on the downgoing and upgoing travel paths as two independent but equal processes, as shown in Figure 5-15.

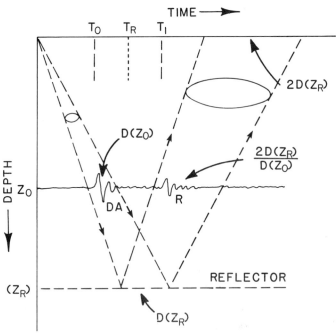

Figure 5-14   This illustration is one way to show how the geometrical spreading of a spherical wavefront reduces VSP amplitudes. D(Z) represents the magnitude of the geometrical divergence of a downgoing circular cone of such a wavefront, which reflects upward from an interface at depth Z_R. DA is the downgoing direct arrival at depth Z_0, and R is the reflection of DA at depth Z_R, which is recorded at depth Z_0.

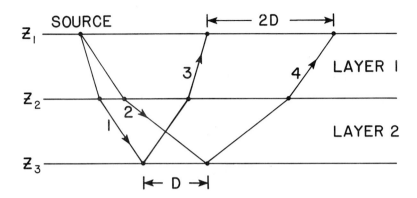

Figure 5-15   Because of the vertical symmetry between downgoing and upgoing rays in horizontally layered media, the geometrical divergence of downgoing rays 1 and 2 between depths Z_1 and Z_3 is the same as the geometrical divergence of upgoing rays 3 and 4 between Z_3 and Z_1.

If the amplitude decay resulting from spherical divergence on the downward path to a flat, horizontal reflector at depth Z_R is $D(Z_R)$, then the concept illustrated in Figure 5-15 says that the two-way path amplitude decay when that primary reflection reaches the surface is $2D(Z_R)$.   If the amplitude decrease is expressed in terms of decibels, such a wavefield behavior means that when the amplitude decay on the downward path to a reflector is 40 db, then the amplitude decay for the two-way path down and then back to the surface is 46 db (not 80 db).

The amplitude decay experienced by a primary reflection recorded at a depth Z_0, which is between the surface and the depth Z_R at which the reflection was generated, is then

$$D = \frac{2D(Z_R)}{D(Z_0)} \tag{9}$$

as shown in Figure 5-14.   This equation simply says that the spherical divergence amplitude decay, D, of a subsurface measured reflection event is the decay, $2D(Z_R)$, that the seismic signal would experience on a two-way path from the surface down to the reflector and back to the surface, followed by a gain restoration, $D(Z_0)$, equivalent to the one-way path divergence from the surface down to the observation depth, Z_0.

In terms of the travel times T_0 and T_1, measured for the downgoing first arrival and the upgoing reflection at depth Z_o (Figure 5-14), the reflector depth, Z_R, occurs at a one-way travel time, T_R, given by

$$T_R = 0.5(T_0 + T_1). \tag{10}$$

This relationship is true only for flat, horizontal reflectors. For any subsurface VSP trace, the gain function, $g(T)$, used to restore the amplitude decay caused by spherical divergence is thus

$$g(T) = G(T) \qquad \text{for } 0 \le T \le T_0, \text{ and} \tag{11}$$
$$g(T) = 2\,G\big((T+T_0)/2\big)/G(T_0) \qquad \text{for } T_0 \le T \le T_{max},$$

where $G(T)$ is the gain function shown in Figure 5-13, T_0 is the first break time for the VSP trace, and T_{max} is the total record time. If $G(T)$ is expressed in decibels, the multiplicative factor, 2, in the first term should be interpreted as an increase of 6 db.

Applying this gain function to the VSP field data shown in Figure 5-10 yields the result shown in Figure 5-16. The downgoing compressional first arrivals show a gradual amplitude decrease with depth after the spherical divergence effect is removed, as they should. The anomalously high amplitude intervals labeled A, B, C, D in Figure 5-12, which occur where the single casing is not adequately coupled to the formation, are more obvious in this wiggle trace display then they are in Figure 5-10. The compressional first arrivals observed inside the doubly cased interval from the surface down to 750 feet are accurately reproduced even though the casings are not cemented together, as can be seen by examining the first 200 milliseconds of record time in this depth interval. The amplitudes of these shallow compressional first arrivals continually decrease as the geophone moves upward in the top few hundred feet of the well because the geophone records only vertical particle motion, which is proportional to $\sin(\emptyset)$ as described in Figure 5-9. The energy source used in this experiment was a compressional vibrator positioned 700 feet from the wellhead. Because of the large source offset, the compressional motion in the top part of the well becomes more horizontal as the recording position approaches the surface.

Although the compressional first arrival amplitudes inside the double casing are correctly recovered, the amplitudes of all later events recorded in this interval are erroneous and appear as extremely high amplitude resonances. There is a dramatic reduction in this high amplitude resonance at a depth of 900 feet, slightly below the surface casing point of 760 feet. The amplitudes of the shallowest three or four traces are so high that they are clipped in this display.

171

uncemented casing Hole
OK for producing useable data.

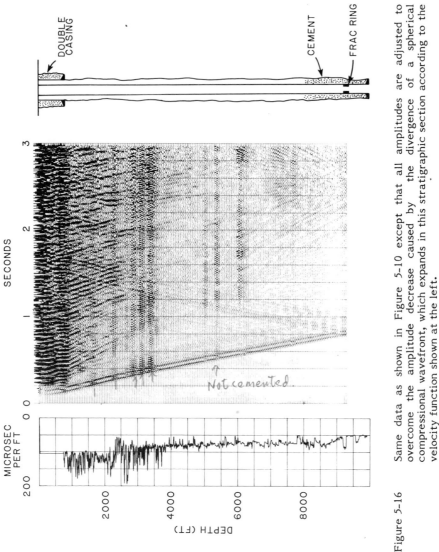

Figure 5-16   Same data as shown in Figure 5-10 except that all amplitudes are adjusted to overcome the amplitude decrease caused by the divergence of a spherical compressional wavefront, which expands in this stratigraphic section according to the velocity function shown at the left.

Meaningful correlations between compressional wavelet amplitude behavior and subsurface geology in the vicinity of this well can now be made. Studies of this type are described in Chapter 7. Similar V^2T amplitude gain functions describing shear wave divergence should be constructed and separately applied to these data in order to analyze shear wave amplitude behavior. If a V^2T function based on actual shear wave velocities and travel times cannot be constructed, one can only assume that the compressional V^2T function also properly restores shear wave amplitudes.

Some VSP data processors use empirical functions of the form

$$g(T) = AT^n, \qquad (12)$$

where T is recording time and A and n are constants, to restore the amplitude decay caused by spherical divergence. This type of gain restoration function is easier to construct than is the type described in Equation 11. Also, it is more similar to the types of gain functions used to process surface-recorded data, and it can accurately account for spherical divergence effects of VSP data if the exponent n is correctly chosen. For example, the amplitude gain function calculated in Figure 5-13 is replotted in logarithmic coordinates in Figure 5-17. Determining the slope and intercept of the plotted points shows that this function can be described by the equation

$$g(T) = (42.1)T^{1.48} \qquad (13)$$

where T is one-way time in seconds. This equation could be used to recover the amplitudes of VSP compressional events in lieu of Equation 11 by defining T as recording time. For North Sea VSP data, n typically is set to a value of about 1.5. This same value of n also applies in the Anadarko Basin of Western Oklahoma and in some offshore areas of West Africa. In some offshore Australia wells, a value of n = 2.0 seems to be more appropriate.

Thus in most wells, if one does not have access to a good quality sonic log, or cannot use VSP first break data so that precise gain functions like that in Figure 5-17 can be calculated, then an empirical function of the form

$$g(T) = AT^n, \quad \text{where } 1.0 \leq n \leq 2.0, \qquad (14)$$

should provide reasonably good amplitude reconstruction of spherically spreading compressional and shear wavefronts. In such a case, one simply has to guess what value of A and n should be used.

The spherical divergence functions considered in this section assume that the reflecting interfaces are flat and horizontal. The use of interval velocities to calculate wavefront divergence through stratigraphic layers having interfaces with arbitrary dip is considered by Hubral and Krey (1980).

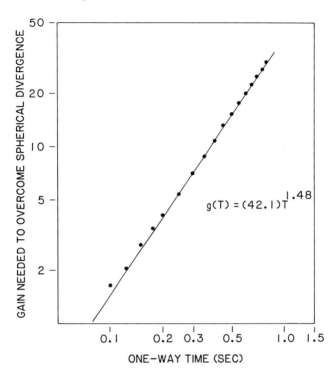

Figure 5-17   The gain function ($V^2_{rms}T$) shown in Figure 5-13, which restores the amplitude decay caused by spherical divergence, can be restated in the form AT^n shown here. This functional form is a convenient way to define and apply spherical divergence corrections.

Removal of Selected Wave Modes

VSP data contain both downgoing and upgoing wave modes which overlay each other in varying degrees of complexity. The analysis of upgoing wave modes is particularly important since these events are the ones recorded by surface seismic measurements. The retrieval of these upgoing modes from VSP data is complicated by the fact that they are considerably weaker in amplitude than are downgoing modes. Thus, numerical procedures that can attenuate downgoing modes without seriously affecting upgoing events are essential in VSP data processing.

In structural interpretations of VSP data, it is sometimes necessary after upgoing modes are separated from downgoing modes, that the upgoing events be further segregated according to the magnitude of their propagation velocities since reflections from flat, horizontal interfaces exhibit a slower upward propagation velocity through a VSP data set than does a reflection from a dipping reflector. Consequently, in structural applications it is important to process VSP data so that upgoing events occurring in a defined velocity range are unaltered, while upgoing events propagating with velocities outside of that range are attenuated.

These types of interpretational objectives indicate how essential it is to selectively suppress specific wave modes recorded in a vertical seismic profile so that more critical, and usually weaker, wave modes can be amplified and studied. Two techniques for removing selected VSP wave modes are discussed in the following sections. The first approach involves the design of velocity filters in frequency-wavenumber (f-k) space. The construction of these filters requires that VSP data be recorded at uniform increments in both time and space. Both the spatial and the temporal sampling intervals must satisfy the Nyquist sampling theorem, which requires that at least two sample points be recorded within the shortest wavelength contained in the data. The second approach is a trace by trace arithmetic subtraction of the estimated wave mode that is to be attenuated. A median filter technique can be used to estimate the wave mode that is to be eliminated, and this estimate is then subtracted from each VSP trace. Median filters are not widely used in seismic data processing, so the mathematical properties of these filters are also briefly discussed.

F-K Velocity Filtering

Velocity filtering is an established data processing procedure for removing unwanted energy modes from digitally recorded seismic data (Embree, et al., 1963; Fail and Grau, 1963; Treitel, et al., 1967; Sengbush and Foster, 1968; Hildebrand, 1982; Dobrin, et al, 1965; Hawes and Gerdes, 1974; Kinkade, 1980). Velocity is a vector quantity; therefore, the propagation velocities of seismic wave modes can differ in the following two fundamental ways:

1. The directions of propagation can be different.

2. The magnitudes of the velocities can be unequal.

A velocity filter can attenuate an unwanted energy mode by taking advantage of either, or both, of these velocity properties.

The fundamental relationships between seismic velocity behavior as expressed in the space-time (Z-T) domain and in the frequency-wavenumber (f-k) domain are shown in Figure 5-18. VSP data are exchanged between these two mathematical domains by for-

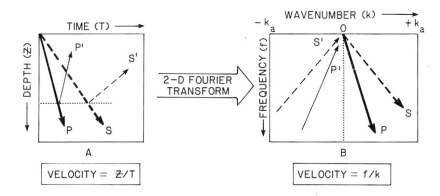

Figure 5-18   The separation of VSP wave modes by velocity filtering can be accomplished by transforming the data from a function of time and space to a function of frequency and wavenumber.   These hypothetical VSP data show how downgoing compressional (P) and slower shear (S) body waves and upgoing compressional (P') and shear (S') reflections transform into f-k space. Downgoing modes appear in the positive wavenumber half plane and upgoing modes in the negative wavenumber half plane.   The width of each line is proportional to the amount of energy contained in that wave mode.

ward and inverse two-dimensional Fourier transforms. In the example shown in Figure 5-18, downgoing energy is arbitrarily defined as having a positive propagation velocity. Thus, the Fourier transform expresses downgoing wave modes in terms of positive wavenumbers.   Upgoing VSP energy modes exhibit negative velocities (by arbitrary definition), and the Fourier transform therefore places all upgoing modes in the negative wavenumber half plane.   Thus, the Fourier transform segregates VSP events into two different half planes of f-k space, depending on the direction that the events travel. Whereas, upgoing and downgoing VSP events always crisscross each other in the space-time (Z-T) domain, they do not overlap in the frequency-wavenumber (f-k) domain if the data are properly sampled.   This separation of VSP wave modes in f-k space provides a convenient way by which downgoing events can be attenuated without suppressing upgoing events.

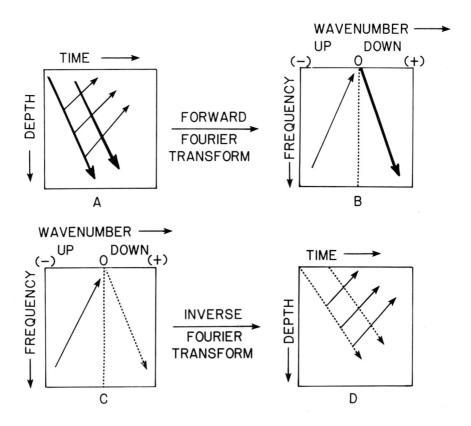

Figure 5-19 The concept of f-k velocity filtering.

A -   Original VSP data. Strong downgoing events shown as heavy arrows. Weak upgoing events shown by light arrows.

B -   Forward two-dimensional Fourier transform of data set A. Downgoing events are in the positive half plane. Upgoing events are in the negative half plane.

C -   Data set B after multiplying entire positive half plane by a small number (e.g. 0.001). Downgoing events are attenuated. Upgoing events are unaffected.

D -   Inverse two-dimensional Fourier transform of data set C. Downgoing events are attenuated. Upgoing events are now stronger than are the downgoing events.

One way to numerically reject downgoing VSP wave modes is diagrammed in Figure 5-19. Once the data are transformed into the f-k domain, multiplying every data point in the positive wavenumber half-plane by a value which is much less than unity will attenuate the amplitudes of the downgoing events. For example, multiplying each data point in the right half plane of data set B in Figure 5-19 by 0.001 would suppress all downgoing modes by 60 decibels. Performing an inverse Fourier transform on these modified f-k data generates a new set of VSP data, D, expressed in space and time, in which the downgoing modes are 60 decibels weaker than the downgoing modes in the original data set, A.

If VSP data contain a wave mode that is traveling in the same direction as a second wave mode which needs to be removed from the data, then the velocity filtering procedure described above must be slightly altered. In addition to using vector property #1 (direction), as in Figure 5-19, to attenuate events which do not occur in the correct half-space of the wavenumber domain, vector property #2 (magnitude) must also be employed so that wave modes in the correct wavenumber half-space traveling with incorrect group velocities are suppressed. This type of filtering problem can be illustrated by the hypothetical data sketched in Figure 5-18, where it is assumed that the vertical seismic profile records downgoing compressional (P) and shear (S) events. If the upgoing compressional wave modes (P') shown in this figure are to be isolated from the total VSP response, then not only must all downgoing events in the right half of the f-k plane be suppressed, but the upgoing shear wave modes (S') traveling in the same direction as the P' events must also be attenuated. A masking function that could be superimposed over the VSP data in the f-k domain in order to suppress all events not traveling in an upward direction with compressional velocity is shown by the dotted region in part A of Figure 5-20. If all f-k data points within this dotted area are significantly reduced in amplitude (say by a factor of 0.01 or 0.001), then all recorded VSP events except those traveling upgoing with compressional velocity are correspondingly attenuated. An inverse Fourier transform of these altered f-k data into the space-time domain is shown as data set B.

Rieber Mixing Created by Velocity Filtering

An important feature of Fourier transform pairs is that a narrow function in one Fourier domain transforms into a wide function in the other Fourier domain. For example, a spike impulse in time transforms into a flat, white frequency spectrum, and spike pairs in the frequency domain transform into continuous monochromatic sine or

178

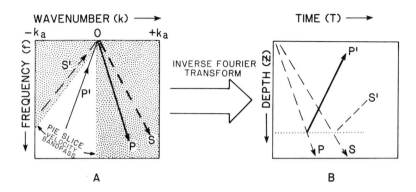

Figure 5-20 The concept of f-k velocity filtering.

A - Same display as data set B from Figure 5-18. All data points within the dotted areas
are multiplied by a small value (e.g. 0.001) in order to reduce the magnitudes of all
energy modes except the upgoing compressional reflections P'.

B - Inverse Fourier transform of data set A. Compare with data set A of Figure 5-18.
All wave modes except upgoing compressional reflections, P', are suppressed.

cosine time functions that extend to infinity. Carrying this principle into the wavenumber
(k) and space (Z) domains means that a function which spans a narrow range of
wavenumbers transforms into a function that is wide in Z space. This feature of Fourier
transforms is particularly important in f-k velocity filtering of VSP data.

Two possible ways that the f-k spectrum of VSP data can be adjusted to reject
downgoing wave modes are illustrated in B and C of Figure 5-21. When the passband of
the velocity filter becomes narrower, as in Part C, the upgoing VSP events occurring
within the passband are extended to cover more and more of the spatial (Z) domain
according to the principle of Fourier transform pairs described in the previous paragraph.
The result is a spatial mixing, or averaging, of the upgoing data along the velocity trend
of each propagating event within the velocity passband. Such spatial compositing of
seismic data is commonly called Rieber mixing (Sheriff, 1973).

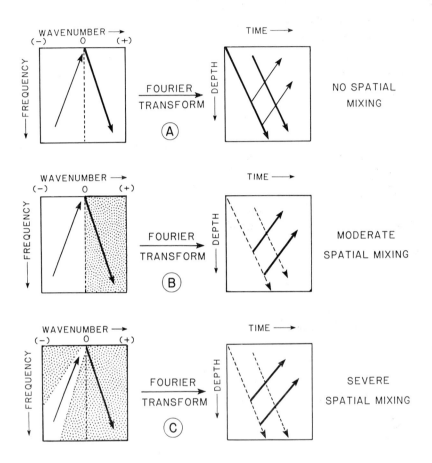

Figure 5-21 Relationship between the width of an f-k velocity bandpass region and the amount of Rieber mixing of VSP data in the time-space domain.

A - Fourier transform of unaltered VSP f-k spectrum.

B - Fourier transform of VSP f-k spectrum with half plane attenuated (attenuated area is dotted).

C - Fourier transform of VSP f-k spectrum reduced to narrow pie slice (attenuated area is dotted).

This spatial mixing of VSP data by velocity filters has both positive and negative aspects. On the positive side is the fact that the waveshapes of events traveling within the velocity passband are smoothed and made more consistent from trace to trace. This result helps an interpreter follow phase relationships through the VSP traces and is particularly important when trying to identify the waveshapes and arrival times of weak upgoing events that emerge at the earth's surface. Consequently, VSP data which are velocity filtered in order to emphasize upgoing events are usually more valuable for interpreting surface-recorded reflection data than are unfiltered VSP data. A negative feature of f-k velocity filtering is the fact that spatial mixing of VSP traces means that the depths at which VSP events start and stop in a stratigraphic column are difficult to define. A simplified example of this type of interpretational problem is given in Figure 5-22. A single seismic reflector is assumed to occur at depth Z_R, and the downgoing first arrival generates a primary reflection at that depth. A downgoing multiple, which is generated in the shallow part of the stratigraphic section above Z_1, is shown following the first arrival. This second downgoing event generates the indicated upgoing multiple reflection at depth Z_R. If these data are velocity filtered in f-k space with a narrow pie-slice filter, such as shown in Part C of Figure 5-21, so that upgoing events are preserved and downgoing events are attenuated, then the resulting altered VSP data after being transformed back to the space-time domain could look like the traces shown on the right of Figure 5-22. All upgoing events are smeared over space so that the upgoing multiple now appears to be generated at depth Z_2 instead of Z_R. Consequently, this event could be easily interpreted as a primary reflection generated by the first arrival at depth Z_2 rather than as a multiple originating at depth Z_R.

If an objective of data processing is to preserve subtle waveform changes from trace to trace, then Rieber mixing of data must be minimized. This objective can be best achieved with f-k filters if narrow reject filters are used instead of narrow pass filters. The contrast between these two types of f-k filter designs is shown in Figure 5-23. In this illustration, it is assumed that all downgoing VSP events are already removed so only the left half of the f-k plane is shown. If the intent is to attenuate all recorded energy except that represented by the upgoing mode P', then the S' energy must be removed. Each of the indicated f-k filters is created by numerically attenuating all data points occurring within the dotted areas of the f-k plane. The narrow pass filter will attenuate S' and preserve P' but only at the expense of severe trace mixing. On the other hand, the narrow reject filter will remove S' and cause minimal Rieber mixing of P' wavelets because the altered f-k passband that is input to the inverse Fourier transform calculation is still quite wide. This approach to f-k filter design in VSP data processing has been discussed by Gaiser and DiSiena (1982a).

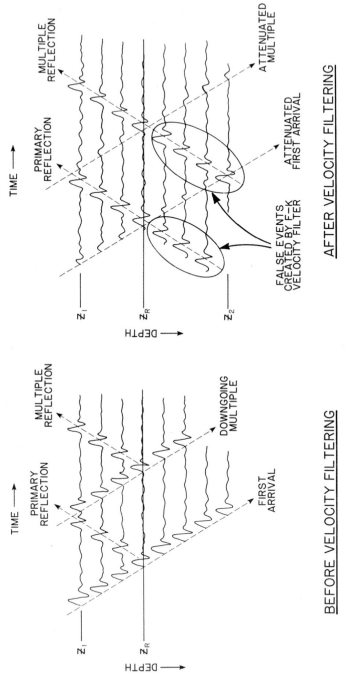

BEFORE VELOCITY FILTERING

AFTER VELOCITY FILTERING

Figure 5-22   These hypothetical VSP data show how f-k velocity filtering can attenuate strong downgoing events and allow weaker upgoing events to be identified and studied. F-K velocity filtering of VSP data results in a Rieber type mixing of the data in the time-space domain. This averaging smooths waveshapes and improves the phase alignments of events inside the velocity passband. However, it also causes these events to extend beyond their actual spatial termination depths.

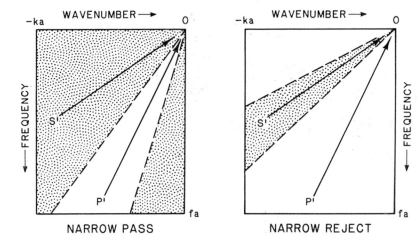

Figure 5-23   Two upward traveling events, P' and S', are shown in the left half of the f-k plane.   Event P' can be preserved and event S' attenuated by either a "narrow pass" or a "narrow reject" type of velocity filter.   These filters are created by numerically attenuating all data points falling within the dotted areas.   Narrow rejection filters are usually preferred because they create less spatial mixing and averaging of event P' when the altered f-k data are converted to the time-space domain.

Median Filters

Median filters were developed as a means of smoothing data in which the signal that needs to be preserved contains abrupt discontinuities.   Initial uses of this filtering concept by such people as Rabiner et al., (1975) and Jayant (1976) were in the area of speech processing where voice data typically contain high amplitude events (spoken words) separated by null responses (pauses in speech).   Frieden (1976) describes how median filters can be used to enhance digital images where one needs to smooth light intensities and color tones yet also honor the abrupt boundaries and edges of imaged objects. Computational algorithms that construct median filters for such purposes are described by Huang et al. (1979) and Evans (1981).   Bednar (1982a) has recently suggested several applications of median filtering in seismic data processing, and Evans (1982) has emphasized how median filters can remove unwanted data spikes.   Many seismic data processors who are aware of the median filter seem to view its capability as limited to that of a data despiker; however, a median filter can also be used as a powerful velocity filter.   Seismograph Service Limited appears to be the first geophysical group to use median filters as multichannel VSP data smoothers and velocity filters, and have submitted a patent application concerning their development (Fitch and Dillon, 1983b).

The term "median" is used here in its correct statistical sense and should not be confused with the terms mean, average value, or weighted mean. If N statistical data samples are arranged in ascending order of magnitude, then the median value is the sample in the $(N+1)/2$ position of the sequence. When N is odd, the median is the middle value of the ordered set of data. If N is even, the median is usually defined as the mean of the two middle terms of the monotonically increasing sequence.

Median filters do a type of data smoothing. Let X_1, X_2, X_3, X_4, X_5 represent a sequence of statistical samples having variable magnitudes. Reordering these values so that they successively increase in magnitude will in general create a different sequence, which might for example be X_3, X_5, X_2, X_4, X_1. The median value of this reordered sequence is X_2. In this instance, the action of a median filter can be represented as:

$$(X_1, X_2, X_3, X_4, X_5) \longrightarrow \boxed{\begin{array}{c}\text{Median}\\ \text{Filter}\end{array}} \longrightarrow X_2. \tag{15}$$

This example illustrates an important mathematical property of a median filter; which is, median filtering is a nonlinear process. The filter output X_2 cannot be represented as a linear combination of filter coefficients convolved with the input data vector. Neither can the output be described as a multiplication of a filter frequency spectrum with the frequency spectrum of the input data vector. If several linear operators $a(t)$, $b(t)$, $c(t)$, ... are applied to a data vector $x(t)$, then the order in which the operations are done is not critical. Thus the repeated convolution

$$y(t) = a(t) * x(t) * b(t) * c(t), \tag{16}$$

where * represents convolution action, is the same as the convolution sequence

$$y(t) = b(t) * a(t) * x(t) * c(t). \tag{17}$$

However, if a nonlinear mathematical process is applied to the data, that process cannot be arbitrarily performed at any step in the processing sequence. If $\emptyset(t)$ is a nonlinear operator, then

$$y_1(t) = \emptyset(t) * a(t) * x(t) \tag{18}$$

184

is not the same result as

$$y_2(t) = a(t) * \emptyset(t) * x(t). \tag{19}$$

Thus, if median filters are used in processing VSP data, it is important to note the point in the processing procedure at which the filtering is done. If two distinct VSP data sets are to be processed with a median filter and compared, the same linear operations must precede and follow the median filtering in both cases; otherwise, one may be comparing nonlinear filter effects and not real subsurface geological characteristics.

Two properties that make median filters attractive in VSP data processing are:

1.   Median filters absolutely reject noise spikes, and

2.   Median filters pass step functions without altering them.

Property 1 is demonstrated in Figure 5-24. Any single data point having a magnitude, B, which is greatly different from that of its neighboring data points, a_n, will always occur at the extreme end of the reordered data vector and can never be the median value. Thus a median filter rejects noise spikes absolutely.

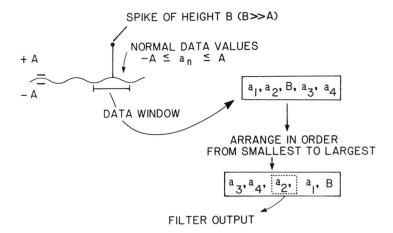

Figure 5-24   One attraction of a median filter is that it absolutely rejects noise spikes. Arranging data values so that they sequentially increase in magnitude will always place the spike value at one end of the rearranged data vector. Thus the median can never be the spike value. A five point median filter, centered on a spike, is illustrated here.

Property 2 is perhaps the more important feature of a median filter because VSP data contain numerous wave modes which exhibit abrupt discontinuities or which originate and terminate at discrete depth points. These types of events can be viewed as "step functions". For example, upgoing reflections are step functions since they do not exist below the impedance interface where they are generated. Likewise, a downgoing intrabed multiple created at depth Z_2 by an upgoing reflection is a step change since it does not exist above Z_2. This latter example of a step function behavior contained in VSP data is shown as feature #2 in Figure 5-25. There are other types of step function phenomena that may occur in VSP data. A second example could be the faulted reflector shown as feature #1 in Figure 5-25. In most VSP applications, it is important that data processing procedures be used which will emphasize these step functions without changing their position in time or space.

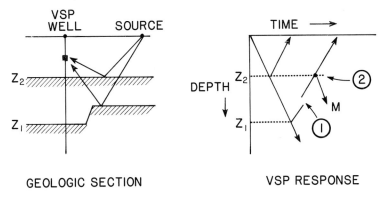

GEOLOGIC SECTION   VSP RESPONSE

Figure 5-25 During data processing, it is important to preserve the time and depth locations of the "step functions" contained in VSP data. The VSP response shown here indicates several wave modes that originate and terminate at discrete points in time and space. Step function 1 marks the fault in the upgoing reflection, Z_1. Step function 2 defines the creation of a downgoing multiple at depth Z_2. Any processing that smears the termination points of these events over time or space adversely affects the interpretation of the data.

The median filtering of a step function is diagrammed in Figure 5-26. This particular step function is intended to represent the behavior of a VSP event sampled vertically as a function of depth at a fixed recording time. Specifically, it could represent the termination of one part of the upgoing VSP event from the faulted unit at depth Z_1 in Figure 5-25 after that event is time shifted to vertically align it along the depth axis. If the source-receiver geometry is such that reflection points traverse the fault as the recording geophone rises in the well, then this upgoing reflection will exhibit an abrupt offset (i.e. a step change) as shown by the fixed-source-offset models calculated

186

in Chapter 6. The filtering action illustrated in Figure 5-26 shows that as a median filter moves downward toward deeper depths along a constant time line passing through a high amplitude part of this vertically aligned upgoing reflection, that it passes through a step reduction in amplitude without repositioning or smearing the step change along the depth axis. Thus, median filters have the admirable qualities of smoothing a data vector yet faithfully preserving all step functions in that vector.

The term "smoothing" has to be used with some caution when describing the filtering action of a median filter. It is not uncommon for the output of a median filter to exhibit small amplitude data spikes at random locations in what is otherwise a reasonably smooth digital function. These small amplitude noise spikes are frequently called "whiskers". These spikes usually do not have a large amplitude, but they can make the output of a median filter have a slightly jagged appearance. Data that have been median filtered should also be bandpass filtered in order to remove these whiskers.

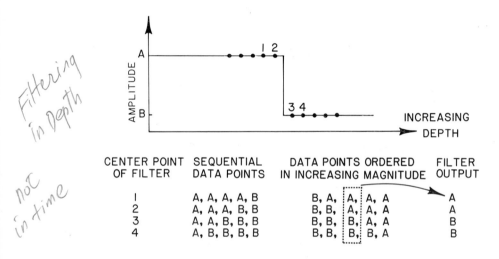

Figure 5-26   A median filter does not smear step functions. This example shows a five point median filter operating on a step function whose magnitude decreases from A to B between sample points 2 and 3. The filtered output is also a step function which decreases from A to B between sample points 2 and 3.

An additional property of a median filter that is important in VSP data processing is the smoothing which results when this type of filter reacts with triangular and boxcar shaped functions. The convolution of a median filter with a triangular pulse is shown in Figure 5-27 and tabulated in Table 5-1. A short (e.g. 3 points) median filter will smooth

Table 5-1   Tabulation of two of the median filter outputs illustrated in Figure 5-27.

CENTER OF FILTER	5 POINT MEDIAN FILTER		7 POINT MEDIAN FILTER	
	DATA SAMPLES IN INCREASING ORDER	FILTER OUTPUT	DATA SAMPLES IN INCREASING ORDER	FILTER OUTPUT
A	1, 1, 1, 1, 1	1	1, 1, 1, 1, 1, 1, 2	1
B	1, 1, 1, 1, 2	1	1, 1, 1, 1, 1, 2, 5	1
C	1, 1, 1, 2, 5	1	1, 1, 1, 1, 2, 5, 8	1
D	1, 1, 2, 5, 8	2	1, 1, 1, 2, 5, 8, 11	2
E	1, 2, 5, 8, 11	5	1, 1, 2, 5, 8, 8, 11	5
F	2, 5, 8, 8, 11	8	1, 2, 5, 5, 8, 8, 11	5
G	5, 5, 8, 8, 11	8	2, 2, 5, 5, 8, 8, 11	5
H	2, 5, 8, 8, 11	8	1, 2, 5, 5, 8, 8, 11	5
I	1, 2, 5, 8, 11	5	1, 1, 2, 5, 8, 8, 11	5
J	1, 1, 2, 5, 8	2	1, 1, 1, 2, 5, 8, 11	2
K	1, 1, 1, 2, 5	1	1, 1, 1, 1, 2, 5, 8	1
L	1, 1, 1, 1, 2	1	1, 1, 1, 1, 1, 2, 5	1
M	1, 1, 1, 1, 1	1	1, 1, 1, 1, 1, 1, 2	1

Table 5-2   Tabulation of the median filter outputs illustrated in Figure 5-28.

CENTER OF FILTER	9 POINT MEDIAN FILTER		11 POINT MEDIAN FILTER	
	DATA SAMPLES IN INCREASING ORDER	FILTER OUTPUT	DATA SAMPLES IN INCREASING ORDER	FILTER OUTPUT
A	2 2 2 2 2 2 2 7 7	2	2 2 2 2 2 2 2 2 7 7 7	2
B	2 2 2 2 2 2 7 7 7	2	2 2 2 2 2 2 2 7 7 7	2
C	2 2 2 2 2 7 7 7 7	2	2 2 2 2 2 2 7 7 7 7	2
D	2 2 2 2 7 7 7 7 7	7	2 2 2 2 2 2 7 7 7 7	2
E	2 2 2 2 7 7 7 7 7	7	2 2 2 2 2 2 7 7 7 7	2
F	2 2 2 2 7 7 7 7 7	7	2 2 2 2 2 2 7 7 7 7	2
G	2 2 2 2 7 7 7 7 7	7	2 2 2 2 2 2 7 7 7 7	2
H	2 2 2 2 7 7 7 7 7	7	2 2 2 2 2 2 7 7 7 7	2
I	2 2 2 2 2 7 7 7 7	2	2 2 2 2 2 2 7 7 7 7	2
J	2 2 2 2 2 2 7 7 7	2	2 2 2 2 2 2 2 7 7 7	2
K	2 2 2 2 2 2 2 7 7	2	2 2 2 2 2 2 2 2 7 7 7	2

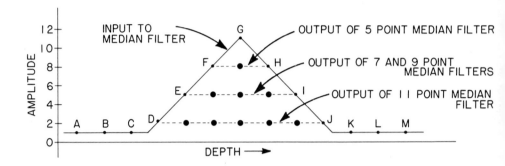

Figure 5-27 The smoothing capability of a median filter is controlled largely by the length of the filter. Some of these filters output values are tabulated in Table 5-1.

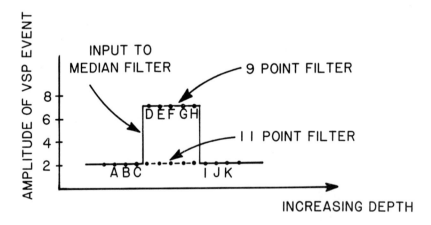

Figure 5-28 A median filter can completely reject a boxcar-shaped amplitude anomaly if the filter is sufficiently long. If N data points occur within the box, the median filter must span at least (2N+1) data points in order to completely reject the high amplitude anomaly. The filter outputs of these 9 and 11 point filters are tabulated in Table 5-2.

only the apex point of the triangle, but the amount of smoothing increases as the filter length increases as shown in the figure. A median filter spanning 15 or more data points would convert this triangular pulse to a perfectly flat trace having an amplitude of 1.

The smoothing behavior that median filters have on a boxcar-shaped function is illustrated in Figure 5-28 and tabulated in Table 5-2. Again, the filter response shows that a square shaped pulse can be completely rejected if the median filter is long enough. Specifically, if the square pulse spans N data points, then a median filter containing (2N+1) or more points will completely reject the amplitude change. The principal point to emphasize is that long median filters can reject VSP amplitude anomalies which appear as pseudo-triangular or pseudo-boxcar shapes extending over depth intervals spanning several VSP traces. This property can be used to attenuate wave modes that cross a VSP data set obliquely rather than vertically. Consequently, one does not have to resort to f-k velocity filtering in order to remove unwanted VSP wave modes.

Removal of Downgoing Wave Modes By Subtraction

If the downgoing VSP wavefield can be accurately estimated, then that estimate can be subtracted from the total VSP response in order to attenuate the strong downgoing modes which conceal and camouflage the weaker upgoing reflections that need to be retrieved from VSP data. This estimation and subtraction technique can be viewed as the four step process illustrated in Figure 5-29. Data set A in this figure represents VSP data displayed in their original field recorded format, where the strong downgoing modes (solid lines) exhibit a down-to-the-right time stepout with depth, and upgoing modes (dashed lines) have an up-to-the-right time stepout. The first processing step is a negative time shift of the data which vertically aligns all downgoing modes. This step is shown as data set B and is accomplished by shifting each trace to the left by an amount equal to its first break time. A real data example of this time shifting procedure is shown in Figure 5-31.

A median filter is now applied to these data in the depth direction along every constant time line. This filtering step, shown as data set C, accomplishes two things:

1.   It severely attenuates all upgoing modes.
2.   It smooths and amplifies all downgoing modes.

190

2 steps.

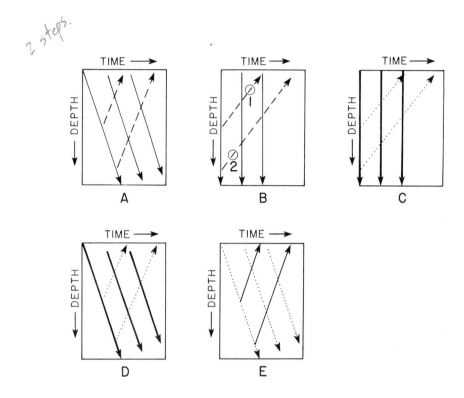

Figure 5-29   The removal of downgoing wave modes by median filtering and
              arithmetic subtraction.

A -   Original VSP data. Downgoing events shown as solid arrows. Upgoing
       events shown as dashed arrows.

B -   VSP data time shifted to vertically align downgoing events.

C -   Data set B after applying a "long" median filter along the depth axis.
       Downgoing events are smoothed and emphasized (heavier solid arrows).
       Upgoing events are strongly attenuated (dotted arrows). A bandpass
       filter may have to be applied in order to remove the "whiskers" created
       by the median filter.

D -   Data set C shifted back to same time alignment as data set A.

E -   Result after subtracting data set D from data set A. Downgoing events
       are heavily attenuated (dotted arrows).   Upgoing events are only
       slightly affected and are now stronger than downgoing events.

These two filtering objectives are achievable only if the basic seismic wavelet has a consistent waveshape from trace to trace. If we assume that an invariant seismic wavelet exists throughout data set B in Figure 5-29, then a median filter applied to these data along fixed time lines will smooth and strengthen all downgoing events, as shown by the heavier lines in C.

The attenuation of upgoing events by a median filter can be explained with the hypothetical data sketched in Figure 5-30. This illustration shows downgoing events vertically aligned and upgoing events with considerable up-to-the-right time stepout, just as occurs in data set B. A median filter operating on an aligned downgoing event, such as along time line T_1, smooths and accentuates the downgoing event since the waveshapes are reasonably constant, and since time line T_1 passes through approximately equal phase positions of the wavelet. It is assumed that whenever wavelets have a constant shape, equal phase points in the wavelets will have approximately the same amplitude. When the median filter passes through a non-vertical event, as it would along time line T_2, the filter rejects the dipping event since that wave mode does not exhibit a uniform vertical phase alignment. The hypothetical situation sketched along time line T_2 corresponds to a median filter operating on a single spike, as is shown in Figure 5-24. If an upgoing event has a steep time stepout, so that an approximate phase alignment of a high amplitude wavelet along time line T_2 occurs over 4 or 5 vertical traces, then the input to the median filter would be more like the triangular function shown in Figure 5-27. As illustrated in that figure, it is essential that the length of the median filter be increased as the width of the triangular amplitude anomaly (i.e. the degree of vertical phase alignment of the dipping event) increases if the filter is to reject the anomaly. When constructing a median filter to convert data set B in Figure 5-29 to data set C, a filter as long as nineteen or more points may be needed in order to achieve complete rejection of some upgoing modes.

The application of this median filter is a critical step in the estimation of the downgoing wave modes because the filtered result (C in Figure 5-29) is assumed to be the downgoing wavefield which is to be subtracted trace by trace from the original data. The subtraction process is accomplished in Figure 5-29 by first time shifting data set C by the opposite amount used to change the data from A to B. This new data set, D, is then subtracted trace by trace from the original data A to yield data set E. This subtraction severely attenuates the downgoing modes in A (the small downgoing residues are shown as weak dashed lines), but it does not affect upgoing modes since no significant upgoing modes occur in D. An alternate subtraction technique for removing downgoing wavefields and preserving upgoing wavefields is described by Mons (1980a).

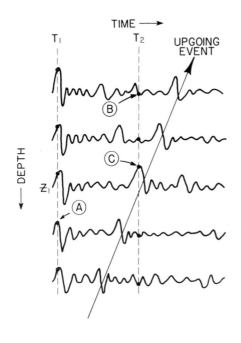

A – OUTPUT OF 5 POINT MEDIAN
     FILTER AT DEPTH Z_i AND
     TIME T_i

B – OUTPUT OF 5 POINT MEDIAN
     FILTER AT DEPTH Z_i AND
     TIME T_2

C – DOMINANT AMPLITUDE OF
     UPGOING EVENT REJECTED
     BY MEDIAN FILTER

Figure 5-30   These hypothetical VSP data are time shifted to vertically align downgoing events. A spatial median filter operating along fixed time lines, such as along T_1, generates smoothed versions of downgoing wavelets at each depth if the wavelets are properly phase aligned and approximately the same shape. Events whose phases are not vertically aligned are attenuated. An example would be the rejection of upgoing event C by a median filter operating along time line T_2. If the upgoing event is partially in phase over several depth traces, the length of the median filter must be increased (as demonstrated in Figure 5-27) in order to reject the triangular-shaped amplitude response of the upgoing event which occurs along a fixed time line.

Although median filtering does not use a Fourier transform technique to separate upgoing and downgoing wavefields, VSP data still should be recorded at vertical spatial increments, ΔZ, so that the Nyquist sampling theorem is satisfied when the preceeding processing procedure is used. Otherwise, the large phase changes in wavelets between widely spaced recording depths, and the attendant large amplitude variations that are associated with these phase changes, invalidate much of what has been stated here and illustrated in Figures 5-29 and 5-30.

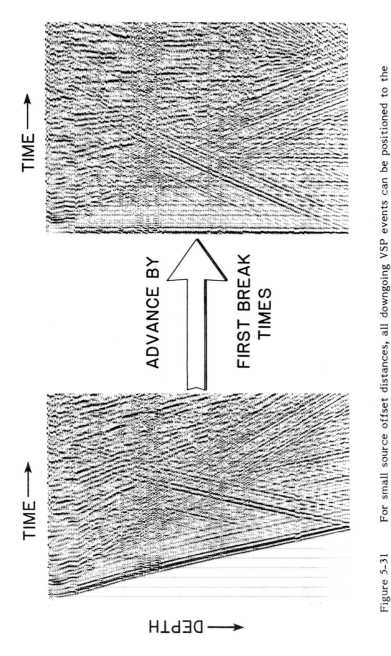

Figure 5-31   For small source offset distances, all downgoing VSP events can be positioned to the two-way times at which they leave the interface where they are generated, irrespective of the depth at which the events are recorded, if each VSP trace is advanced in time by an amount equal to its first break time. Some of these traces need to be better adjusted in time in order to optimally align the first breaks.

One point that should be emphasized regarding any wavefield subtraction process is that first break times must be measured with extreme care and should be accurate to within 0.5 ms if possible. Otherwise, the static time shifts that create data set B in Figure 5-29 will not be precise, and downgoing events will not be vertically aligned with optimum phase agreement. Median filtering of poorly aligned data, such as is shown in Figure 5-31, generates excessive noise. Likewise, when wavefields A and D are subtracted, they must be in perfect time registry relative to each other, or the difference, E, will be dominated by noise. An example of upgoing events retrieved from VSP data by median filtering and wavefield subtraction is shown in Figure 5-44.

Designing Deconvolution Operators from Downgoing VSP Wavefields

Surface geophones cannot distinguish between downward reflected seismic wavelets leaving the earth's surface and upgoing wavelets arriving at the surface. However, downgoing events are easily separable from upgoing events in vertical seismic profiling data since the direction of VSP geophone deployment creates opposite time-depth stepouts for upward and downward traveling wavelets. By appropriate velocity filtering, a vertical seismic profile can be decomposed into a downgoing wavefield and an upgoing wavefield. The ability to retrieve the complete downgoing seismic wavefield from VSP data allows one to calculate robust deconvolution operators which will separate upgoing multiples from upgoing primary reflections. The concept of using a downgoing wavefield to design deconvolution operators for removing multiples from an upgoing wavefield is described in patents assigned to Seismograph Service (England) Limited (Anstey, 1980; Fitch and Dillon, 1983c).

An advantage of using downgoing VSP events to calculate deconvolution operators is that the operators are determined from a wavefield (i.e. the downgoing events) whose signal strength is 20 db to 40 db greater than that of the usual wavefield used to calculate deconvolution operators (i.e. weak upgoing events recorded at the surface). The calculation is therefore based on the best possible description of the multiple relationships that exist in the stratigraphic section near a VSP well, and the influence of noise on the calculation is minimized. It is necessary that all downgoing events be analyzed in terms of their proper two-way times relative to each other in order to calculate the deconvolution operators needed in this procedure. The positioning of downgoing events to their correct two-way travel times is accomplished by advancing each VSP trace by its

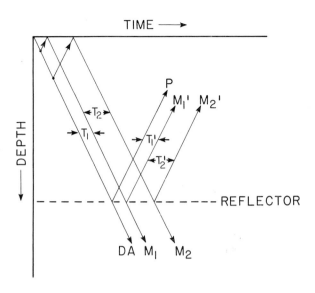

Figure 5-32   The time delays T_1' and T_2' by which upgoing multiples M_1' and M_2' follow each primary reflection, P, are the same as the time delays T_1 and T_2 by which the downgoing multiples M_1 and M_2 follow the direct arrival, DA. Thus numerical operators that preserve DA and attenuate M_1 and M_2 will also preserve P and attenuate M_1' and M_2'.

first break time, as illustrated in Figure 5-31. Each first arrival then occurs at time zero, and all subsequent downgoing events following the first breaks are at their proper two-way times.   Both downgoing surface multiples and downgoing subsurface intrabed multiples are properly positioned as a function of two-way time by these static time shifts. The assumptions of small source offset distance and flat, horizontal reflectors must be kept in mind when performing this time shifting, as is emphasized in the text describing the features of Figure 5-3. A single trace estimate of the total downgoing wavefield can be made by vertically summing all, or a selected subset, of the aligned traces shown on the right side of Figure 5-31.   This summing process is called downstacking (Anstey, 1980).

There is only one downgoing event which generates primary reflections, that being the downgoing first arrival at each subsurface interface. This first arrival is labeled DA (direct arrival) in Figure 5-32, and the upgoing primary reflection generated by it is labeled P.   All later downgoing events, such as near-surface multiples M_1 and M_2, generate upgoing multiples, which are labeled M_1' and M_2'. The time delays, T_1' and T_2', between these successive upgoing events are the same as the time delays, T_1 and T_2, between the successive downgoing events that created them. Therefore, deconvolution operators that cancel the downgoing events occurring after the first arrival will also cancel upgoing multiples that follow each upgoing primary reflection generated by the first arrival. This concept is the key to using the timing relationships of downgoing events to remove upgoing multiples.

The calculation and application of the deconvolution operators needed in this procedure are illustrated in Figures 5-33 and 5-34. The data in panel A of these two figures are, respectively, the downgoing and upgoing wavefields obtained by applying f-k velocity filters to the same VSP data set. The data in Figure 5-33 are filtered to attenuate upgoing events, thus these data represent the total downgoing wavefield, which consists of the direct arrival followed by numerous downgoing multiples. The data in Figure 5-34 are filtered to attenuate downgoing events and thus represent the total upgoing wavefield, which consists of primary reflections and several multiples which follow each primary reflection. In addition, the data in Figure 5-34 are time shifted to vertically align upgoing events. The multiple pattern in a vertical seismic profile varies with depth because the number of downgoing intrabed multiples changes with depth; therefore, a distinct deconvolution operator may have to be calculated for each VSP trace in Figure 5-33. Upgoing intrabed multiples generated below a geophone's recording depth are usually not properly attenuated in that geophone's deconvolved upgoing response, unless multiples with equivalent polarity, amplitude behavior, and timing delays are already contained in the downgoing wavefield at that geophone level. One should try to achieve adequate deconvolution results with a single deconvolution operator if possible, particularly if the intent is to use this VSP derived operator to remove multiples in surface-recorded data. The data in panel B of Figure 5-33 show the result of applying a distinct deconvolution operator to each trace in panel A. For each such test, a processor must make a subjective judgment as to whether or not the amount of multiple attenuation is adequate, and whether or not the final waveshape of the first arrival is acceptable.

The result of applying these same operators to the upgoing wavefield is shown in Figure 5-34. The effect of the deconvolution operators on several features of the upgoing wavefield is described in the figure label. The deconvolved set of upgoing events in Figure 5-34 can now be vertically summed (upstacked), and the composite trace will contain only a small amount of contamination due to multiples. Such a trace is a valuable interpretational tool that can be used to specify whether a particular surface-recorded event near this well site is a primary reflection or a multiple.

One interesting aspect of this type of approach to calculating deconvolution operators is that multiples having time delays up to a certain value can be attenuated without affecting multiples with longer time delays. Thus, the origin of groups of multiples can be studied throughout a stratigraphic section without the attendant

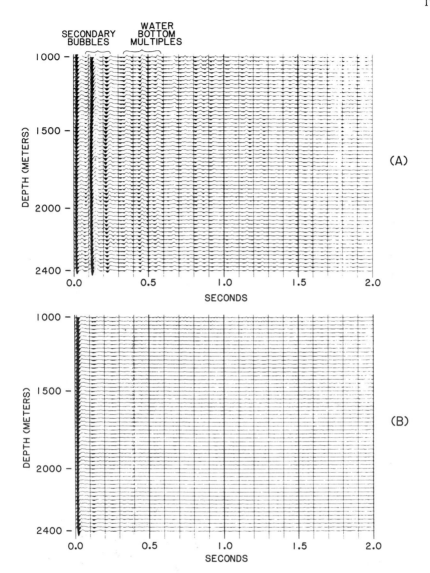

Figure 5-33

A -   Downgoing wavefield extracted from a VSP recorded in 250 meters of water with a single 200 in³ airgun. The secondary bubble oscillations and some water bottom multiples are labeled on the top trace.

B -   Same downgoing wavefield after predictive deconvolution operators, 100 ms long with prediction distances of 40 ms, are calculated and applied to each trace of "A". A gain function of the form KT^n is applied to each trace of panels "A" and "B" in order to overcome the amplitude decay caused by spherical divergence.

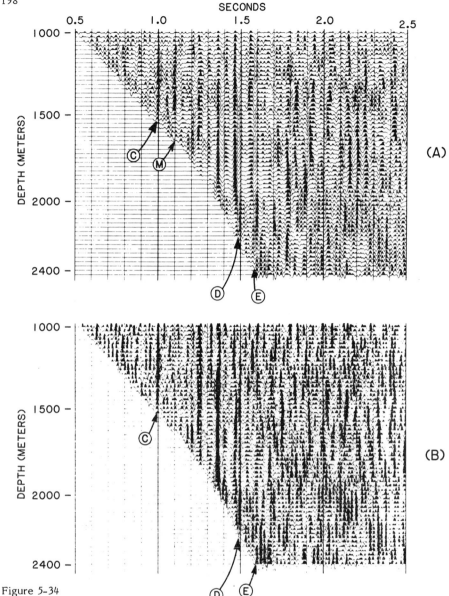

Figure 5-34

A - Upgoing wavefield obtained from the same VSP data that yielded the downgoing wavefield in panel A of Figure 5-33. Event M is a secondary bubble image of primary reflection C. Primary reflection event E has a time delay relative to primary reflection D which is unfortunately the same as the time delay of the secondary bubble oscillation.

B - Upgoing wavefield after the predictive deconvolution operators determined in Figure 5-33 are applied to the data in panel A above. Primary reflection C is preserved, but multiple M is suppressed. Event D is resolved into a doublet. Event E is severely attenuated even though it is not a multiple.

confusion of all multiples being present. Some geophysicists are of the opinion that this insight is the only way that deep, late-arriving intrabed multiples can be successfully removed from surface-recorded data, since surface data usually do not contain enough repetitions of these events to allow statistical deconvolution techniques to succeed.

An example of a VSP derived deconvolution operator being applied to surface-recorded data is shown in Figure 5-35 (Hubbard, 1979; Seismograph Service Ltd, 1980b). The surface-recorded data in the top panel have been subjected to state of the art processing, which includes far-field signature deconvolution applied both before and after stack, a statistical deconvolution, and a 2-D migration. The resulting seismic section is split at the location of a VSP well, and a panel of six repeated traces labeled, "Deconvolved Upgoing VSP Wavefield", is inserted. The deconvolution operator applied to the upgoing wavefield was calculated from the downgoing VSP wavefield and represents an operator that (a) satisfactorily attenuates all downgoing events following the first arrival, and (b) converts the VSP first arrival wavelet to a preferred waveshape. The VSP traces inserted into this display could be chosen as any of the following:

1. A repeated set of the single VSP trace created by vertically summing (upstacking) all of the traces in the deconvolved upgoing VSP wavefield.

2. A repeated set of a single representative VSP trace chosen from the deconvolved upgoing VSP wavefield at a depth above a major zone of interest.

3. Six consecutive VSP depth traces chosen from the deconvolved upgoing VSP wavefield above a selected zone of interest.

In any of these cases, the deconvolved upgoing VSP data are a good estimate of the primary reflection events that should be recorded at the surface near the well site. Comparing the conventionally processed surface data with the deconvolved upgoing VSP data inserted into the display shows that some VSP reflections are not accurately portrayed in the surface data, and that some surface-recorded events have no counterparts in the VSP data.

The bottom data panel shows the same surface data, except that the deconvolution operator used to remove multiples from the VSP data is now also applied to the surface-recorded events before stacking them. No other deconvolution is performed in the processing. Hubbard (1979) outlines the following process by which this deconvolution operation could be performed:

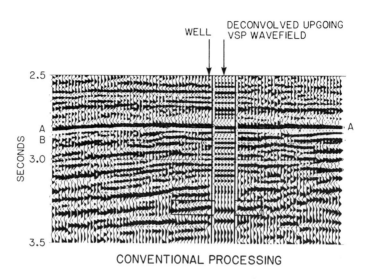

CONVENTIONAL PROCESSING

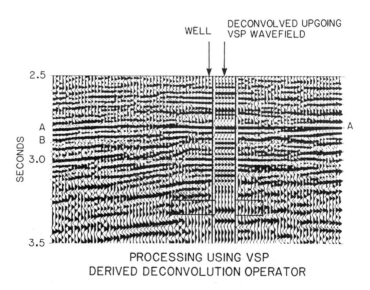

PROCESSING USING VSP
DERIVED DECONVOLUTION OPERATOR

Figure 5-35   A comparison of surface-recorded data processed by conventional deconvolution techniques and by a deconvolution operator determined from the downgoing VSP wavefield (After Hubbard, 1979, and Seismograph Service Ltd., 1980b).

1. Select a VSP trace from the downgoing wavefield just above some zone of interest and determine its autocorrelation function. This VSP trace will define all multiples that enter the stratigraphic interval immediately below its recording depth. These VSP data should be processed to preserve correct reflection amplitudes in order to construct optimum deconvolution operators.

2. From this autocorrelation, calculate a predictive deconvolution operator that removes the seabed reverberation from the VSP trace. (If the VSP data were recorded in a marine environment).

3. Apply this predictive deconvolution operator to the downgoing VSP traces and recalculate the autocorrelation function of the selected VSP trace. This VSP trace now defines all multiples, except the seabed reverberations, which enter the zone of interest.

4. Derive an inverse operator that removes the remaining multiples left in the VSP trace.

5. Apply the operator in step 4 to the surface-recorded data.

6. Apply a predictive deconvolution to the surface-recorded data to remove the seabed reverberation. (If marine data are involved.)

These steps do not necessarily define a unique or optimum approach to deconvolution design; other options are possible. For example, the two deconvolution operators in steps 2 and 4 could be combined into a single operator. However, in Figure 5-35, the agreement between the VSP data and the surface data processed in this manner is much improved compared to the agreement obtained when the surface data are processed with conventional deconvolution techniques. Note particularly that the unconformity between A and B is easier to interpret in the bottom panel. Likewise, several multiple events in the conventionally processed data, such as the large amplitude event in the center of the rectangular area outlined at 3.3 seconds, are removed by the VSP derived deconvolution operator. The low frequency VSP event between 3.3 and 3.4 seconds can also be seen extending to the right of the well after the VSP derived deconvolution operator is applied to the surface data.

Wavelet Processing

The interpretive value of VSP data is greatly increased if every seismic shot emits the same basic source wavelet, so that wavelet variations recorded downhole can be confidently assumed to be manifestations of geological conditions and not artifacts of a variable energy source. Consequently, great care should be exercised in the field to insure that a uniform physical environment is maintained around the energy source used to generate downgoing wavelets, that the coupling between the energy source and earth does not change during the course of a VSP experiment, and that the same amount of energy is created by the source each time it is activated. Even when careful field procedures are followed, VSP energy sources still create variable wavelets. Numerical wavelet processing is thus required in order to convert all source wavelets to a standard waveshape.

In order to perform wavelet processing of VSP data, the wavelet propagating away from the source must be recorded independently of the downhole VSP data. This practice of recording source monitor wavelets in vertical seismic profiling is described in Chapter 4, and several illustrations of VSP energy sources in Chapter 3 show monitoring geophones or hydrophones placed close to the source position.

The concept of wavelet processing of VSP data is illustrated in Figure 5-36. The near-field wavelets shown in Column A represent the response recorded by near-field monitor geophones, such as shown in Figures 2-9, 2-11, and 2-13. A rather severe, but nonetheless realistic, change between the Nth and Mth source output wavelets is indicated. Neither wavelet resembles the desired shot wavelet shown in column C; consequently, wavelet shaping operators must be designed to convert these actual output wavelets to the desired waveshape. If an operator successfully converts a near-field wavelet to the selected standard input wavelet, then that operator is also applied to the downhole data recorded for that same seismic shot. If an operator cannot be designed that converts the source wavelet to the standard input wavelet, then the downhole trace recorded for that seismic shot must be carefully inspected to determine if it needs to be eliminated from further analysis.

Horizontal Stacking

When a surface energy source is offset a horizontal distance, H, from a vertical borehole, then reflection signals received by a geophone positioned in the well originate

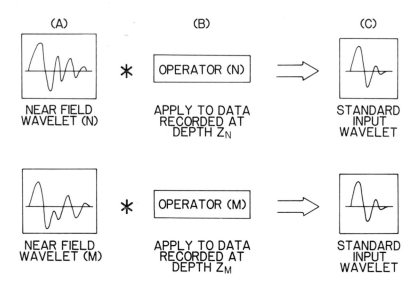

Figure 5-36   Wavelet processing of VSP data by means of wavelets recorded in the near field of a VSP energy source. The symbol * implies convolution.

from subsurface reflection points which are also laterally offset from the borehole. If a reflector is flat and horizontal, and if the borehole geophone is not far below the surface, then a reflection point can be laterally offset as far as a distance, H/2, from the wellbore. A succession of reflection points extending laterally away from a VSP well can be created by (a) keeping a surface energy source at a fixed offset distance and varying the depth of the borehole geophone, (b) keeping a geophone at a fixed depth and varying the offset distance of the source, or (c) varying both the depth of the geophone and the offset distance of the energy source. Recording geometries (a) and (b) are illustrated in Figure 8-8 and will be discussed in Chapter 8.

Whenever any type of seismic data are recorded so that subsurface responses from a succession of horizontal reflection points are preserved, then the recorded data can be arranged so that they represent a horizontal profile of the earth. Thus the ability to convert vertical seismic profiling data into a horizontal profile of the subsurface exists if the VSP shooting geometry creates reflection points that extend laterally away from a study well. Any numerical procedure that creates horizontal profile data from VSP data will be called VSP horizontal stacking. Outside of the Soviet Union, the first publicly reported horizontal stacking technique that allows VSP data to be recorded with diverse field geometries appears to be that developed by Wyatt and Wyatt (1981b).

In order to illustrate how VSP data create a horizontal image of the subsurface extending laterally away from a well, a hypothetical vertical seismic profile is diagrammed in Figures 5-37 and 5-38. Four flat, horizontal reflectors, penetrated by a vertical borehole, are shown. Raypaths extend from a surface energy source to each reflector and back upward to a borehole geophone at depth Z_G. The raypaths are drawn as straight lines simply for convenience. In reality, these rays would refract at interfaces A, B, and C so that the surface locations of reflection points X_B, X_C, X_D would be farther away from the well than drawn. The principle fact to note from this raypath picture is that primary reflection events recorded at depth Z_G originate from subsurface points that move farther away from the wellbore as recording time increases (i.e. as reflector depth increases).

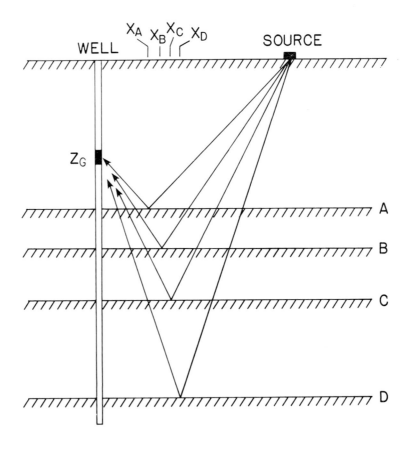

Figure 5-37   An illustration showing how reflection points, X_n, move farther away from a vertical borehole as recording time (reflector depth) increases. For simplicity, raypaths are drawn as straight lines even though distinct velocity layers exist in this hypothetical stratigraphic section.

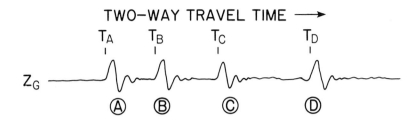

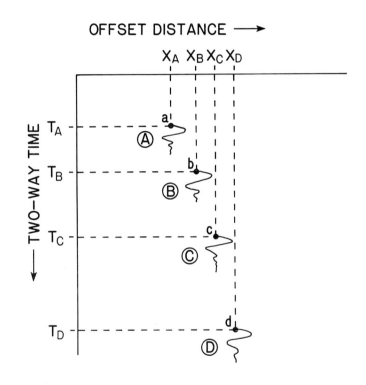

Figure 5-38   The top trace shows the response of the geophone at depth Z_G in Figure 5-37 after downgoing events are attenuated, and after the four upgoing reflections, A, B, C, D, are delayed so that they occur at their correct two-way travel times, T_A, T_B, T_C, T_D. In the lower illustration, these four reflections are positioned in time at the horizontal offset distances, X_A, X_B, X_C, X_D, where they are generated. Points a, b, c, d define a "VSP NMO curve".

The VSP response recorded by the geophone at depth Z_G is shown at the top of Figure 5-38. It is assumed that all events except upgoing primary reflections are numerically removed from the trace in this display, and that the four reflection events A, B, C, D are time shifted to their surface arrival times T_A, T_B, T_C, and T_D by properly accounting for the effects of source offset and curved raypaths. Note that these four reflections are created at increasing offset distances X_A, X_B, X_C, X_D in Figure 5-37. Consequently, in the lower illustration of Figure 5-38, each reflection is positioned in time at its horizontal point of origin. The computational procedure of separating the VSP trace at the top of this figure into several data windows, and then correctly positioning these data windows in the two-dimensional space, reflection time vs. horizontal offset distance, as shown at the bottom of the figure, is the basis of VSP horizontal stacking. Numerical techniques which perform this function are considered proprietary by those companies that have developed this data processing capability, and explicit schemes have not been published. The procedures are built upon a type of NMO correction that is appropriate to VSP source-receiver geometry, and which can be applied to data recorded at all geophone depths.

The preceding analysis shows that a single VSP trace, recorded as a function of time at a fixed geophone depth, can be broken into several data segments, and these segments can then be repositioned at their proper two-way travel times at various horizontal distances from the VSP well. Each data segment must also be compressed or stretched according to the VSP field geometry. This single trace concept now needs to be extended to several VSP traces in order to establish the principle of VSP horizontal stacking. Figure 5-39 shows rays reflecting from two interfaces, A and B, into four different geophone positions Z_1, Z_2, Z_3 and Z_4. Again, the raypaths are drawn as straight lines rather than as realistic refracted paths simply for convenience. The geophone depths are chosen so that all four raypaths reflect from subsurface points directly under surface points X_1 and X_2. The primary reflections recorded at these four depths are shown in the traces drawn at the top of Figure 5-40. For simplicity, only the four primary reflection events are shown. All downgoing events, upgoing multiples, and noise modes are assumed to have been removed. These four reflections are labeled 1, 2, 3, 4 as they are in Figure 5-39, and they are also time-shifted to their surface arrival times, T_A and T_B. In the lower illustration of Figure 5-40, the same process used to position the events in Figure 5-38 at their proper offset distances has now positioned these four reflections at their correct offset locations, X_1 and X_2. Thus from just those VSP traces recorded between Z_4 to Z_1, a horizontal stack can be created that defines the horizontal reflection character of reflectors A and B between offset distances X_1 and X_2. Recording data at other geophone positions between Z_4 and Z_1 will cause additional wavelets to reflect from interfaces A and B between points X_1 and X_2 and improve the quality of the horizontal stack. For example, a raypath from the source to reflector A in Figure 5-39, and then to

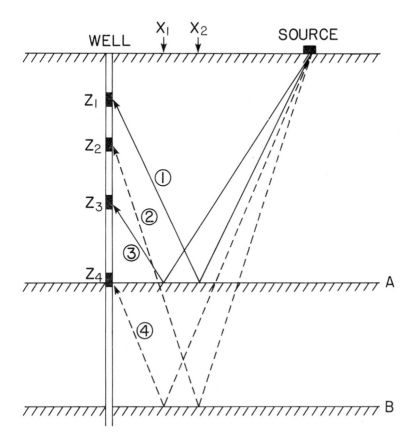

Figure 5-39   An illustration showing how reflection points along a horizontal interface move closer to the wellbore as a borehole geophone approaches that interface.   Ray paths are drawn as straight lines only to simplify the illustration.

geophone Z_2, would create a wavelet located at time T_A between wavelets 1 and 3 in the bottom display of Figure 5-40. If a well penetrates a sequence of flat, horizontal re-flectors, VSP data recorded over the entire depth interval of the well can create a horizontal stack extending from the wellbore to half the distance to the surface source. Real data examples of horizontal stacks created from VSP data have been published by Wyatt and Wyatt (1981b, 1982a, 1982b). A field procedure that raises and lowers seismic sources and receivers in adjacent boreholes so that reflection points move along geological interfaces between the boreholes is described by Hawkins (1981).   This technique is also a type of horizontal stacking.

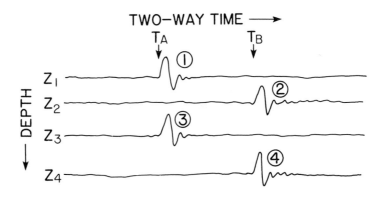

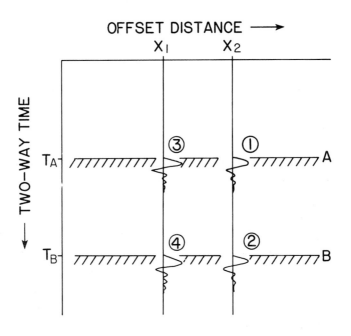

Figure 5-40   The traces at the top show the geophone responses in Figure 5-39 after all upgoing multiples and all downgoing events are attenuated and after the remaining primary reflections are shifted to their correct two-way travel times. In the lower illustration, these same reflection events are plotted at the offset distances X_1 and X_2 where they were created to show how VSP data can be converted into a horizontal stack.

Migration

Only flat, horizontal reflectors are considered in the horizontal stacking procedure described in the preceding section. Whenever horizontally stacked data are created from VSP data recorded in wells where dipping reflectors exist, the data should be migrated either before or after stacking so that reflections occur at their proper spatial positions and dip attitudes.

VSP migration differs from ordinary seismic migration of surface-recorded data in that the spatial coordinate axes defining the shot and geophone positions are orthogonal rather than parallel. Consequently, a unique kind of migration procedure is required for VSP data. Some geophysical research groups are developing f-k VSP migration procedures, but their progress has not been publicized. Wyatt and Wyatt (1981b, 1982a, 1982b) show some examples of VSP data migrated with the VSPCDP technique.

Figure 5-41 illustrates a VSP well penetrating three reflectors, R_1, R_2, and R_3, which have increasing dips. The well penetrates these reflectors at depths A, B, and C. The VSP responses drawn at the right show how these reflections should appear after the data are delayed by their first break times, assuming that the source remained at a fixed position while the data were recorded. If the offset distance to the source were small, and if the reflectors were flat and horizontal, the upgoing reflection events would lie along the vertical dashed lines. Instead, they curve toward the downgoing first arrival because of their dip. Examples of synthetic VSP data representing responses created by dipping reflectors will be shown in Chapter 6. Each reflection curvature originates at depth A, B, or C. The local curvature diminishes as the geophone recording distance above A, B, and C increases.

When migrating seismic reflections to their proper attitudes and spatial positions, one advantage of migrating VSP data, rather than surface-recorded data, is that VSP reflections are always recorded as properly migrated events along the downgoing first break line. That is, reflection R_1 is properly migrated at depth A when it is recorded since the wellbore penetrates that reflector, and no migration adjustment is ever needed at its point of origin. This same observation is true for reflection R_2 at depth B and for reflection R_3 at depth C. There are no such control points when migrating surface-recorded reflections. It must be emphasized, though, that VSP migration procedures must adjust the upgoing reflection events, R_1, R_2, and R_3, at all other points along their trajectories except at A, B, and C.

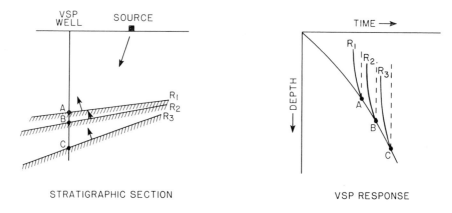

STRATIGRAPHIC SECTION   VSP RESPONSE

Figure 5-41   The unmigrated appearance of VSP reflection events created by dipping interfaces. Any migration technique utilizing VSP data has the advantage that VSP reflections are recorded as perfectly migrated data along the first break curve ABC.

Maximum Coherency Filtering

When VSP data are recorded in a well which penetrates reflecting interfaces having variable dips, or in a well which passes near subsurface diffractors, the upgoing wavefield contains events that have a large range of vertical velocities. Such a wavefield is often too complicated to interpret because some dipping reflections are weak and difficult to see, some diffractions are so strong that they overwhelm other features, and some diffractions are hidden by stronger events. For a completely rigorous interpretation of such data, it is helpful to segregate these upgoing events according to their vertical propagation velocities, and then recombine these velocity filtered events into a single display in a way that makes the amplitudes of all of the events better equalized than they were in the original data. The concept of maximum coherency filtering has been used with some success to produce this type of wavefield display (Balch, et al., 1980b, Parrott, 1980).

The basic procedures involved in maximum coherency filtering are diagrammed in Figure 5-42. As a first step, the upgoing VSP wavefield is usually separated from the downgoing wavefield by f-k filtering; i.e., all data positioned in the right half of the f-k

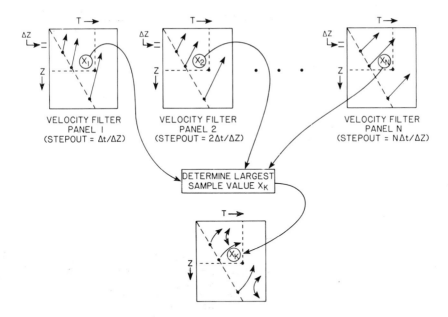

Figure 5-42   The concept of maximum coherency filtering.   A complicated upgoing wavefield containing events spanning a large range of velocities due to dipping reflectors, diffractors, etc., is separated into several simpler wavefields by filters that pass only those events traveling with a velocity close to a specified $\Delta T/\Delta Z$ time-depth stepout value. These velocity filtered versions of the original wavefield are shown in the top row. Each sample point, $X(Z,T)$, in the final maximum coherency wavefield, shown at the bottom, represents the maximum amplitude sample value, $X_K(Z,T)$, occurring at position (Z,T) in this panel of N velocity filtered wavefields.

plane are attenuated, as was shown in Figure 5-21B.   An alternative technique for separating the upgoing and downgoing wavefields would be the median filtering and wavefield subtraction approach described in Figure 5-29. The resulting upgoing wavefield is then further separated into several simpler wavefields by applying multichannel velocity filters that attenuate events whose propagation velocity does not occur in a narrow velocity passband centered about

$$V = \Delta Z/\Delta T. \tag{20}$$

These simplified wavefields are represented by the top row of N velocity filter panels in Figure 5-42.   The vertical distance, ΔZ, in Equation 20 is arbitrary, but is commonly

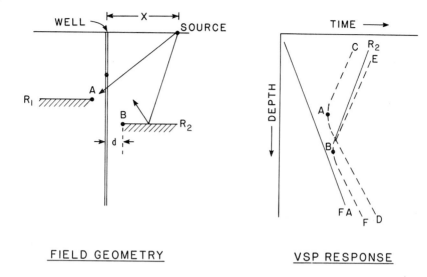

FIELD GEOMETRY   VSP RESPONSE

Figure 5-43   The appearance of compressional diffraction events in vertical seismic profiles. If diffraction leg AC is recognizable over an appreciable depth interval, it will travel approximately parallel to compressional reflection R_2 unless the source offset and diffractor offset distances are large. Diffraction legs AD and BF parallel the compressional first break line FA if the source and diffractor offsets are zero. They increasingly diverge from FA as these offsets increase (Courtesy S. B. Wyatt, Phillips Petroleum Company).

defined as the vertical separation between successive VSP traces. The time interval, ΔT, is adjusted over a user selected range

$$\Delta T = \pm \Delta t, \pm 2\Delta t, \ldots, \pm N\Delta t, \tag{21}$$

where Δt is an arbitrary time increment. It is particularly important to allow both positive and negative vertical velocity values when designing the multichannel velocity filters if the VSP data contain diffraction events. To clarify this point, consider the simple stratigraphic situation containing two diffractors, A and B, shown on the left in Figure 5-43. The appearance of these subsurface structural features in a vertical seismic profile is shown at the right. The event labeled FA indicates the downgoing first arrival. Diffraction event B, as drawn, almost blends into reflection event R_2. If R_2 is a compressional reflection and both the source offset distance, X, and the diffractor offset distance, d, are small, then this behavior means that event B is a P-wave diffraction, since a shear wave diffraction would have an apparent velocity much slower than R_2. If the source offset distance, X, and the diffractor offset distance, d, are zero, then the P-wave diffraction leg BE would blend smoothly into R_2, and leg BF would parallel FA. If these offset distances are large, then the legs of a P-wave diffraction exhibit a slower apparent velocity than do upgoing and downgoing compressional body wave events. In such cases, the upper leg of a P-wave diffraction will not parallel upgoing P-wave reflections, and neither will the downgoing diffraction leg parallel FA.

Since reflector R_1 cannot be imaged because it lies to the left of both the source and the borehole geophone, diffraction A is an isolated event that is not associated with a measured reflection. Diffraction curve CAD is also drawn as a compressional diffraction event. The asymptote of the lower leg of this compressional diffraction event will parallel the downgoing compressional first arrival, FA, if the source offset, X, and diffractor offset, d, are zero. Also, when both of these offsets are zero, the asymptote of the upper leg will parallel the time-depth stepout trend of an upgoing compressional reflection. For large source offsets and modest diffractor offsets, both diffraction legs of event A will have slower apparent velocities and will not travel parallel to upgoing and downgoing P-wave events.

If a downgoing diffraction leg is defined as having an apparent vertical velocity, $-\Delta Z/\Delta T$, then the upgoing leg has an apparent vertical velocity of $+\Delta Z/\Delta T$. Consequently, both positive and negative velocity filtered data are needed to preserve a complete diffraction event. The retrieval of downward extending diffraction legs from data that have already been f-k filtered in order to attenuate downward traveling events may not always be successful, but often there is sufficient diffraction energy left after the f-k filtering so that complete diffraction events can be recreated with carefully designed multichannel velocity filters.

Once a complete suite of velocity filtered data panels is constructed, as shown in Figure 5-42, a different fixed gain can be applied to each panel, if desired, so that the maximum amplitude in each panel is the same. A new display, unifying the major features of each panel into a single picture, can then be constructed by defining each sample point in the final display to have the maximum amplitude of those sample values found at that same space-time coordinate in the N panels of velocity filtered data. If $X_k(Z_i,T_j)$ represents the sample value at depth Z_i and recording time T_j in velocity panel k, then the sample value $Y(Z_i,T_j)$, at that space-time coordinate in the final display is

$$Y(Z_i,T_j) = \text{sgn}\big(\text{Max}\,|X_k(Z_i,T_j)|\big), \quad k = 1, 2, \text{---} N, \tag{22}$$

where sgn is +1 if $X_k(Z_i,T_j)$ is positive and -1 if $X_k(Z_i,T_j)$ is negative. Adjusting the maximum amplitude of each velocity filtered panel to the same value before performing the data sorting defined by Equation 22 allows upgoing events spanning all allowed vertical velocities to have equal amplitude weights in the final display, which is a property that does not exist in the original data. Some interpreters do not want any amplitude adjustments to be made on the velocity filtered panels because these

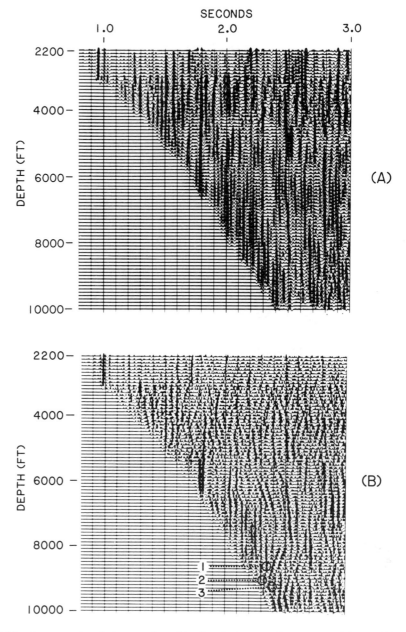

Figure 5-44   An example of the stratigraphic detail that can be extracted from VSP data by appropriate velocity filtering techniques. The upgoing wavefield in panel A was obtained by a wavefield subtraction process. Median filtering of this wavefield yields the results in panel B. (After Kennett and Ireson, 1981).

adjustments give false impressions about the magnitudes of reflection coefficients and the strengths of diffraction events. They simply create a suite of velocity filtered panels, as in Figure 5-42, and then search for the maximum amplitude occurring at each time-space coordinate in accordance with Equation 22. Either philosophy of amplitude adjustment is acceptable, depending upon the objectives sought in interpreting the data.

An example of the detail that can be extracted from VSP data by this type of procedure is shown by the data in Figure 5-44 (Kennett and Ireson, 1981). The processing steps followed in creating these data differ in some respects from those used to describe the technique portrayed by Figure 5-42, but the general concept remains the same. That is, the data are filtered so that all events in the upgoing wavefield which occur in a broad velocity range are amplified, even if some of the events are subtle or heavily camouflaged in the unfiltered data. The processing procedure uses a proprietary, and patented, median filtering concept (Fitch and Dillon, 1983b). The data in panel A have been time shifted so

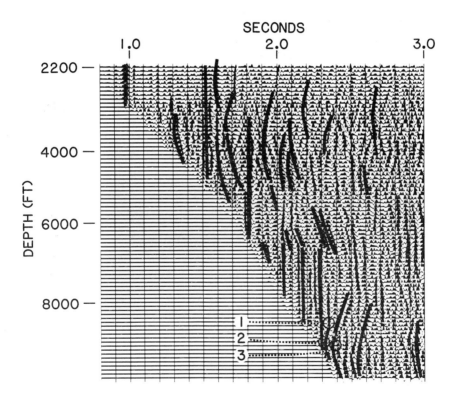

Figure 5-45   Same data as in Part B of Figure 5-44. Some selected dipping reflectors and diffraction events are emphasized by heavy lines.

that upgoing reflections from flat, horizontal reflectors are vertically aligned, and downgoing events have been attenuated by wavefield subtraction (Figure 5-29). The upgoing wavefield in panel A is then filtered so that events having velocities between -12 ms per trace spacing and +12 ms per trace spacing are preserved. The filtered result is shown in panel B. One reflection termination (labeled 1), and two diffraction apices (labeled 2 and 3), are circled because they will be used in an interpretation example discussed in Chapter 8 (Figure 8-33). Several other diffractors and reflector terminations which cannot be seen in data set A are obvious in the remainder of filtered data set B. These same filtered data are displayed in Figure 5-45 with some of the major subsurface features revealed by the filtering emphasized with heavy lines.

It should be emphasized that all diffraction events contained in this display are shear wave diffractions, not compressional diffractions, because none of them terminate smoothly into a compressional reflection event. Instead, each one has a curvature that cuts across the compressional reflections. Since the data were recorded with vertically oriented geophones in a vertical borehole, these diffraction events are assumed to be SV modes. A vertically oriented velocity sensitive geophone will not record a compressional diffraction event when it is at the same depth as the diffractor because the particle velocity created by the diffraction arrival is perpendicular to the geophone case. Thus the high amplitude vertex of a compressional diffraction is not recorded by a vertically oriented geophone. As a result, the weaker compressional diffraction tails, which are recorded since they arrive at recording depths where there is a vertical component to the particle velocity vector, cannot be confidently identified.

The detail that can be recovered from VSP data by separating the data into several distinct data sets by restrictive velocity filters, balancing the amplitudes of events occurring in selected velocity ranges, and then recombining the data into a unified display is sometimes amazing. Certainly the data in Part B of Figure 5-44 allow much more detailed insight into the subsurface than do the data in Part A. There are many ways that the velocity filtering can be done. One may use statistical median filters, mutichannel velocity filters, filters designed in f-k space, filters with a broad velocity passband, several filters with narrow passbands, balance data amplitudes after filtering, avoid any type of amplitude balancing, or use a host of other options. Each specific VSP data case has to be considered individually in order to select the best processing sequence.

Interpolation and Extrapolation of Data Across Skipped Zones

VSP data are usually not recorded in all depths intervals of a drilled borehole. For example, poor seismic coupling between multiple casings may not allow usable data to be

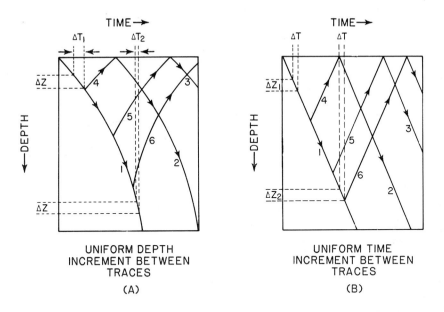

TIME→   TIME→

UNIFORM DEPTH
INCREMENT BETWEEN
TRACES

(A)

UNIFORM TIME
INCREMENT BETWEEN
TRACES

(B)

Figure 5-46   VSP data can be recorded in the field, or processed after recording, so that the downgoing first breaks are uniformly incremented in space or uniformly incremented in time. Propagating events follow curved trajectories when the first arrivals are plotted as a constant depth increment, but travel in straight line trajectories when the first breaks are displayed so that they are separated by a constant time increment. Equivalent wave modes are numbered in each drawing.

recorded in the shallow part of a well. Deeper in a study well, washouts or uncemented casing may not allow adequate geophone-to-formation coupling so that additional zones have to be skipped. Thus, VSP data tend to be sampled in an inconsistent way in the depth domain, and any numerical technique that processes VSP data as a function of depth is detrimentally affected by these data gaps. In order to improve any data processing that involves spatial filtering, and also to clarify the behavior of wavelet reflection and transmission throughout a drilled stratigraphic section, there may be instances where it is advantageous to interpolate or extrapolate VSP data across skipped recording intervals.

Since seismic propagation velocity increases with depth, downgoing and upgoing events travel across VSP traces in curved paths when the data are displayed so that there is a constant depth increment between geophone levels. This behavior is shown in a schematic form in Part A of Figure 5-46. The time increments, ΔT_1 and ΔT_2, between first break times spanning a fixed depth interval, ΔZ, decrease with depth because of this velocity increase. If VSP data are recorded at non-uniform spatial intervals which are carefully chosen so that there is a constant time increment between the first breaks of successive traces, then these curved travel paths convert into straight line paths as shown in Part B. Equivalent wave modes are numbered in these two displays. When VSP data are in the form shown in Part B, a fixed time increment, ΔT, between the same phase point of an upgoing or downgoing wavelet spans a spatial distance that increases with depth. Thus ΔZ_2 in Part B is larger than ΔZ_1. If the first breaks are caused by a downgoing compressional wave, then any recorded shear wave events will not be positioned in straight line paths in a constant-time-increment format. Thus, in this discussion we will assume that only one wave mode, either a compressional wave or a shear wave, but not both, is being interpolated or extrapolated across skipped recording intervals.

Digital data processing is less expensive and easier to implement whenever the necessary mathematical manipulations can be satisfactorily performed by simple linear functions and linear arithmetic. This observation is particularly true when interpolating or extrapolating seismic trace data. Partially for this reason, Chun et al. (1982) have emphasized that extrapolation or interpolation of VSP data across skipped depth intervals is best achieved if the data are first converted to a constant-time-increment form. Since all downgoing events travel in straight line trajectories in this format, and so do all upgoing events if the reflecting interfaces are flat and horizontal, then only linear functions are required for interpolation and extrapolation of both upgoing and downgoing events, as shown in Figure 5-47, Part B. If the data are in the usual constant-depth-increment format, more complicated curvilinear functions have to be used in order to extend the data into skipped intervals.

There are two ways by which constant-time-increment VSP data can be created. First, the data can be recorded in a borehole at non-uniform spatial increments so that there is a constant difference between the first break times of adjacent traces. This option is the preferred way by which constant-time-increment data should be obtained. In order to select the correct depth positions where a borehole geophone should be locked in order to achieve this timing behavior, one must have a sonic log already recorded in the well, or some equivalent detailed foreknowledge of the velocity behavior through the drilled section. The second option is to record VSP data at regularly spaced depth intervals, and then later numerically interpolate and extrapolate between these field traces to create a combination of pseudo and real field traces at constant time increments. Chun et al. (1982) point out that if either (or both) of these options is followed, and constant-time-increment data are created between the earth's surface and some point below the deepest recording depth, then better wave mode separation and superior velocity filtering result because spatial aliasing is avoided and boundary effects on spatial filters are minimized. In addition, upgoing and downgoing wavefields can be

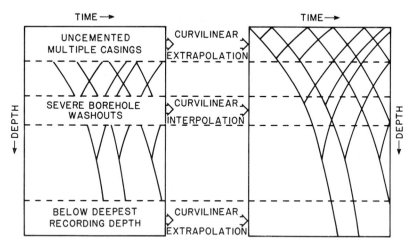

(A) – CONSTANT DEPTH INCREMENT DATA

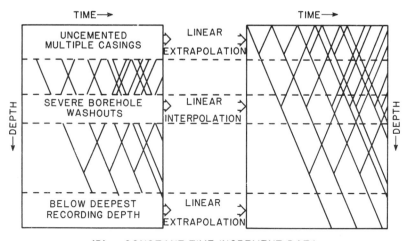

(B) – CONSTANT TIME INCREMENT DATA

Figure 5-47   Upgoing and downgoing VSP events can be numerically extrapolated into, or interpolated across, depth intervals where they were not recorded. These data manipulations are easier to do, and are usually more accurate, if they are performed on VSP data expressed in a constant-time-increment format because only linear processes are required.

better interpreted because more insight into primary and multiple reflections in the skipped intervals is achieved. Two key concepts in this data extrapolation and interpolation procedure are that upgoing reflections generate downgoing waves at the earth's surface, and that all downgoing waves create upgoing events at the same depths where primary reflections are generated.

If VSP data are recorded in a constant-time-increment format in several depth intervals so that extensive portions of the total downgoing and upgoing wavefields are captured, then the wavefields in the skipped depth intervals can be reconstructed by continuing each recorded wavelet along its linear trajectory, accounting for polarity reversals of reflections at interfaces where the acoustic impedance increases, and adjusting for amplitude decay of the propagating wavelets. After velocity filtering this combined set of pseudo and real VSP trace data, or after performing any numerical process that will benefit from trace data existing in a larger spatial region than it originally did, the interpolated synthetic traces can be eliminated from the data set so that further analyses deal only with the original field recorded traces.

CHAPTER 6

STRUCTURAL AND STRATIGRAPHIC MODELING
OF VERTICAL SEISMIC PROFILES

Some simple, yet important, VSP principles can be demonstrated by means of ray trace modeling. Ray tracing is an established procedure for investigating seismic reflection and transmission processes, yet few examples of ray trace methodology being applied to vertical seismic profiling problems can be found in public geophysical literature. Most seismic data processing centers have some type of ray tracing software which can be used to synthesize the behavior of surface reflection data recorded over mathematically defined structural and stratigraphic models of the subsurface. Altering this software so that it permits geophone locations to be at arbitrary points in the subsurface, rather than being restricted to the earth's surface, will provide a capability for calculating synthetic vertical seismic profiles. Types of questions pertaining to vertical seismic profiling that can be addressed by ray trace modeling are the following:

1./ How can the dip of a seismic reflector be determined from VSP data?

2. How much subsurface coverage is achieved in vertical seismic profiling for a given source-reflector-wellbore geometry?

3. What are some of the ways to detect a fault by vertical seismic profiling?

4. What magnitude of fault throw can be determined with VSP data?

5. How do angular unconformities appear in VSP data?

6.   What frequency content and timing accuracies need to be achieved with VSP data in order to detect a specific stratigraphic anomaly near a wellbore?

7.   When should multiple source offsets be used in a VSP field experiment instead of a single source offset?

One report illustrating VSP ray trace modeling of some of these exploration problems has been published (Seismograph Service, Ltd., 1980a). In addition to demonstrating the structural and stratigraphic messages contained in VSP data, ray trace modeling is also invaluable for determining the optimum VSP field geometry that can address a given structural or stratigraphic problem. Constructing ray trace models before starting a VSP experiment will result in fewer instances where VSP sources and geophones are positioned so that they record data which are incapable of achieving a given exploration objective. Ray trace modeling also provides a framework for an economic analysis during the planning stage of a VSP field experiment. For example, modeling can prohibit the recording of more data than are needed to achieve an exploration objective. It can also indicate how much data need to be recorded in order to properly image a subsurface anomaly, and thus allows an explorationist to make a reliable estimate of how much time and money will be required to record the data. Ray trace modeling can even serve as a quality control mechanism for an on-site QC observer during a VSP field experiment. If reasonable estimates of the subsurface conditions are known before a VSP survey is conducted, then synthetic data made from VSP models based on these estimates can tell QC personnel where they should look in the real VSP field data in order to judge if the data quality is sufficient to achieve the desired exploration objectives.

VSP Ray Tracing Assumptions

Ray trace modeling assumes that seismic energy propagates along mathematically straight line segments (i.e. rays), and that reflection and transmission processes occur at a mathematical point when one of these straight lines intersects a boundary between two different rock units. This intersection of an acoustic impedance boundary and a seismic ray is called a specular reflection point if Snell's law of reflection and refraction defines the direction that the reflected and transmitted rays take when they leave the point (see Figure 6-1). Actually, seismic energy propagates through the earth as spherically

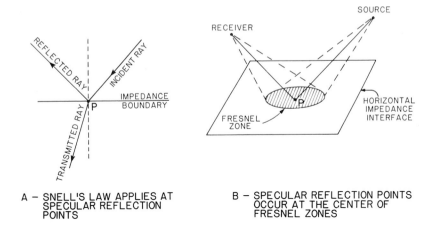

A — SNELL'S LAW APPLIES AT
SPECULAR REFLECTION
POINTS

B — SPECULAR REFLECTION POINTS
OCCUR AT THE CENTER OF
FRESNEL ZONES

Figure 6-1   A specular reflection point indicates where an upgoing VSP reflected ray originates. By definition, a specular reflection occurs only if Snell's law defines the directions that the reflected and transmitted rays travel. A specular reflection point such as P can be viewed as the center of the first order Fresnel zone, which is the actual subsurface area from which reflected energy arrives in phase at a VSP geophone.

spreading wavefronts, and reflection and transmission processes occur not at mathematical points but in large circular (or elliptical) Fresnel zones. Each specular reflection point shown in the ray trace models in this text marks only the center of the Fresnel zone corresponding to a reflected raypath. The complete Fresnel zone that surrounds each of these specular reflection points can be quite large, and this topic will be discussed in some detail in Chapter 7. The physical dimensions of these Fresnel zones must be kept in mind when interpreting ray trace models since a Fresnel zone defines the size of the reflector area which contributes to each reflection raypath. Particular attention should be given to situations where a Fresnel zone may be larger than the lateral dimensions of a structural or stratigraphic feature of an earth model. In such cases, if a raypath can exist mathematically, it generally yields the proper timing for the reflected event, but the amplitude of the synthetic reflection will be incorrect since in the real earth some of the Fresnel zone would extend past one or more edges of the feature.

Diffracted energy can be defined as all energy which propagates along raypaths other than those given by Snell's law of reflection and refraction. In real earth seismology, diffracted energy can, in some instances, be the dominant type of response recorded by geophones. However, since all energy propagation in ray trace modeling obeys Snell's law, the calculated results shown in this chapter disregard diffracted energy. Diffraction effects can be ignored in VSP modeling as long as the wavelengths contained in the seismic pulse are much less than the radius of curvature of any part of a reflecting surface occurring in the model. Thus, ray trace modeling in the vicinity of sharp curvature changes in an impedance boundary should be viewed with caution. Several models of this nature will be considered; specifically, sharp corners occur in the fault and unconformity models that will be shown. All models that are presented in this chapter have these additional assumptions, or restrictions, involved in the calculations.

1. The subsurface stratigraphy can be completely described by a two-dimensional cross section of the earth. (Three-dimensional modeling will be illustrated in a later section of the chapter after several two-dimensional models are studied.)

2. The borehole in which VSP geophones are located is vertical.

3. There is a single layer having a constant seismic propagation velocity of 8000 ft/sec between the earth's surface and the structural or stratigraphic anomaly to be modeled.

4. No noise is added to the calculated results.

5. The modeling software calculates pressure wavefields, not particle velocity responses. This pressure signal is not decomposed into three-component motion, thus the calculated amplitudes do not accurately represent the output of a vertically oriented VSP geophone.

6. The VSP geophone, if it is moved, is positioned at successive depth decrements of 50 feet.

7. For fixed-offset source models, wiggle trace plots will show every geophone response (i.e., wiggle traces will occur at vertical sample intervals of 50 feet), but raypath plots will show only every sixth ray path (vertical sample interval of 300 feet). All raypaths and wiggle trace responses will be plotted when walkaway source models are calculated.

8. The propagating wavelet is a 10 to 50 Hertz zero-phase bandpass function.

Dipping Reflector Models

Many seismic reflectors are not parallel to the earth's surface but exhibit some dip relative to horizontal. Determination of reflector dip is critical to exploration since dip affects the magnitude of structural closure, defines basin geometry, implies tectonic history, and controls depositional processes. Consequently, it is important to demonstrate how reflector dip appears in VSP data and to determine what resolution limits must be achieved in VSP measurements in order to satisfactorily measure dip.

A simplified model of a flat reflector, with its resultant VSP response for a fixed offset source location, is illustrated in Figure 6-2. A vertical dashed line passing through the reflection origin at a depth of 8000 feet is drawn on the wiggle trace display in order to emphasize the symmetry of the time-depth stepout of the downgoing and upgoing events. At any position above the reflector, the time interval, T_A, between this bisecting line and the downgoing first arrival is identical to the time interval, T_B, between the bisector and the upgoing reflection if the source is located at the wellhead. T_B is larger than T_A by a negligible amount for small source offset distances. A key VSP interpretational rule illustrated by this model is:

Interpretation Rule 1

For small source offset distances, if the downgoing VSP direct arrival and an upgoing reflection exhibit the same amount of time stepout with depth, then the reflector has zero dip.

When the source offset distance approaches, or exceeds, the magnitude of the reflector depth, there is considerable asymmetry between downgoing and upgoing VSP events for a flat reflector. The time-depth appearance of data recorded for a large offset shooting distance is illustrated by the models shown in Figures 6-3 and 6-4. In this type of shooting geometry, the time interval T_B, defined in Figure 6-2, is greater than time interval T_A, and the ratio, T_B/T_A, becomes progressively larger as the source offset distance increases. Refraction arrivals would be observed at some geophone depths in a real vertical seismic profile which used such large source offsets, but refraction events are not included in these calculations. In general, Interpretation Rule 1 will be valid in most VSP recording conditions.

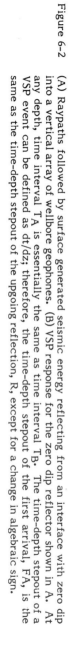

(A)

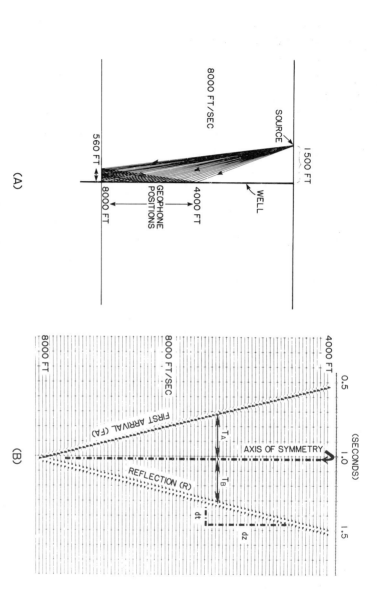

(B)

Figure 6-2 (A) Raypaths followed by surface generated seismic energy reflecting from an interface with zero dip into a vertical array of wellbore geophones. (B) VSP response for the zero dip reflector shown in A. At any depth, time interval TA is essentially the same as time interval TB. The time-depth stepout of a VSP event can be defined as dt/dz; therefore, the time-depth stepout of the first arrival, FA, is the same as the time-depth stepout of the upgoing reflection, R, except for a change in algebraic sign.

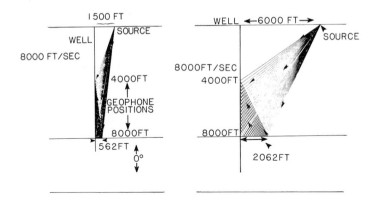

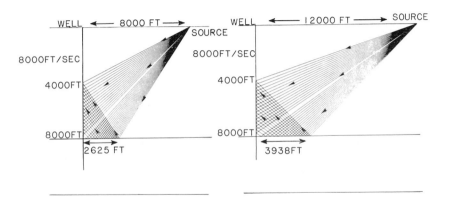

Figure 6-3   Raypaths calculated for a flat reflector and large source offset distances. An offset distance of 1500 feet is not considered large for a reflector depth of 8000 feet and is included only to provide a reference with which the other models can be compared.

The effect of reflector dip in VSP data is illustrated by the models shown in Figures 6-5 through 6-8, where the VSP responses for reflectors having dips of 0, 10, 20, and 30 degrees are calculated. The source offset distance is held constant at 1500 feet in all of these models. The source is positioned downdip from the hypothetical VSP well to generate the model responses in Figures 6-5 and 6-6, and updip from the well in the models shown in Figures 6-7 and 6-8. Examination of the wiggle trace responses in Figures 6-6 and 6-8 shows that the time-depth stepout of an upgoing reflection from a

228

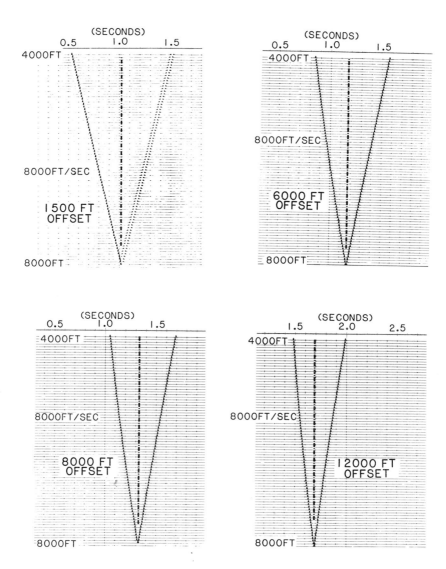

Figure 6-4   Synthetic VSP seismograms corresponding to the flat reflector, large offset models shown in Figure 6-3. The vertical dashed line passes through the point of origin of each reflection. Note how the upgoing reflection on each trace lies farther from this dashed line than does the downgoing event that created the reflection.

dipping interface is steeper than the time-depth stepout of the downgoing direct arrival, regardless of whether the source is updip or downdip from the well. The time-depth stepout behavior of the upgoing reflection events in all of these models is summarized in Figure 6-9. A fact illustrated by these modeling exercises is that formation dip can be recognized more readily in VSP data recorded with a source at a fixed offset distance if the source is located updip from the well rather than downdip. This statement is true because, for a given reflector dip, an updip source location creates a greater time-depth asymmetry between upgoing and downgoing events. These reflection comparisons are best illustrated by the summary chart shown in Figure 6-9, Part B. These model calculations lead to the following interpretation rules for data recorded with small source offset distances:

Interpretation Rule 2

If an upgoing VSP reflection event has a slope that is greater than the slope of the downgoing direct arrival in the same depth interval, then the reflector is not horizontal.

Interpretation Rule 3

As reflector dip increases, the slope of the upgoing VSP reflection becomes steeper.

It is important to note the difference in the amount of subsurface coverage achieved for updip VSP shooting versus downdip shooting. At least fifty percent more of a dipping reflector's surface is imaged in these models when the VSP source is updip from the wellbore than when the source is downdip (cf. Figures 6-5 and 6-7). If the intent of a VSP experiment is to look as far as possible laterally away from a borehole in order to acquire information about stratigraphy, structure, or rock properties, then updip source placement is recommended. If the intent is to get optimum concentration of reflection signal close to the borehole, then a downdip source location is better.

The slope of the downgoing direct arrival in Figures 6-6 and 6-8 is essentially the same at every depth since the formation velocity down to the reflector is constant, and the source offset is relatively small. However, if a reflecting interface is dipping, the slope of the upgoing reflection continuously increases as the reflection is followed downward toward the reflector, even though the formation velocity in the layer does not

230

change. This decrease in time stepout, dt/dz, with depth is difficult to see at these plot scales, but readers with excellent eyesight may detect a change in reflection slope by carefully inspecting the 20 and 30 degree dip cases in Figures 6-6 and 6-8. The presence of this increase in reflection event curvature is another property of VSP data that implies reflector dip.

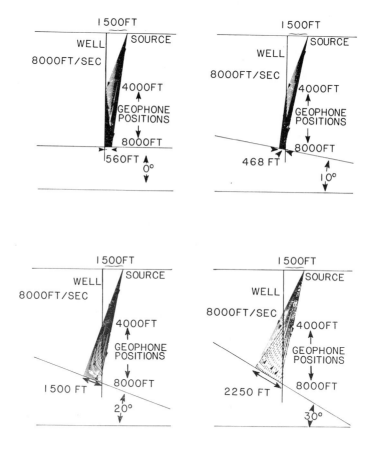

Figure 6-5   Raypaths calculated for a downdip source, which is offset a distance of 1500 feet, and various reflector dips. The footage value written below each dipping reflector defines the distance from the wellbore to the farthest reflection point and is measured along the reflecting interface.

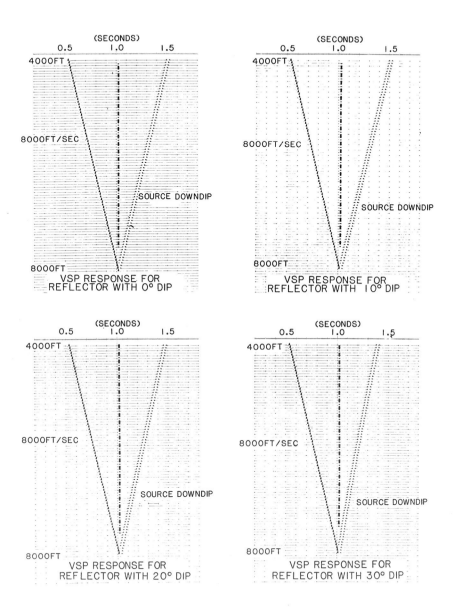

Figure 6-6   Synthetic VSP seismograms calculated for the dipping reflector models shown in Figure 6-5. The source is downdip from the study well in these models.

232

In real VSP data, where the seismic propagation velocity increases as depth increases, a downgoing direct arrival will exhibit increasing slope with depth, and not a constant slope as shown in Figures 6-6 and 6-8. In such a case, an upgoing reflection from a flat, horizontal interface will exhibit a mirror image of the same curvature expressed by the downgoing direct arrival. However, an upgoing reflection from a dipping interface will exhibit an increase in its slope, as a function of depth, that is greater than the slope of the downgoing direct arrival. Thus, the time-depth stepout asymmetry between downgoing and upgoing events in a constant velocity model is a valid generalization of the asymmetry that would be observed in real earth situations where velocity increases with depth.

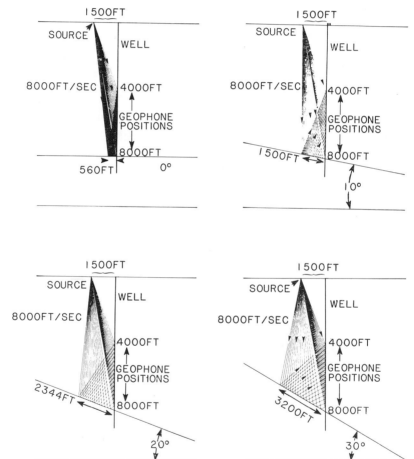

Figure 6-7   Raypaths calculated for an updip source location distance of 1500 feet and various reflector dips. The footage value written below each dipping reflector defines the distance from the wellbore to the farthest reflection point and is measured along the reflecting surface.

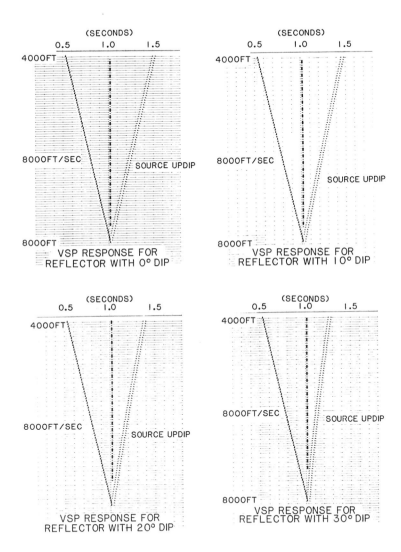

Figure 6-8   Synthetic VSP seismograms calculated for the dipping reflector models shown in Figure 6-7. The source is updip from the study well in these models.

234

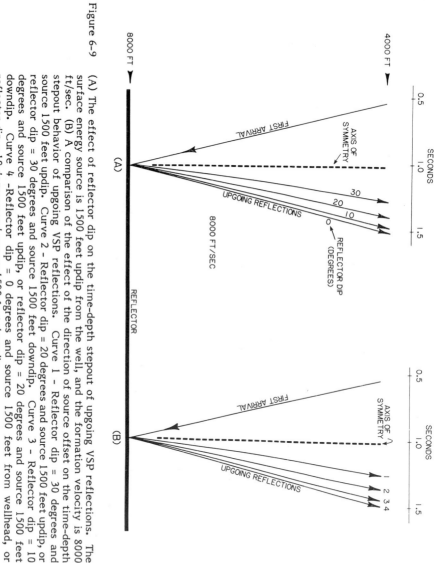

Figure 6-9   (A) The effect of reflector dip on the time-depth stepout of upgoing VSP reflections. The surface energy source is 1500 feet updip from the well, and the formation velocity is 8000 ft/sec.   (B) A comparison of the effect of the direction of source offset on the time-depth stepout behavior of upgoing VSP reflections.   Curve 1 - Reflector dip = 30 degrees and source 1500 feet updip.   Curve 2 - Reflector dip = 20 degrees and source 1500 feet updip, or reflector dip = 30 degrees and source 1500 feet downdip.   Curve 3 - Reflector dip = 10 degrees and source 1500 feet updip, or reflector dip = 20 degrees and source 1500 feet downdip.   Curve 4 -Reflector dip = 0 degrees and source 1500 feet updip, or reflector dip = 10 degrees and source 1500 feet from wellhead, or reflector dip = 10 degrees and source 1500 feet downdip.

A shortcoming of VSP data recorded with the field geometry described by these models is that one cannot infer the strike direction of a dipping reflector as long as the data are recorded with a single component geophone, as is assumed in these calculations. This ambiguity is built into the mathematical calculations because azimuthal dependence is eliminated by the assumption of a two-dimensional earth. In real VSP data, however, the same ambiguity exists if only a vertically oriented geophone is used, since such a geophone records the same response irrespective of the azimuthal direction from which an upgoing signal arrives at the subsurface geophone position. This inability to define the azimuthal orientation of a reflector should be kept in mind when examining all 2-D models in this chapter and when interpreting any real VSP data shown in the text which have been recorded with single-component vertically oriented geophones. This important characteristic of VSP data leads to the following interpretational rule:

Interpretation Rule 4

The strike direction of a dipping reflector cannot be determined from VSP data generated by one surface source at a fixed offset location and recorded downhole by only vertically oriented geophones.

The strike direction of a dipping reflector can be determined via three-component VSP measurements if the orientation of the downhole triaxial geophone system is known. Also, the time-depth stepout behavior of an upgoing reflection from a flat dipping interface that lies obliquely to the vertical plane containing a surface source and a wellbore can be modeled by three-dimensional ray tracing and compared with real VSP data in order to determine strike. An example of such a model will be shown later in this chapter.

If each synthetic trace in Figure 6-6 or 6-8 is delayed by its first break time, then the upgoing reflection from the flat, non-dipping reflector shown in these figures would occur at the same recording time at each geophone depth, in accordance with the static time correction described in Equation 1 of Chapter 5. In other words, delaying all traces by their respective first break times causes the reflection wavelet from a flat, horizontal interface to vertically align in phase on all traces. Since an upgoing reflection from a dipping interface occurs at earlier recording times than does a reflection from a horizontal boundary (Figure 6-9, part A), then delaying all traces by their first break times causes a reflection from a dipping interface to be a curved event that lies to the left (i.e. at earlier record time) of the vertical trajectory followed by an event reflecting from a horizontal boundary. An example of this behavior is shown in Figure 5-44.

The curved trend of black peaks originating at 1.37 seconds at a depth of 4500 feet in this figure are the upgoing events from a dipping interface, after each trace is delayed by its first break time.

Analytical Determination of VSP Reflector Dip

The preceding models demonstrate the effect of reflector dip on VSP data. Although these models are informative, the critical need, from the viewpoint of a seismic interpreter, is the ability to analyze VSP reflection behavior in real data and assign actual dip angle values to the interfaces which generated the observed reflections. One way to determine formation dip is to create as realistic a geological model as possible of the subsurface imaged by the VSP data, calculate a series of synthetic ray models for a variety of formation dips, using the same source and geophone positions which existed when recording the field data, and determine which dip values create synthetic VSP data most like the real data. This approach is a useful interpretational tool if the geological model is reasonably accurate.

A second approach is to determine an analytical equation which expresses reflector dip in terms of quantities that can be measured from real VSP data. With this analytical capability, an interpreter can calculate reflector dip without having to do extensive ray trace modeling. The following unpublished analytical development by S. B. Wyatt of Phillips Petroleum Company is one convenient formulation that relates known time and depth coordinates of a VSP reflection event to the dip of that reflector.

The raypath followed by a seismic wavefront which travels downward from a surface source positioned updip from a VSP well, then reflects upward from a single interface dipping at angle α, and arrives at a subsurface geophone above this reflector is shown in Figure 6-10. For simplicity, a constant velocity medium and straight raypaths are assumed. In this figure, L is the horizontal distance between the wellhead and the source position, Z is the vertical depth to the VSP geophone, and H is the depth down the vertical wellbore to the dipping interface. All of these distances should be known in a real vertical seismic profile. Some dashed lines are also shown in the figure which define the following additional distances:

$$a_1 = H \cos (\alpha) \tag{1}$$
$$a_2 = (H-Z) \cos (\alpha) \tag{2}$$
$$a_3 = a_1 - L \sin (\alpha) \tag{3}$$
$$a_4{}^2 = (L^2 + Z^2) - (a_3 - a_2)^2 \tag{4}$$
$$(d_1 + d_2)^2 = a_4{}^2 + (a_3 + a_2)^2 \tag{5}$$

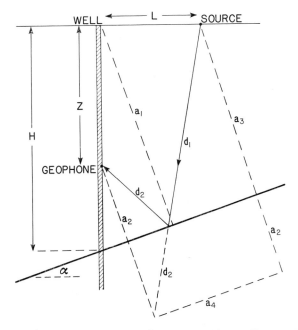

Figure 6-10   Raypath geometry for an updip source, offset a distance L from a VSP well, and imaging a single dipping reflector. The ray travels a distance d_1 to the reflector and a distance d_2 from the reflector to a borehole geophone. The direct ray from the source to the geophone is not shown. (Courtesy S. B. Wyatt, Phillips Petroleum Company)

If the average velocity in the layer above the interface is V, and the reflection arrival time at the geophone is t, then

$$(d_1 + d_2) = Vt \qquad (6)$$

Using Equations 1, 2, 3, 4 and 6 to change the parameters in Equation 5 yields the expression

$$H\cos^2(\alpha) - L\sin(\alpha)\cos(\alpha) = \frac{V^2 t^2 - (L^2 + Z^2)}{4(H-Z)} \qquad (7)$$

or

$$(H^2 + L^2)\cos^4(\alpha) - (2Hk + L^2)\cos^2(\alpha) + k^2 = 0 \qquad (8)$$

238

where

$$k = \frac{V^2t^2 - (L^2 + Z^2)}{4(H-Z)} \tag{9}$$

Since the average velocity, V, can be determined from the first break time of the direct arrival recorded at depth H, then all quantities in Equation 8, except α, are known, and formation dip can be calculated. Alternative analytical expressions relating reflector dip to known or measureable VSP geometrical distances and travel times have been published by Kennett et al. (1980), Balch et al. (1980b), Demidenko (1969), and Michon and Omnes (1978).

Fault Models

Detection of subsurface faults is one valuable application of vertical seismic profiling. Ray trace modeling can demonstrate how faults are expressed in real VSP data, and how VSP field experiments should be designed to best reveal them. The models that will be investigated can be segregated into two categories:

1. Models employing a single surface source, located at a fixed offset distance from a well, with a downhole geophone positioned at successive depth levels that extend over a sizeable vertical interval of the wellbore.

2. Models employing several surface sources, located at successively increasing offset distances from the well, with a downhole geophone at a single, fixed depth. These models are called walkaway vertical seismic profiles because of the nature of the source movement involved.

Most VSP experiments are performed with a single source (or source array) positioned at a fixed distance from the wellhead; consequently, it is essential to know how a faulted reflector appears in VSP data recorded in this manner. Some structural models using this type of shooting geometry, and their respective synthetic VSP responses, are shown in Figures 6-11 and 6-12. A constant velocity of 8000 ft/sec is assumed in the single model layer between the surface and the faulted reflector. The velocity in the bottom layer is 12000 ft/sec.

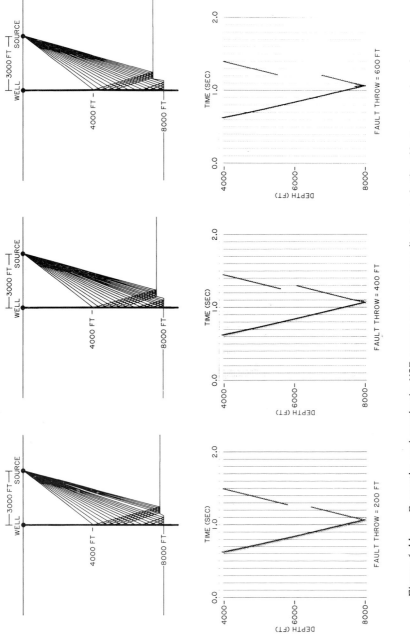

Figure 6-11   Raypaths and synthetic VSP responses corresponding to a single offset source imaging a
normal fault when the source is on the upthrown side of the fault.

240

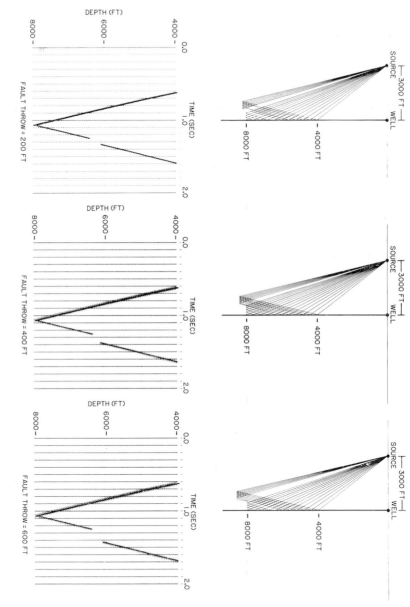

Figure 6-12   Raypaths and synthetic VSP responses corresponding to a single offset source imaging a normal fault when the source is on the downthrown side of the fault.

Fault throws of 200, 400, and 600 feet are modeled, but the horizontal distance along the interface from the vertical wellbore to the fault plane is kept fixed at 500 feet in every case. The models in Figure 6-11 assume that the source is on the upthrown side of the fault, and the models in Figure 6-12 place the source on the downthrown side. The sharp corners of this normally faulted reflector are diffraction points, but the mathematics of the ray tracing procedure used to construct these models ignores diffraction effects. In real VSP data, a diffraction event would trail off of both segments of the upgoing reflection. In every model, the upgoing reflection undergoes an abrupt discontinuity when the geophone depth is such that the reflection points begin to traverse the fault plane. The upper portion of the reflection event is displaced toward earlier recording times if the source is on the upthrown side of the fault and toward later recording times if the source is on the downthrown side. The amount of reflection discontinuity depends on the height of the fault throw.

The principal features of these VSP models are summarized in Figure 6-13. The upgoing dashed line represents the continuation of reflection R if that event were generated by a horizontal, unfaulted interface. Note that the vertical discontinuity in the reflection is larger when the source is on the upthrown side of a normal fault (events A, B,

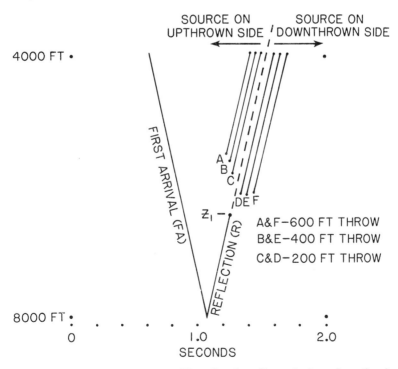

Figure 6-13 Effect of fault throw on VSP reflection discontinuity when the fault is imaged by a single offset source.

C) because a larger fault shadow is created compared to when the source is on the downthrown side (events D, E, F). For a single source located at a fixed offset during VSP recording, these synthetic responses lead to the following interpretational rule:

Interpretation Rule 5

An upgoing VSP reflection, which is broken into segments having abrupt offsets in a time-depth plot, is probably created by specular reflection points spanning one or more faults.

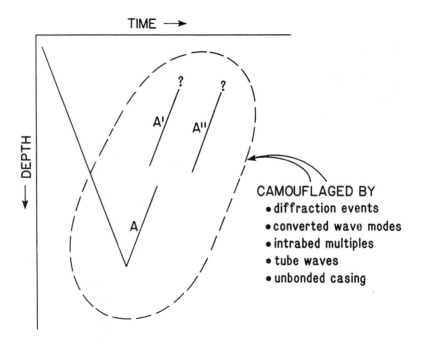

Figure 6-14   The interpretation of a faulted reflector, A, is complicated by the types of noises listed here, as well as by the fact that more than one upgoing event (e.g., A' and A") could be assumed to be the continuation of A.

In real VSP data, it can often be difficult to apply this interpretational guideline because data quality may make it impossible to decide where a weak reflection event terminates, and in addition, there are usually several recorded events which could be selected as the upper reflection segment of a faulted reflector. For example, in Figure 6-14, should event A' or A" be chosen as the extension of the faulted reflector A? Some of the factors that exist in real VSP data which make it difficult to choose between A' and A" are listed in this illustration. One of the more serious problems to consider is to decide if A' and A" are both primary reflections, or whether one or both of them is a multiple. In other cases, diffraction events and converted wave modes complicate the interpretation. In fortunate instances, the faulted reflector creates a unique reflection waveshape which is easily distinguishable regardless of the direction and magnitude of the displacement of the shallower reflection segment. Even though noise contamination may make it difficult to answer specific questions about a faulted stratigraphic unit, such as, "Is the well on the downthrown or upthrown side of a fault?", or "What is the amount of fault throw?", Interpretation Rule 5 is still a valuable principle of VSP data behavior that should be used when analyzing a VSP data set since it confirms, at least, the existence of a fault near the wellbore.

Models describing the VSP responses of walkaway source vertical seismic profiles which image a faulted reflector are shown in Figures 6-15 and 6-16. In Figure 6-15, the geophone remains fixed at a depth of 4000 feet while the source traverses from 4700 feet to the right of the well in lateral increments of 300 feet to a final position 3000 feet to the left of the well. Fault throws of 200, 400, and 600 feet are modeled. The model parameters are the same in Figure 6-16, except the geophone is at a depth of 6000 feet. No downgoing multiples, which can overlay the upgoing reflections and confuse the interpretation, are calculated in these simple single layer models.

The shallowest event, occurring at 0.5 seconds at the well location in Figure 6-15 and at 0.75 seconds in Figure 6-16, is the direct arrival. The lateral distance between reflection points on the faulted interface is 100 feet when the geophone is 4000 feet above it; thus, the discontinuity for the 600 ft throw begins on the fifth trace to the right of the well location in Figure 6-15. When the geophone is moved 2000 feet closer to the fault, the distance between the reflection points decreases to 55 feet. Consequently, the 600 ft throw discontinuity begins on the eighth trace from the well in Figure 6-16. Obviously, the lateral resolution of a fault's location improves as a VSP geophone moves closer to a faulted unit. Also, it should be noted that the source may have to be moved farther away from the well as the geophone approaches the fault so that a sufficient number of reflection points lie on both the downthrown and upthrown sides. In real VSP data, it would be difficult to properly interpret the location and throw of a fault if only four or five reflection events were recorded on the upthrown side, as shown in Figure 6-16.

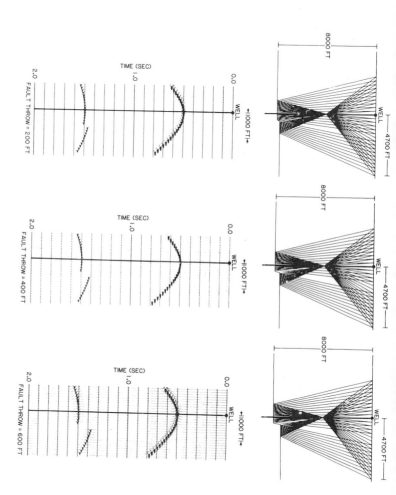

Figure 6-15   Raypaths and synthetic VSP responses corresponding to a walkaway source imaging a
normal fault when the borehole geophone is 4000 feet above the fault. The distance
between source positions is 300 feet, but the trace spacing is 100 feet since that is the
calculated distance between reflection points along the faulted interface. The shallower
event is the direct arrival.

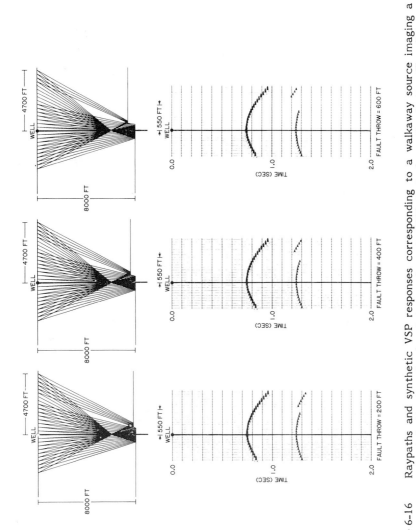

Figure 6-16   Raypaths and synthetic VSP responses corresponding to a walkaway source imaging a normal fault when the borehole geophone is 2000 feet above the fault. The distance between source positions is 300 feet, but the trace spacing is 55 feet since that is the calculated distance between reflection points along the faulted interface. The shallower event is the direct arrival.

This type of model response allows an explorationist to decide what frequency bandwidth would have to be recorded in order to detect a certain magnitude of fault throw. For example, the vertical displacement in reflection time when a fault throw of 200 feet is imaged is approximately equal to the time span of the 10 to 50 Hertz wavelet used in these calculations. It is advantageous, for interpretational purposes, if the fault time displacement is of the same order as the time extent of the propagating wavelet, or larger.

It is difficult to convert the time interval spanned by the reflection discontinuity into an accurate estimate of the fault throw as long as the data are left in the recording format shown in Figures 6-15 and 6-16. The vertical time differences for fault throws of 200, 400, and 600 feet would be 50, 100, and 150 ms, respectively, but the measured time differences are approximately 40, 85, and 130 ms when the geophone is 4000 feet below the surface (Figure 6-15), and 15, 60, and 80 ms when the geophone is moved down to 6000 feet (Figure 6-16). These time differences are not obvious functions of the vertical depth displacement of the faulted unit because they are determined from two oblique travel paths originating at widely separated source positions. For example, inspection of Figure 6-16 shows that the source must be moved laterally 1800 feet in order for the raypaths to traverse the shadow zone of a fault with 600 feet of throw.

For interpretational purposes, it is better to remove the normal moveout associated with each data trace so that the reflection events appear as if the source were always directly above each reflection point. It should be noted that the horizontal spacing between reflection points varies with reflector depth in VSP recording geometry. Thus, one can say that NMO-corrected VSP data have a fixed trace spacing along a single horizontal interface, but it is not correct to assume that this same trace spacing is valid at other reflector depths. This behavior is an important distinction between VSP data, which are recorded with a geophone located below the source, and surface-recorded data, where the geophone stays at the same depth level as the source. The lateral distance between reflection points for surface-recorded data is the same at all depths if the reflectors are flat and horizontal. Timing differences measured between any two reflection events in an NMO-corrected upgoing wavefield can be assumed to represent vertical travel times and should be directly proportional to fault throw.

Angular Unconformity Models

Many oil and gas fields are located at angular unconformity surfaces where source rocks, reservoir facies, and reservoir seals are commonly juxtaposed. Even if an angular

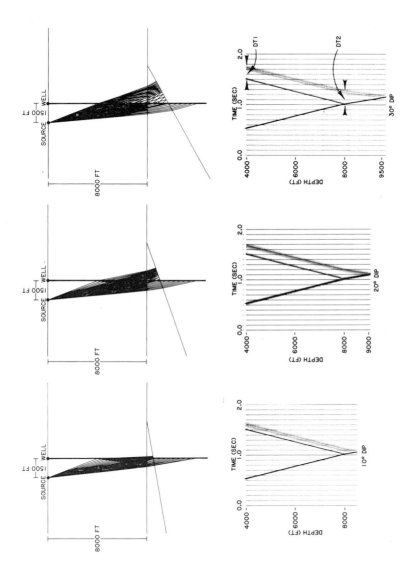

Figure 6-17   Raypaths and synthetic VSP responses corresponding to a single offset source, positioned downdip from a well, and imaging an angular unconformity. The velocity in the wedge shaped truncated unit is 12000 ft/sec. The velocity above and below the wedge is 8000 ft/sec. Several rays are traced to geophone positions below the wedgeout.

248

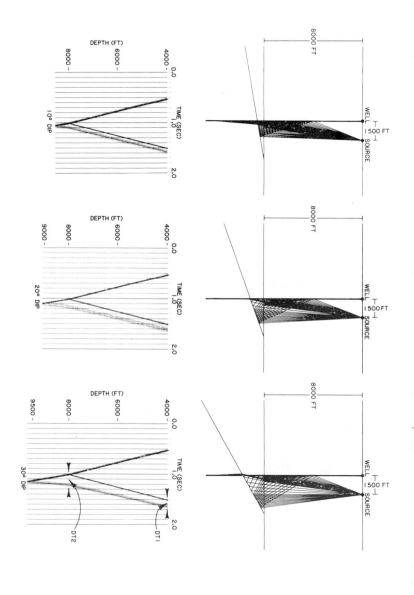

Figure 6-18   Raypaths and synthetic VSP responses corresponding to a single offset source, positioned updip from a well, and imaging an angular unconformity. The velocity inside the wedge is 12000 ft/sec. The velocity in the layers above and below the wedge is 8000 ft/sec. Several rays are traced to geophone positions below the wedgeout.

unconformity is not prospective for hydrocarbons, it is still necessary to recognize these stratigraphic boundaries because they are so critical to understanding the tectonic and depositional history of an area. Surface-recorded seismic reflection data comprise one of the most widely used and successful data bases for identifying and mapping these types of unconformities. Since VSP data are commonly used to aid the interpretation of surface-recorded seismic data, it is important to recognize how angular unconformities appear in VSP data.

The simple models in Figures 6-17 and 6-18 will be used to demonstrate the appearance of unconformable reflectors in a vertical seismic profile. The angle at which the bottom interface intersects the top horizontal interface is set at 10, 20, and 30 degrees, respectively. In each model, a vertical borehole penetrates both interfaces at a horizontal distance of 3000 feet from the angular pinchout. The vertical distance between the two interfaces is not small enough to allow the two reflection wavelets to interfere with each other. Only one source is positioned 1500 feet away from the wellhead. Geophone responses are calculated at vertical intervals of 50 feet from 4000 to 9000 feet below the surface.

The velocity in the wedge-shaped, truncated unit is 12000 ft/sec. The velocity above and below this unit is 8000 ft/sec. These models could also be viewed as representing a high velocity, thinning sand encased in a low velocity shale, but the view taken in this exercise is that each model represents a dipping section of alternating high and low velocity beds that has been eroded and overlain by low velocity sediment.

When the source is downdip from the well (Figure 6-17), the reflecting points on the dipping interface lie on the opposite side of the well from the reflection points along the horizontal interface. When the source is updip from the well (Figure 6-18), reflection points along both interfaces occur on the same side of the well, but even in this case, the reflection points on the dipping interface do not lie vertically beneath the reflection points on the horizontal boundary. Such behavior implies that this type of subsurface structure can be correctly interpreted only after the VSP data are migrated. The necessity for applying migration procedures that properly position VSP reflection events will be more obvious in the walkaway source models considered next. In addition, the reflection points along the dipping interface are not uniformly spaced. They are widely separated in the downdip direction, and the spacing decreases as they climb updip.

In order to interpret the presence of an unconformity, the two upgoing reflections must exhibit some difference in their time-depth stepout behaviors. One indicator of a difference in reflector dip is a comparison of the time interval between the two reflections measured at two different recording depths. Two reference depths, at 4000 feet and at 7800 feet below the surface, are identified in the 30 degree dip model in Figures 6-17 and 6-18. The time interval between the two reflection events at 4000 feet is labeled DT1, and the time difference at 7800 feet is labeled DT2. Measured values of DT1 and DT2 for each model are listed in Table 6-1. The fact that DT1 is smaller than DT2 implies that the two reflection events are imaging interfaces that have different dips. These two time intervals would differ even if the top interface were not horizontal, as long as its dip were not the same as the dip of the lower interface. The difference, (DT2 - DT1), is considerably larger when the source is placed updip from the well. Thus, one could conclude that the existence of reflectors having different amounts of dip (i.e. angular unconformities) can be recognized more easily if updip source placement is used in the field.

Table 6-1

Time Measurements Indicating a Difference in the Dip of the
Unconformable Reflectors Modeled in Figures 6-17 and 6-18.

Source Position	Dip (Degrees)	DT1 (MS)	DT2 (MS)	DT2-DT1 (MS)
Downdip	10	90	90	0
	20	165	170	5
	30	215	245	30
Updip	10	65	75	10
	20	100	140	40
	30	125	195	70

These model results suggest an additional VSP interpretational rule that should be noted; viz,

<u>Interpretation Rule 6</u>

Upgoing reflection events which converge as the
recording depth decreases are usually mapping
an angular unconformity.

These same unconformity models are shown imaged by a walkaway vertical seismic profile in Figure 6-19. The source is "walked" at horizontal increments of 300 feet from 5700 feet to the right of the well to 5700 feet to the left while the geophone remains fixed at a depth of 4000 feet. Raypaths are shown reflecting from both interfaces, and the reflection responses are plotted in the wiggle trace displays starting at 1.5 seconds at the well location. The shallow event occurring at 0.5 seconds at the well trace is the direct arrival.

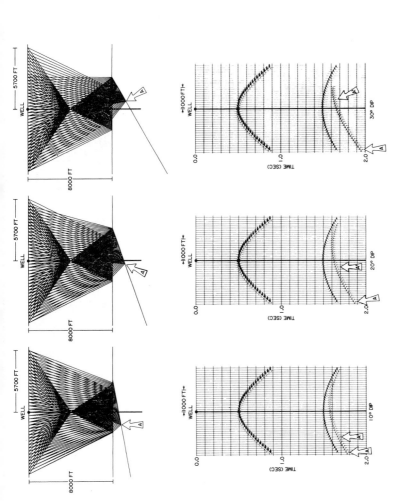

Figure 6-19   Raypaths and synthetic VSP responses corresponding to a walkaway source imaging an angular unconformity. The geophone is 4000 feet above the unconformity surface. The velocity in the wedge is 12000 ft/sec. The velocity above and below the wedge is 8000 ft/sec. If the data were migrated, event A would move laterally to A'. The shallow event between 0.5 and 0.9 seconds is the direct arrival. The trace spacing represents the distance between reflection points only on the horizontal interface. A smaller trace spacing should be used to image the dipping interface.

252

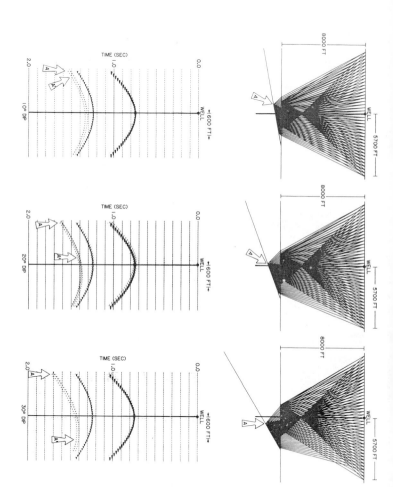

Figure 6-20

Raypaths and synthetic VSP responses corresponding to a walkaway source imaging an angular unconformity. The geophone is 2000 feet above the unconformity surface. The velocity in the wedge is 12000 ft/sec. The velocity above and below the wedge is 8000 ft/sec. If the data were migrated, event A would move laterally to A'. The shallow event between 0.75 and 1.05 seconds is the direct arrival. The trace spacing represents the distance between reflection points only on the horizontal interface. A smaller trace spacing must be used when imaging the deeper, dipping reflector.

In this model, it should be noted that the lateral distance between successive reflection points on the flat, horizontal interface is 100 feet, so this horizontal scale is used in plotting the VSP traces. The reflection points along the dipping reflector are separated by 97 feet when the dip is 10 degrees, but this separation reduces to 90 and 85 feet, respectively, when the dip increases to 20 and 30 degrees. Thus, the indicated plot scale of 100 feet per trace spacing, although correct for the horizontal interface, is not accurate for the reflections returning from the dipping interface. The horizontal interface is imaged laterally a distance of 1900 feet from the wellbore, but more importantly, the area of the dipping reflector that is imaged does not lie directly below that part of the horizontal reflector which is illuminated. The reflection point occurring farthest to the left on the dipping interface is labeled A in the ray models. The reflection event associated with this reflection point is also labeled A in the wiggle trace plots. If these trace data were migrated, reflection event A would move laterally to position A'. The point A' does not show the correct vertical location of the migrated position of A. Migration would move A' higher than its plotted position in these wiggle trace displays. These models demonstrate the critical need for migrating VSP data recorded with a walkaway source and form a basis for making the following interpretational rule:

Interpretation Rule 7

If an angular unconformity is imaged by a walkwaway type of VSP shooting, the reflection events that return from dipping interfaces must be moved updip.

These same angular unconformities are modeled in Figure 6-20 except the geophone is now at a depth of 6000 feet, and is thus closer to the unconformity. The distance between reflection points on the horizontal interface is now 60 feet. Consequently, the area around the borehole is seen with a better horizontal resolution compared to when the geophone was 4000 feet above the unconformity, but a smaller area is imaged. The last reflection point on the horizontal interface is only 1140 feet from the wellbore. The recording time position of the reflection event created at depth point A on the dipping interface is again labeled A in the wiggle trace data, and the correct horizontal location of that reflection event after migration is shown by the label A'. Note that all shots, except for the last three on the right, contribute to the image of the reflector dipping at 30 degrees; whereas, nine shots fail to contribute to the image of this reflector in Figure 6-19 when the geophone is 4000 feet above the unconformity.

An interpretation of any reflecting interface below the horizontal boundary will be difficult if a large reflection coefficient exists at the top of this unconformable unit and if several downgoing multiples are generated in the shallow part of the stratigraphic model. The resulting multiple images of the horizontal interface would overlay primary reflections occurring below this boundary and could severely distort their reflection waveshapes, or even completely mask them. This situation is common in real earth seismology, and is one of the principal reasons why it is difficult to construct seismic images below some unconformities. An example of real VSP data being used to look below an angular unconformity will be described in Chapter 8 (Figure 8-36).

Three-Dimensional VSP Ray Tracing

Many subsurface imaging problems cannot be accurately represented by a two-dimensional model of the earth. In structurally complex areas, such as salt dome flanks, overthrust zones, or continental rift margins, seismic reflectors may change dip and strike attitudes over rather small lateral distances in any horizontal direction. For such situations, seismic raypaths calculated under the assumption that reflector geometry changes in only one horizontal direction can lead one to false conclusions concerning where to locate a VSP source and a borehole receiver in order to image geological anomalies, and can also lead to false interpretations of recorded VSP data. VSP ray tracing in three dimensions should definitely be used to analyze these types of problems.

Calculating three-dimensional raypaths through contorted multilayered media is conceptually simple, but difficult to perform even with modern computers. It is perhaps even more difficult to display the calculated three-dimensional raypaths accurately, quickly, and in an easily interpretable format. Currently, there is widespread emphasis on developing schemes for recording closely spaced surface seismic reflection data in two horizontal directions so that accurate three-dimensional images of the subsurface can be created. Consequently, several industrial groups are in various stages of providing three-dimensional ray tracing programs and three-dimensional display devices to the geophysical industry. Once these computer software and hardware technologies are available, then coding the computer programs to accept subsurface geophone positions, rather than restricting receiver locations to just the earth surface, will permit rigorous three-dimensional VSP ray trace modeling.

Until a versatile three-dimensional ray tracing capability is available, simple three-dimensional earth models using simple geometrical assumptions about subsurface interfaces are quite valuable. For example, allowing only one velocity layer to exist above a surface that needs to be analyzed, and restricting this reflecting surface so that it is a planar interface, provides important insights into some basic three-dimensional VSP raypath behaviors. Consider the earth model shown in Figure 6-21 as an illustration. This three-dimensional model consists of only two velocity layers separated by a planar reflecting interface whose strike and dip directions can be arbitrarily defined. In this illustration, the dip of the interface is held constant at 10 degrees, but the strike direction differs by 90 degrees for the two cases. The source remains in the same position, 1000 feet east of the well.

The problem addressed by this simple model is one that is not uncommon in VSP surveys; i.e., a VSP energy source cannot always be located at the best possible position for conducting an experiment. For instance, in farm and ranch areas, a landowner may refuse to allow VSP field equipment on his property. In federal and state parks, vehicles must often stay on established roads. In other areas, severe topographical features may restrict source locations. An example of this type of field situation is shown in Figure 6-22. At this site, the surface source could be located only on the access road leading to the well site, which is shown on the ridge crest. A question that should be answered before committing to a vertical seismic profile in such a well is, "Can reflected signals from the subsurface feature that needs to be imaged be recorded in the well if the source is located at this spot?". If the subsurface structure and velocity can be reasonably estimated, then VSP ray tracing is a valid way to determine whether or not the desired imaging objective can be achieved. The analysis is more precise if the calculations are performed in three-dimensional space.

Questions that could be answered by such an analysis are suggested by the models in Figure 6-21. For example, if the source position is forced to lie in a certain azimuthal direction (e.g., due east of the wellhead in this case), where does this restriction cause subsurface reflection points to fall? For a reflector having a specified dip and strike, can a target anomaly be properly illuminated with this restricted source placement?

A stratigraphic trap type of anomaly, such as a point bar sand, is shown lying on the hypothetical planar reflecting surface in this illustration. These three-dimensional ray tracing solutions show that the reflection points traverse the sand body when the source and the wellbore are aligned along the dip direction of the inclined reflector. However, if the reflector dips to the south by 10 degrees or more, then the target cannot be imaged unless the source position can be changed. Simply placing a VSP energy source and a VSP geophone on opposite sides of a target to be imaged does not insure that reflection points will fall on the target if dipping beds are involved. Thus, even when earth models of simple geometry are used, three-dimensional ray tracing results can be of great value both for planning proper field geometry before recording VSP data, and for assisting the interpretation of the data after recording and processing.

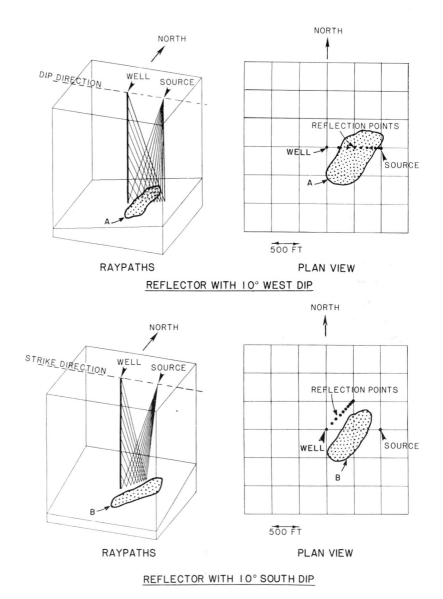

NORTH

DIP DIRECTION WELL SOURCE

A

RAYPATHS

NORTH

REFLECTION POINTS

WELL

SOURCE

A

500 FT

PLAN VIEW

REFLECTOR WITH 10° WEST DIP

NORTH

STRIKE DIRECTION WELL SOURCE

B

RAYPATHS

NORTH

REFLECTION POINTS

WELL

SOURCE

B

500 FT

PLAN VIEW

REFLECTOR WITH 10° SOUTH DIP

Figure 6-21   An example of interactive computer displays showing how subsurface reflection points on dipping interfaces can be located by 3-dimensional VSP ray tracing (Courtesy K. D. Wyatt and C. R. Stricklin, Phillips Petroleum Company). A and B are hand drafted to represent targets that need to be imaged by the VSP raypaths.

Figure 6-22   A typical problem regarding proper source positioning in VSP field work. Government regulations sometimes prevent vehicles from entering forested areas, so energy sources must be located only on access roads to a well site. This particular source is a Bolt Model LSS-IT land airgun. The drill rig on the ridge crest marks the well site (Courtesy F. B. Bodholt, Phillips Petroleum Company).

A slightly more complex three-dimensional ray tracing analysis involving two dipping interfaces is shown in Figure 6-23. The shallow reflector, A, dips to the south at 10 degrees. The deeper reflector, B, dips to the south at 20 degrees. Both interfaces are illuminated by sources located at three different surface locations.   One source is positioned to the west, along strike, one is located updip to the northeast, and one downdip to the southeast. In the oblique view, a single ray from each source is followed down to reflector B and back upward to a borehole geophone in the vertical well. The ray from Source 1 penetrates reflector A at point 1A and reflects upward from interface B at point 1B. Similarly, the ray from Source 2 penetrates interface A at point 2A and reaches interface B at point 2B.

The locations of a few reflection points on each interface, when a geophone is positioned at a depth of 1000 feet, are shown in the plan view at the bottom.   The notation "1 (20)" denotes where a shot wavelet created by Source 1 reflects from the

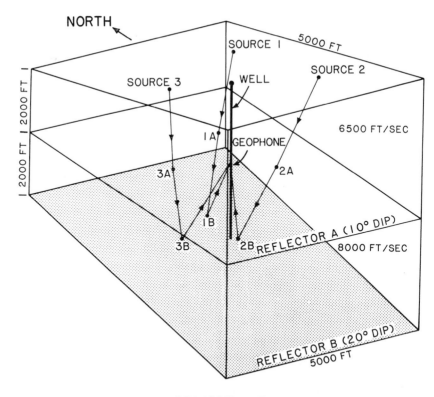

OBLIQUE VIEW

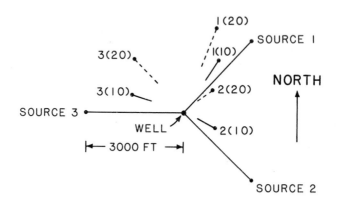

PLAN VIEW

Figure 6-23 Location of reflection points by 3-D ray tracing when VSP data are recorded in a multilayered earth having reflecting interfaces that are not horizontal. (Courtesy of D. B. Neff, Phillips Petroleum Company).

deeper interface which has a dip of 20 degrees. The point "3 (10)" denotes where energy leaving Source 3 reflects from the shallow interface which has a dip of 10 degrees.

For each source, all reflection points climb updip from the vertical plane passing through that source and the wellbore. Inspection of the figure shows that the amount of updip shift of each reflection point is a function of the depth to the reflector, the dip of the reflector, and the angle between the azimuth direction from the well to the source and the strike direction of the reflector. Reflection points for geophone depths greater than 1000 feet will lie along slightly arcuate paths between the plotted reflection points and the wellbore. The tracks of these reflection points along the 10 degree dip interface are approximated by the short solid line segments that extend from each of the three reflection points on that interface toward the well. The dashed lines approximate the track taken by reflection points moving along the 20 degree dip interface toward the wellbore as the recording geophone is lowered toward that interface.

Although the two preceding models are highly simplified descriptions of a three-dimensional earth, the calculated results emphasize that a person interpreting VSP data should consider the following interpretation guideline:

Interpretation Rule 8

> If VSP data are recorded in a well that penetrates complicated subsurface structure and stratigraphy, an analysis of the data should not be considered complete until a three-dimensional ray trace modeling study is made.

Synthetic VSP Calculations

The VSP ray models discussed in this chapter are simplified by allowing the earth to consist of only two or three homogeneous layers, and by not allowing multiple reflections to be created. A more realistic scheme for calculating synthetic VSP responses, which allows many reflecting interfaces and any order of multiples to be included, has been reported by Wyatt (1981a). Several simplifying assumptions are involved in this calculation process:

1. All reflecting interfaces must be flat and horizontal.

2.   Downgoing and upgoing seismic wavelets must travel normally to all interfaces. This assumption prohibits offset source geometry.

3.   An impedance boundary must occur at equal vertical time increments of magnitude Δt. Typically, Δt is chosen to be one millisecond.

The only input data needed in the calculation are the acoustic impedance value in each synthetic earth layer and an initial downgoing shot wavelet. The impedance values are usually provided by sonic and density log data. The shot wavelet is either assumed or calculated from real surface seismic data or VSP data.

A comparison of part of a real vertical seismic profile with synthetic VSP data calculated by Wyatt's technique is shown in Figure 6-24. A sonic log recorded in the VSP well was used to create the synthetic impedance layers, and an average of several of the actual VSP direct arrivals was used as the shot wavelet. The comparison between the real data and the synthetic data is quite good because the sonic log in this depth interval is a close estimate of the actual seismic impedance sequence. In situations where the near surface section of the earth generates many downgoing multiples, the synthetic data will not match a real vertical seismic profile as well as in this example, since sonic and density logs are not recorded in the first few hundred feet of a borehole. Thus, the shallow impedance behavior that creates these multiples can only be approximated in the synthetic calculation.

An alternate theoretical approach to constructing synthetic VSP data is described by Temme and Muller (1982). The mathematics of their modeling approach incorporates all surface and internal multiples, adjusts amplitudes for spherical divergence, and allows anelastic absorption. A point source at a finite offset distance from a vertical array of subsurface receivers can be imitated in addition to an acoustical, plane wave, zero-offset model similar to Wyatt's. These finite offset source models are also limited to just the acoustic mode of wave propagation.

Summary

The models shown in this chapter are far too simple to represent real earth conditions. Nonetheless, they do offer clues and suggestions as to what subsurface geological conditions can be associated with certain characteristics of VSP reflection events. Thus, numerical modeling plays a valuable role in helping explorationists understand the stratigraphic and structural messages contained in recorded VSP data.

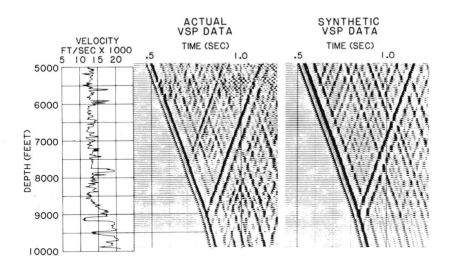

Figure 6-24 Comparison between actual VSP data and synthetic VSP data constructed from sonic log data recorded in the VSP well. (After Wyatt, 1981a)

Perhaps more importantly, modeling is a tremendous aid in designing VSP field experiments so that specific subsurface targets are appropriately illuminated by the wavefields arriving at preselected borehole geophone positions. This latter application of modeling has not been emphasized as much in this chapter as have the interpretational aspects of modeling. However, modeling anticipated VSP responses before commencing a VSP field experiment is highly recommended, particularly if the data must be recorded under the duress of either a limited time or a limited money budget.

The hazards of modeling VSP problems in only two dimensions cannot be overstressed. The differences between raypaths in three-dimensional models versus raypaths in two-dimensional models is sometimes startling. Without question, some VSP field experiments have failed or been marginally successful because explorationists did not study the anticipated structural and stratigraphic situation in three dimensions before commencing the field measurements.

One obvious need in VSP ray trace modeling is that the borehole must be allowed to take an arbitrary path through a hypothetical stratigraphic section so that VSP recording in a deviated well can be imitated. The restriction of a vertical borehole, which is used in the illustrations in this chapter, does not allow some important VSP problems to be studied.

The intent of this chapter has been to illustrate a few basic behaviors of VSP data before concentrating on interpretations and applications of real VSP data in the next two chapters. Consequently, only primary reflection events have been emphasized. A modeling feature that should be added when doing more rigorous studies would be that one or more shallow reverberating layers should be added so that downgoing multiples can be created. Some of the characteristics of upgoing VSP reflections which have been pointed out in this chapter are greatly altered and camouflaged when downgoing multiples overlay the upgoing events. Obviously, diffraction effects should also be included in the model calculations, but this type of modeling extends beyond ray trace theory and encompasses wave equation theory. An example of a VSP model constructed with a finite difference wave theory algorithm has been published by Kelly et al. (1982).

Chapter 7

EXPLORATION APPLICATIONS OF VERTICAL SEISMIC PROFILING

Properly recorded VSP data should define all upgoing and downgoing seismic events that are generated in a stratigraphic section near a study well. Therefore, VSP measurements allow an explorationist to answer the following questions about an upgoing seismic event arriving at the ground surface:

1. Is the event a primary reflection or a multiple?
2. What is the depth at which the event was generated?
3. What is the time at which the event arrives at the surface?
4. What is the reflection waveshape describing the stratigraphic situation where the event was generated?

These questions address the fundamental principles of seismic stratigraphy, so they are crucial issues that must be considered in order to achieve an optimum interpretation of surface-recorded seismic data. Whether or not these questions can be definitively answered in a particular instance depends on the quality of the recorded VSP data. Unfortunately, there are experiments in which VSP data quality is adversely affected by energy sources that are inadequate because they have low signal strength, poor repeatability, and restricted bandwidth; by recording instruments which create phase and amplitude distortions and have insufficient dynamic range; by seismic noise contamination; and by poor field procedures.

Reflection Coefficients

      Seismic waves are reflected at interfaces which separate rock layers having different acoustic impedances, and the polarity, amplitude, and phase characteristics of these reflected wavelets are determined by the reflection coefficients that exist at these interfaces. Since upgoing reflected wavefields contain much of the subsurface information that is sought in vertical seismic profiling, it is essential that the physics of seismic reflection coefficients be understood before discussing exploration and production applications of VSP data. Mathematical expressions describing the reflection coefficients of seismic waves propagating in the earth can be found in many geophysical papers, but these definitions are often misleading or confusing if the author does not specify what seismic parameter - pressure, particle velocity, or particle displacement - is being considered and in what direction the wave is traveling when it arrives at the reflecting interface. This information must be known in order to correctly calculate a reflection coefficient for a seismic wavelet.

      A correct understanding of the mathematics of a seismic wave reflection coefficient is essential when interpreting vertical seismic profiles because VSP data record both upgoing and downgoing waves, which reflect from both the top and bottom sides of interfaces throughout a stratigraphic section. The polarity of a reflected wavelet created at one of these impedance interfaces, as well as the amplitude of its associated transmitted wavelet, depends upon the direction in which the incident wavelet is traveling when it arrives at this impedance change. Also, in marine vertical seismic profiles, a reflection event measured in the subsurface by a velocity-sensitive geophone must often be correlated with its equivalent reflection measured near the surface by a pressure-sensitive hydrophone. Consequently, it is important to understand the differences between the reflection coefficients which define reflected particle velocity wavefields and reflected pressure wavefields. These problems seldom confront an explorationist interpreting only surface-recorded reflection data.

      A single interface separating two rock layers having different acoustic impedances is shown in Figure 7-1. In this illustration, V_n and ρ_n represent the seismic propagation velocity and rock bulk density in layer n, which lies below layer (n-1). There are three types of detectors that are commonly used to record seismic field data; viz, geophones (which measure particle velocity), accelerometers (which measure particle acceleration), and hydrophones (which measure pressure). For a plane seismic wave, normally incident on the interface drawn in Figure 7-1, the correct definitions for the reflection coefficient of these three commonly measured seismic parameters at this impedance change are:

UPWARD DIRECTION

shown

$$R = \frac{\rho_{N-1} V_{N-1} - \rho_N V_N}{\rho_{N-1} V_{N-1} + \rho_N V_N}$$

Figure 7-1   Definition of reflection coefficient for normally incident particle velocity
pulse traveling downward through layer (n-1) into layer (n).

Downgoing particle velocity wave
Downgoing particle acceleration wave $\Big\}$ $R = \left[\dfrac{\rho_{n-1} V_{n-1} - \rho_n V_n}{\rho_{n-1} V_{n-1} + \rho_n V_n} \right]$ (1)
Upgoing pressure wave

Upgoing particle velocity wave
Upgoing particle acceleration wave $\Big\}$ $R = \left[\dfrac{\rho_n V_n - \rho_{n-1} V_{n-1}}{\rho_n V_n + \rho_{n-1} V_{n-1}} \right]$ (2)
Downgoing pressure wave

These reflection coefficient expressions will be particularly valuable when making
stratigraphic and lithological interpretations of VSP data discussed later in this chapter.

Simple normal incidence models that confirm the correctness of the mathematical
expressions in Equations 1 and 2 are shown in Figure 7-2. The model on the left shows a
rigid interface where a downgoing seismic wavefield encounters a second medium having
an impedance infinitely larger than the medium in which the incident wavefield exists.
The incident wavefield cannot move the second medium because of its tremendous

impedance, thus the boundary is rigid. By definition, the particle velocity must be zero at such an interface. The incident, reflected, and transmitted particle velocity wavefields are labeled v_i, $R_v v_i$, and $T_v v_i$, respectively, where R_v and T_v are the reflection and transmission coefficients of the incident particle velocity wavefield at this impedance boundary. If the boundary does not move then

$$v_i(Z_R) + R_v v_i(Z_R) = T_v v_i(Z_R) = 0 \ . \qquad (3)$$

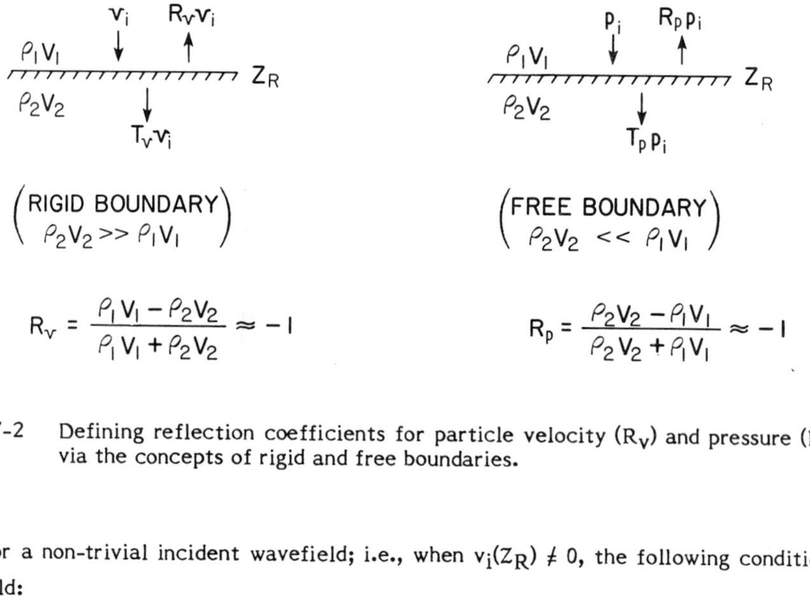

downgoing Wave

shown

$$R_v = \frac{\rho_1 V_1 - \rho_2 V_2}{\rho_1 V_1 + \rho_2 V_2} \approx -1 \qquad \qquad R_p = \frac{\rho_2 V_2 - \rho_1 V_1}{\rho_2 V_2 + \rho_1 V_1} \approx -1$$

Figure 7-2   Defining reflection coefficients for particle velocity (R_v) and pressure (R_p) via the concepts of rigid and free boundaries.

Thus, for a non-trivial incident wavefield; i.e., when $v_i(Z_R) \neq 0$, the following conditions must hold:

$$R_v = -1 \text{ and } T_v = 0. \qquad (4)$$

If Equation 4 is true, and since $\rho_2 v_2 \gg \rho_1 v_1$ as shown in Figure 7-2, then, necessarily, the reflection coefficient for particle velocity is

$$R_v = \frac{\rho_1 V_1 - \rho_2 V_2}{\rho_1 V_1 + \rho_2 V_2} \ . \qquad (5)$$

This expression has the same mathematical form as Equation 1.

The model on the right shows a free interface where a downgoing seismic wavefield encounters a second medium having an impedance infinitely smaller than the medium in which the incident wavefield exists. Because of the tremendous impedance decrease below the boundary, the incident wavefield encounters no effective resistance at the interface. The absence of any resistance to movement is the same as saying that the pressure at Z_R is zero. The incident, reflected, and transmitted pressure wavefields are respectively, p_i, $R_p p_i$, and $T_p p_i$, where R_p and T_p are the reflection and transmission coefficients of a downgoing incident pressure wavefield at this impedance boundary. If the interfacial pressure is zero, then

$$p_i(Z_R) + R_p p_i(Z_R) = T_p p_i(Z_R) = 0 \tag{6}$$

For a non-trivial wavefield; i.e., when $p_i(Z_R) \neq 0$, the following conditions must be imposed:

$$R_p = -1 \text{ and } T_p = 0. \tag{7}$$

If Equation 7 is true, and since $\rho_1 v_1 >> \rho_2 v_2$, then the reflection coefficient for pressure must be

$$R_p = \frac{\rho_2 v_2 - \rho_1 v_1}{\rho_2 v_2 + \rho_1 v_1} \ . \tag{8}$$

This is the same mathematical form as Equation 2. Thus, the mathematical expressions for R_v and R_p differ in algebraic sign, and at any arbitrary impedance boundary

$$R_p = -R_v. \tag{9}$$

The functional relationship between reflection and transmission coefficients is the same irrespective of whether pressure or particle velocity behavior is being described; i.e.,

$$T_v = 1 + R_v, \text{ and} \tag{10}$$

$$T_p = 1 + R_p.$$

Equation 10 leads to the following mathematical forms for the transmission coefficient, T, of the three commonly measured dynamic variables associated with plane seismic waves, normally incident on the boundary between layers n and n-1 in Figure 7.1:

Downgoing particle velocity wave
Downgoing particle acceleration wave $\Bigg\}$ $T = \dfrac{2\rho_{n-1}V_{n-1}}{\rho_{n-1}V_{n-1} + \rho_n V_n}$ (11)
Upgoing pressure wave

Upgoing particle velocity wave
Upgoing particle acceleration wave $\Bigg\}$ $T = \dfrac{2\rho_n V_n}{\rho_{n-1} V_{n-1} + \rho_n V_n}$ (12)
Downgoing pressure wave

In this formulation, earth layer n lies below layer (n-1), as is also the condition implied in Equations 1 and 2.

The expressions in Equations 1 and 2 will be used to determine reflection coefficient values in the "Seismic Amplitude Studies" section of this chapter. The data examples in Figures 7-19 and 7-20 are particularly pertinent to this discussion. Appropriate expressions for the reflection and transmission coefficients of plane waves that impinge on an interface at non-normal incidence are given by Brekhovskikh (1960) and Tooley, et al. (1965). Thorough, but easily understood, derivations of reflection and transmission coefficients can also be found in Kinsler and Frey (1950) and in Ingard and Kraushaar (1960). The discussions by Kinsler and Frey are particularly helpful in that they illustrate the phase relationships among pressure, particle velocity, strain, displacement, and dilatation waves for both positive and negative directions of propagation (see their Figure 5.2).

Identification of Seismic Reflectors

Surface-recorded seismic reflection data are essential to hydrocarbon exploration. A conscientious interpreter of these data attempts to determine what each reflection event implies about subsurface stratigraphy and depositional facies. When doing such interpretation, a person should keep in mind that good quality VSP data are capable of defining the depth at which each upgoing primary reflection is created in a stratigraphic section near a study well, and also the time at which each event should arrive at the earth's surface. Thus, by vertical seismic profiling, the correct stratigraphic interpretation to attach to each surface-recorded reflection event can be established by a physical measurement, and does not have to be based on assumptions or synthetic calculations.

Using VSP data with high signal-to-noise properties, an interpreter can answer such questions as:

1. Is a reflection generated at a lithostratigraphic boundary or at a chronostratigraphic boundary?
2. Which lithological boundaries can be seen with seismic data and which cannot?
3. How reliably do synthetic seismograms made from well log data identify primary and multiple reflections?

These questions involve some of the fundamental principles of seismic stratigraphy as proposed by Vail, et al. (1977), as well as the interpretational benefits of synthetic seismogram modeling.

An example of VSP data recorded in a well where the stratigraphic and lithological conditions that create seismic reflections can be identified is shown in Figure 7-3. These data were recorded in a vertical borehole with vertically oriented velocity-sensitive geophones in the downhole geophone package. The lithological control in this VSP well consists of continuous mud log cutting samples, 160 sidewall cores, and a comprehensive set of commercial well logs. Therefore, the lithological description is about as complete as one can achieve from borehole data.

The basic seismic wavelet ingrained within the rather extensive downgoing VSP first arrival is a symmetrical, zero-phase wavelet generated by a compressional vibrator. At any depth the one-way travel time for the direct arrival is the time coordinate of the apex of the first trough of the first arrival wavelet measured at that particular depth. Each upgoing reflection event should also be a symmetrical wavelet, with its travel time being the time coordinate of the apex of the central leg of the reflection wavelet. Using these wavelet criteria for measuring timing relationships, four reflection events, labeled A, B, C, and D, are traced downward to the depths A', B', C', and D' at which they originate. The depth A' is defined as the subsurface position where the projected trend of the black peaks, A, intersects the apex of the first trough of the downgoing first arrival. Depths B', C', and D' are defined by similar projections of the black peaks B, C, and D. The sonic and density logs are shown to verify that acoustical impedance contrasts do occur at depths A', B', C', and D', and that these contrasts can be defined in terms of negative reflection coefficients for a downgoing particle velocity pulse.

270

Figure 7-3

An example of the reliability with which VSP data can often identify primary seismic reflectors. Four upgoing primary reflections are shown by the lineup of black peaks labeled A, B, C, D. The subsurface depth of the interface(s) that generated each reflection can be defined by extrapolating the apices of the black peaks downward until they intersect the first break loci of the downgoing compressional event. These depths are labeled A', B', C', D'. These are raw field data. No processing has been done other than a numerical AGC function has been applied to equalize all amplitudes.

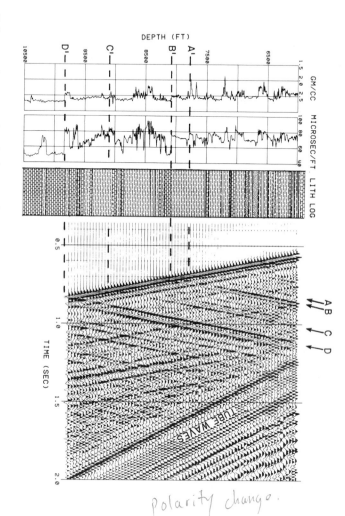

DEPTH (FT)

GM/CC   MICROSEC/FT   LITH LOG

TIME (SEC)

TUBE WAVES

A B   C   D

Polarity change.

Since the central leg of the four marked reflection events is a black peak, but the central leg of the incident wavelet is a trough, all four reflections must be generated where there is a negative reflection coefficient for a downgoing velocity pulse in order for the wavelet polarity to be reversed. This interpretation of the algebraic sign of the four reflection coefficients from the polarities of the downgoing and upgoing wavelets is supported by the analysis of particle velocity vectors shown in Figure 7-4. A velocity sensitive VSP geophone is shown at depth Z_1, which is a distance, $M\lambda$, above a stratigraphic interface that generates an upgoing reflection. The distance, $M\lambda$, is chosen so that the seismic wavelet experiences M cycles of 360 degrees of phase change as it propagates from Z_1 to the interface at depth Z_R (M being an arbitrary integer). Thus, whatever phase orientation the particle velocity vector has at the interface, it has that same orientation at depth Z_1, and vice versa. If the particle velocity vector that creates the central leg of the geophone output is oriented downward at Z_1, and the geophone output is a trough, then that same particle velocity orientation and geophone polarity exist

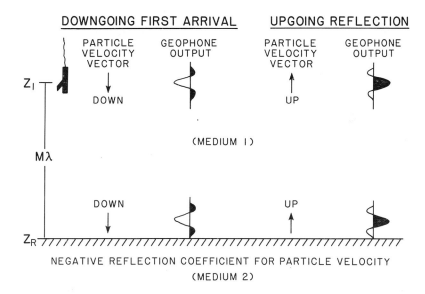

Figure 7-4   Interpreting the algebraic sign of a reflection coefficient from the polarities of downgoing and upgoing events recorded by a velocity sensitive geophone.

for the downgoing wavelet at the interface. A negative reflection coefficient is assumed at the interface, so the particle velocity vector of the upgoing reflected wavelet at depth Z_R has a phase shift of 180 degrees relative to the incident particle velocity vector. Consequently, the voltage output of a velocity sensitive geophone responding to this reflection event would also shift by 180 degrees and appear as a peak, rather than a trough. The upgoing reflection event has the same particle velocity orientation and geophone polarity at Z_1 as it has at the interface, since the propagation distance from Z_R to Z_1 is $M\lambda$. Thus, if an upgoing reflection event recorded by a velocity-sensitive geophone has a polarity opposite to that of the downgoing first arrival, then that reflection is generated by an interface having a negative reflection coefficient for a downgoing particle velocity pulse. The acoustic impedance increases below such an interface.

A key observation that can be made from the data in Figure 7-3 is that the lithological log, which is constructed from continuous cutting samples and a drilling rate log, shows that events B and D clearly mark lithostratigraphic boundaries between shale and limestone. However, events A and C appear to originate within thick shale units. If this lithological description of the subsurface is true, then reflections A and C would support the concept proposed by Vail et al. (1977), and others, that seismic reflectors are not necessarily lithostratigraphic markers. Close examination of additional lithological control in this well shows that silty and limy components appear in the shale units at depths A' and C'. A detailed illustration of all the lithologically sensitive subsurface

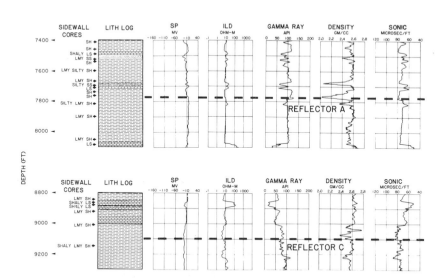

Figure 7-5   Subsurface data describing the physical rock properties creating seismic reflections A and C in Figure 7-3.

information recorded in the well at these reflector depths is shown in Figure 7-5. A sidewall core was obtained at each depth marked with a solid arrowhead. The lithology of each core sample is written beside the sampled depth. The depths A' and C', identified from the VSP data as the depths at which reflections A and C originate, are shown by the two dashed lines. These subsurface data imply that some seismic reflections in the vicinity of this well should be mapped as interfaces that are better characterized by slight changes in sediment grain size, or variations in cementation, than as boundaries where there is a change in basic lithology, since for both reflectors A and C, the basic lithology is shale. In stratigraphic trap exploration, it is essential to know in this type of detail exactly what stratigraphic and lithological interpretations can be assigned to key seismic reflection events. Mapping a reflection event as the boundary between a limestone and a shale, when it actually marks only a slight difference in grain size or a calcareous interval in a shale, will lead to numerous dry holes. It should be noted that although the slight change in grain size at depth A' is a mild alteration in lithology, the log data show that it is a facies change that is marked by significant changes in velocity, density, and cementation, and thus is a strong seismic reflector. This data example shows that a properly executed vertical seismic profile can accurately identify the depths and surface arrival times of critical seismic reflectors penetrated by a VSP well, and is therefore an invaluable interpretive aid that indicates the geological significance of surface-recorded reflection events.

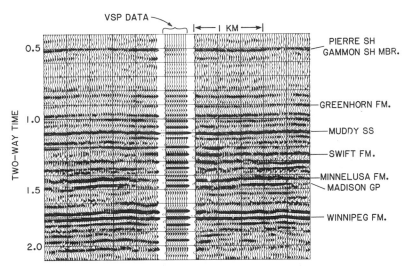

Figure 7-6   Comparison between surface-recorded reflection data and processed VSP data at the USGS Madison Limestone Test Well No. 2. (Altered from Balch, et al., 1981b).

An example of VSP data which identify stratigraphic markers throughout an extensive geologic section is shown in Figure 7-6. These data were recorded, processed, and interpreted by Balch, et al. (1981b) for the U. S. Geological Survey. Upgoing primary reflections were retrieved from the borehole geophone responses by applying several of the data processing techniques described in Chapter 5. A wavelet shaping process creates an excellent waveform character match between the surface-recorded data, which were generated by compressional vibrators, and the VSP data, which were generated by a land airgun. The depth and stratigraphic significance of each reflecting interface were identified in the vertical seismic profile by an interpretational procedure similar to that described in Figure 7-3. The quality of these VSP data, together with the excellent correlation between the processed VSP waveshapes and the surface-recorded reflection waveshapes, insures that subsurface geology will be accurately transferred to the surface-recorded seismic data.

Another example which illustrates how VSP data identify stratigraphic interfaces in surface-recorded reflection data is given in Figure 7-7 (Wyatt and Wyatt, 1982b). This illustration accumulates in a single montage the pertinent subsurface data that can assist in a stratigraphic interpretation of surface reflection data recorded over a VSP study well. The VSP data are filtered to attenuate downgoing events, deconvolved to remove upgoing multiples, and time shifted to vertically align upgoing events reflected from flat, horizontal reflectors. These data are then vertically summed into a single composite trace containing all upgoing VSP primary events. This composite trace is repeated ten times and inserted between the VSP data and the stacked surface-recorded data in order to facilitate correlating upgoing VSP reflections with events recorded at the surface. Different depth datums were used when processing the surface-recorded data and the VSP data. Consequently, the two data sets must be shifted 40 ms relative to each other in order to align equivalent events.

Six of the more prominent upgoing VSP primary reflections which correlate with surface-recorded events are followed down to the depths at which they originate, and then traced laterally along the dashed lines to the sonic log to identify the stratigraphic feature creating each event. The fourth dashed line from the top (at a depth of 6450 feet) marks a positive reflection coefficient for a downgoing particle velocity pulse; all other lines mark negative reflection coefficients. These negative reflectors correlate with black peaks in the VSP data and in the surface seismic traces. Many stratigraphers, who have to use surface seismic data in order to interpret the subsurface, are beginning to use VSP data in this manner in order to tie subsurface stratigraphy to the surface-recorded data.

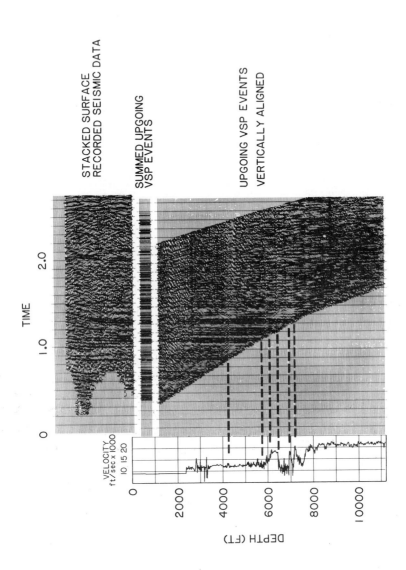

STACKED SURFACE
RECORDED SEISMIC DATA

SUMMED UPGOING
VSP EVENTS

UPGOING VSP EVENTS
VERTICALLY ALIGNED

TIME

VELOCITY
ft/sec × 1000
10 15 20

DEPTH (FT)

Figure 7-7   An example of correlating subsurface geological information (i.e. a well log) with surface-recorded seismic reflections by means of a vertical seismic profile (From Wyatt and Wyatt, 1982b).

Comparison of VSP Data with Synthetic Seismograms

The traditional tool that has been used to establish a correspondence between subsurface stratigraphy and surface-measured seismic data has been the synthetic seismogram. The preceding section illustrates how a vertical seismic profile can also be used to identify primary seismic reflectors in a drilled stratigraphic section. Logic would tell one that VSP data should be a better, more accurate, and more powerful tool for interpreting surface seismic data than is a synthetic seismogram because VSP data are real seismic measurements. In contrast, a synthetic seismogram is exactly what it says it is, a synthetic representation of seismic measurements. In a vertical seismic profile, one can use the same energy source, the same type of geophone, and the same recording instrumentation that was used when acquiring surface seismic data. In a synthetic seismogram calculation, one can only approximate these aspects of the total seismic recording process.

Of particular importance is the fact that a VSP measurement responds to lateral impedance changes in the subsurface which are of the same size as a first order Fresnel zone, whereas a synthetic seismogram does not. To illustrate this important difference between a synthetic seismogram's definition of upgoing seismic reflections, and the upgoing events measured by a vertical seismic profile, consider the stratigraphic situation shown in Figure 7-8. Here a well penetrates a sand body which has a lateral extent equal to the dominant wavelength, λ , contained in a seismic wavelet propagating through the stratigraphic section. Sonic and density log data recorded in this well will indicate a change in acoustic impedance at the top and bottom of the sand. Consequently, a synthetic seismogram calculated from these log data will create reflection events at these two lithological boundaries. However, actual surface seismic data recorded across this feature will not show any reflection event because the lateral dimension of the sand is too small to create a reflected wavefront. The sand body is a diffractor, not a reflector, as far as a propagating seismic wavefront is concerned. In some situations, depending on the overall seismic noise conditions, one might see the diffraction pattern of the sand expressed in an unmigrated stacked seismic section, but the sand body would not be identifiable in a migrated seismic section. The principal point to be emphasized by this example is that the synthetic seismogram indicates a reflection, but the actual seismic data do not. This difference exists, even though the log data are correct and the synthetic seismogram calculation is accurate, simply because log data measure only those rock properties that occur very close to a borehole.

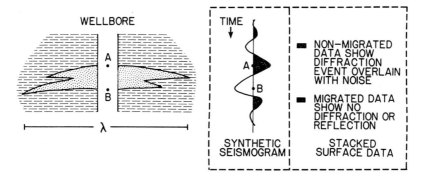

Figure 7-8   A common stratigraphic situation illustrating why a synthetic seismogram response may not look like surface-recorded reflection data. The width of the sand body is equal to the dominant wavelength, λ, of the propagating seismic wavelet, thus the sand is a point diffractor as far as surface measurements are concerned, and is not a reflector. However, sonic and density log data recorded in the well are insensitive to the lateral extent of the sand, so the synthetic seismogram made from these log data indicates strong reflections at the top and bottom of the sand.

The reverse of this situation can also occur; i.e., a synthetic seismogram can indicate no reflection event at a depth where actual surface seismic data contain a reflection. A stratigraphic condition that could create this discrepancy is shown in Figure 7-9. Here a well passes between two laterally extensive sands without penetrating either

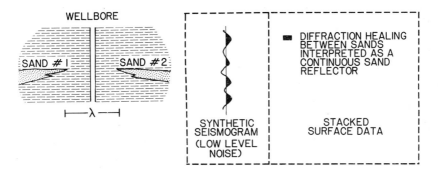

Figure 7-9   A second stratigraphic situation illustrating why a synthetic seismogram response does not always look like surface-recorded surface data. The gap between the sand bodies is equal to the dominant wavelength, λ, of the propagating seismic wavelet; thus, strong diffraction tails from the two sand reflections will extend across the gap, creating the appearance of a continuous reflection event. However, sonic and density log data recorded in the well will not detect the sands, so the synthetic seismogram made from the log data will not contain a sand-shale reflection.

of them. Accurate log data recorded in the well will indicate no acoustic impedance change, so a synthetic seismogram made from the log data will contain no reflection at the depth of the sand. However, when a seismic wavefront propagates through this interval, diffraction tails will extend away from the ends of the two sand bodies and overlap at the well position. Since the well is less than a first order Fresnel zone radius from each sand, the surface-recorded data appear as a continuous sand reflection.

When the sonic and density log data recorded in a well are an accurate representation of the seismic impedance sequence seen by a propagating seismic wavelet, then a synthetic seismogram made from these data will be a good representation of what surface reflection data at that well site should be. Because of stratigraphic situations like the two conditions just described, a nagging and often unanswered question is, "How faithfully do log data represent a seismic impedance sequence?" A paper by Goetz et al. (1979) that discusses other reasons why sonic log data do not represent correct seismic velocity changes in a stratigraphic section is recommended reading.

Data recorded in two closely spaced VSP wells will be used to compare the technical merits of interpreting surface-recorded seismic reflection data by means of VSP measurements and also by means of synthetic seismograms. A complete suite of well logs were recorded in these two study wells immediately after the boreholes reached total depth. In addition, a borehole gravity meter survey was performed after each well was cased so that two density logs; i.e., a standard compensated density log and the borehole gravity meter density values, exist for each well. It should be emphasized that the boreholes were not optimum for recording log data because numerous washouts occurred. The wells were intended to be marginally economic gas wells, so they were drilled in the cheapest, quickest possible way in order to enhance the economics. Water was used as the drilling fluid for the top one-third or more of each hole, and then a mud program was started for the bottom portion of the well. The resulting severe borehole rugosity makes all log data suspect, but this type of borehole is typical of hundreds of wells drilled in the United States each year. For this reason, it seems important to compare synthetic seismograms calculated from log data recorded in such a borehole with VSP data recorded in the same well.

The well log data, together with the synthetic seismograms calculated from the log data, are shown in Figures 7-10 and 7-11 as functions as seismic travel time. The sonic logs are reasonable estimates of the seismic propagation velocity through the drilled interval because they had to be corrected only 1 to 2 percent in order for the integrated times to agree with the check shot travel times. A sonic log can, of course, err in its fine detail and still satisfy this constraint on its integrated one-way time.

The most questionable log data that enter into the synthetic seismogram calculation are the bulk density values provided by the borehole compensated density curve. These data are questionable, not because the log instrumentation is faulty, but because the severe borehole rugosity prohibits the density tool from functioning properly. Inspection of the figures shows that there is a significant difference between this log curve and the density values calculated from the borehole gravity meter data. The following observations should be made about these density estimates:

1. Many details of the borehole compensated density curve replicate the detailed character of the integrated sonic log. Both curves should look alike since decreases in rock porosity would create increases simultaneously in each curve. If there were no other rock density estimate available, this general correlative behavior between the compensated density curve and the sonic log velocity would be taken as a confirmation that these density data could be used in synthetic seismogram construction even though the density values in some intervals are questionable.

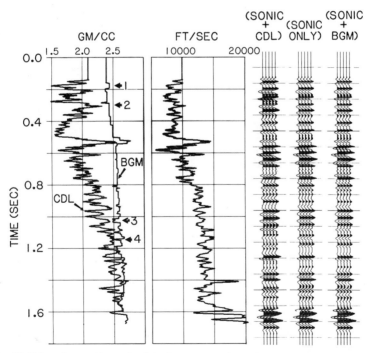

Figure 7-10   Well log data and synthetic seismograms describing the stratigraphic section at VSP well "P". The log curves labeled "CDL" and "BGM" identify a compensated formation density log and a borehole gravity meter profile, respectively. The numbered arrows define depths where the two density curves have different slopes and could create different reflection coefficients. The synthetic seismograms are numerically leveled and do not have true relative amplitudes.

2.   Borehole gravity meter (BGM) readings are costly, time consuming measurements; consequently, these data are collected at as few depth locations as can be justified. The gravity meter readings yield a constant density value between successive gravity meter stations, so the BGM curves in the figures have a blocky appearance. Detailed density behavior cannot be revealed by these gravity meter data because of the large depth increments between the recording levels. However, the general trend, and magnitude, of the BGM curve should be a correct picture of the vertical profile of the bulk density of the rocks penetrated by this well. There is an obvious, fundamental difference between these borehole gravity meter results and the density estimates provided by the borehole compensated density curve, particularly in the upper half of the

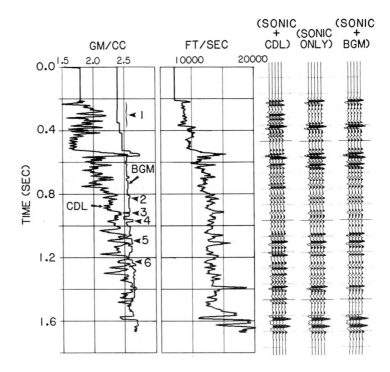

Figure 7-11   Well log data and synthetic seismogram describing the stratigraphic section at VSP well "Z". The log curves labeled "CDL" and "BGM" identify a compensated formation density log and a borehole gravity meter profile, respectively. The numbered arrows define depths where the two density curves have different slopes and could create different reflection coefficients. The synthetic seismograms are numerically leveled and do not have true relative amplitudes.

well. The abrupt density changes that occur at each gravity meter station level are smoothed before the BGM data are used to calculate the reflection coefficients used in constructing synthetic seismograms.

3. Approximately 150 sidewall cores were collected throughout the stratigraphic section penetrated by each well, and the grain densities and porosity values were determined in several samples taken from these cores. These laboratory measurements confirm that, in these wells, the borehole gravity meter data are a more reliable indication of rock bulk density than are the borehole compensated density curves. This finding should obviously not be extrapolated to other wells which have different borehole conditions, and which penetrate different geological environments.

Synthetic seismograms calculated from these well log data are also shown in Figures 7-10 and 7-11. These seismograms contain no multiple reflections, and thus are synthetic representations of what surface-recorded primary reflections should be at each well site. The wavelet used to generate the seismograms in each figure is an average of 20 of the VSP direct arrivals recorded at various depths in each well. Thus, the same basic wavelets are contained in the synthetic seismograms and in the VSP traces, which is essential for a fair comparison of the synthetic and real data. The amplitudes of the synthetic seismograms are "leveled" by a numerical automatic gain control function calculated over successive 800 ms data windows.

It is important to note that there is essentially no difference in the synthetic reflection response calculated with and without density data included. This behavior is one that is often observed in synthetic seismogram construction. The inclusion of density data in a synthetic seismogram calculation generally manifests itself by altering the amplitudes of events; often it does not make a radical change in reflection waveshape nor create or eliminate reflection events. This behavior is particularly common when a borehole compensated density curve and an integrated sonic log agree in as much detail as they do in these two wells. Any amplitude effects created by the inclusion of density data in the seismogram calculation are generally camouflaged by the numerical gain function applied to the data in these displays. In this analysis, relative reflection amplitudes will be ignored when comparing the synthetic seismograms and the VSP data with the surface seismic data crossing the wells. Instead, the emphasis is on whether or not the synthetic seismogram and the VSP data show reflection events at the same arrival times as the

282

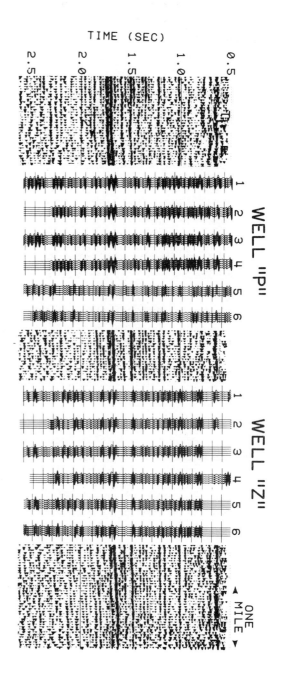

Figure 7-12   Comparison between upgoing VSP events and a surface seismic line crossing study wells "P" and "Z". The numbered VSP data panels were generated in the following ways - (1) vertical sum of unfiltered VSP data, (2) restricted vertical sum of unfiltered VSP data, (3) vertical sum of VSP data after velocity filtering to attenuate downgoing events, (4) restricted vertical sum of VSP data after velocity filtering to attenuate downgoing events, (5) vertical sum of VSP data after predictive deconvolution of upgoing wavefield, and (6) restricted vertical sum of VSP data after predictive deconvolution of upgoing wavefield. VSP data were recorded all the way up to the surface in well P. Only a few levels were recorded in well Z above a two-way time of 750 ms.

283

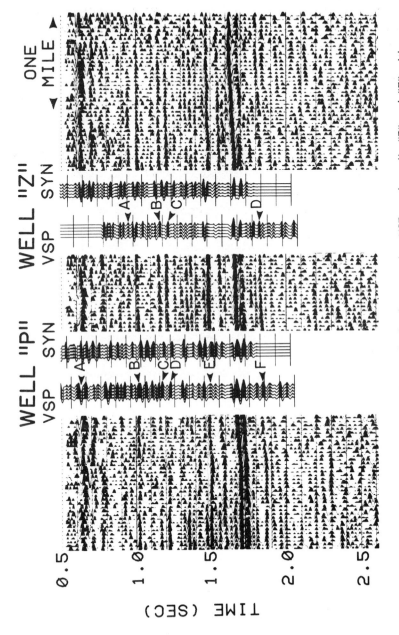

Figure 7-13   Comparison of surface seismic data crossing VSP study wells "P" and "Z" with synthetic seismograms and VSP data recorded in the wells. The lettered arrowheads show where the VSP data are a better match to the surface data than are the synthetic seismogram data.

surface-recorded reflections, and whether or not the waveshapes are like those observed at the surface. The synthetic seismogram constructed from the borehole gravity meter data is also shown in each figure. Surprisingly, it apparently makes no important difference which density estimate, the borehole compensated density curve or the borehole gravity meter data, is used to generate the synthetic seismic response. Some depth intervals where the two density curves disagree are numbered on the BGM curve in each figure, but the synthetic responses in these intervals are essentially the same. Because all three versions of the synthetic seismograms made from the log data in each well are so much alike, only the synthetic seismogram made using the sonic log data will be used in the remaining analysis.

Several versions of the upgoing events that can be created from the VSP data recorded in each well are compared with surface seismic data crossing the wells in Figure 7-12. The data processing techniques referred to in the short descriptive phases in the figure label are discussed in Chapter 5. The term "vertical sum" used to describe some of the processing procedures means that each VSP trace is delayed by its first break time, as in Figure 5-4, and all traces are then summed. Only a few geophone levels were recorded in well Z in the top 3500 feet of the stratigraphic section; consequently, except for a few isolated events, the VSP responses shown for that well start at approximately 750 ms two-way time. Data panels 1 and 3 should contain both upgoing primary and multiple reflections since all upgoing events are included in these vertical sums. The remaining data panels should contain only primary reflections since they are created by vertically summing upgoing events occurring in a short data window immediately following first break times, or by vertically summing upgoing events after a multiple suppressing deconvolution operator has been applied to the upgoing wavefield. Upgoing multiples are not a severe problem in these wells.

At each well, numerical correlations were made between that well's synthetic trace and several surface-recorded traces spanning the well site in order to estimate how satisfactorily the synthetic trace matches the surface data. Numerical correlation calculations were also made between the VSP traces in Figure 7-12 and the surface-recorded data. These correlation values show that the VSP derived traces are a better match to the surface reflection data than are the synthetic seismogram traces. This conclusion can be visually confirmed by the illustration in Figure 7-13, where one VSP data panel from each well is shown in a side-by-side comparison with its corresponding synthetic seismogram and the surface-recorded data. VSP data panel 3 is chosen for well P and data panel 5 for well Z. Some surface-recorded events which the VSP data match more accurately than do the synthetic seismograms are labeled A, B, C, ... Both wells terminate at the base of the strong reflector at approximately 1.7 seconds; consequently, the synthetic seismic traces end at that time. Note, however, that a VSP measurement looks below the bottom of a well and records events, such as D and F, which occur several hundred feet below the total depth of each well. This capability of looking below a drilled depth is one great advantage of VSP data over synthetic seismograms which will be further emphasized in Chapter 8. The VSP traces are arbitrarily terminated below 2.5 seconds.

The comparison between the VSP data and the surface-recorded data in this instance is certainly not optimum. Better agreements between VSP data and surface-recorded data have been demonstrated elsewhere. However, this example, together with the data analyses shown in Figures 7-3, 7-6, and 7-7, emphasize an important reason for the interest in vertical seismic profiling, that being:

> VSP data, properly recorded, are the best tool that an explorationist can use to define primary seismic reflections arriving at the earth's surface. In particular, VSP data, properly recorded, provide a more reliable estimate of surface-arriving primary reflections than does a synthetic seismogram.

This statement will be true as long as the qualification, "properly recorded", is satisfied. Unfortunately, in some wells, the quality of VSP data is adversely affected by unbonded casing, dominating tube waves, poor field procedures, or a host of other problems to such an extent that a VSP's definition of upgoing events can be inferior to that provided by a synthetic seismogram.

In addition, one important characteristic of the earth which becomes obvious as VSP data are examined is that most downgoing multiples occur immediately after the first arrival wavelet. This fact means that the shallow part of a stratigraphic section is the dominant multiple-generating part of the earth. However, log data are usually not recorded in the shallowest parts of a well. In those few instances when sonic and density logs are recorded in the topmost part of a well, these data are often unreliable because the hole diameter is too large or too rugous for logging. Consequently, it is difficult for a synthetic seismogram to accurately portray the total seismic response of a stratigraphic section since the input data needed to describe the reflection and transmission characteristics of the critical, shallow part of the earth are unavailable. This handicap does not exist in vertical seismic profiling, and VSP data contain all multiples generated in the shallow parts of a geological column.

In summary, the thrust of this section has been to illustrate the potential superiority of VSP data over synthetic seismograms for exploration purposes, and not to propose the abolishment of synthetic seismogram usage. Quite the contrary, synthetic seismograms should be more widely used than they are. The synthetic seismogram is, and will remain, one of the most valuable interpretive tools that an explorationist can have. One must simply keep in mind that the log data used to generate a synthetic seismogram have an extremely limited radius of investigation, and in some instances may not represent the actual seismic impedance sequence which a propagating seismic wavefront encounters as it travels through the stratigraphic section penetrated by a well.

Fresnel Zones and VSP Horizontal Resolution

A first order Fresnel zone defines the smallest lateral dimension of a subsurface anomaly which can be resolved by surface-recorded seismic data. VSP data provide a better lateral resolution of the subsurface than do surface reflection measurements

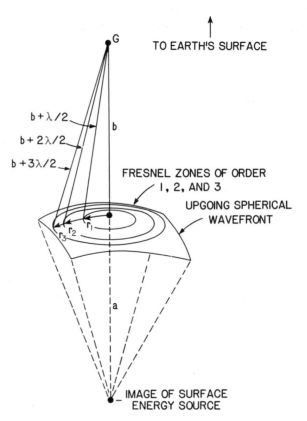

Figure 7-14   The ability of seismic measurements to resolve subsurface geological features in a horizontal sense is controlled by the first order Fresnel zone. This illustration shows a VSP geophone, G, approaching a subsurface reflector, which is being illuminated by a spherical wavefront created by a surface source above G (i.e. a > b). The raypath picture is simplified by using virtual raypaths from the source to the reflector.

because the Fresnel zones involved in VSP measurements are smaller than the Fresnel zones associated with surface positioned geophones. The following analysis that supports this statement assumes that a VSP energy source on the earth's surface creates a spherically spreading wavefront, which travels down to a horizontal impedance interface located at depth "a", and then reflects upward to a VSP geophone located below the earth's surface. These assumptions result in the raypath and wavefront geometry shown in Figure 7-14. In this illustration, the surface source and the raypaths from the source to the reflector are shown as virtual images positioned below the reflector. A VSP geophone G is shown at a distance "b" above the reflecting interface. Suppose that we restrict the propagating wavefront so that it contains only one wavelength, λ, and then freeze the wavefront in space as shown in Figure 7-14. Whenever raypaths from geophone G to any two spatial points located on a constant phase surface of this wavefront differ by less than $\lambda/2$, those two spatial points are defined to be within the same first order Fresnel zone on that wavefront surface. Thus, the radius, r_1, of the first order Fresnel zone whose center is a distance, b, directly below geophone G, has a value such that the travel path from the geophone to the circumference of the zone is $(b + \lambda/2)$. The first, second, and third order Fresnel zones are indicated in the figure.

A cross-sectional view of a possible source-to-reflector-to-geophone raypath is shown in Figure 7-15. Again, the surface source is shown as a virtual image on the opposite side of the reflector from its true position. The extra length PQ for the non-normal raypath determines the radii of the Fresnel zones. The basic assumption involved

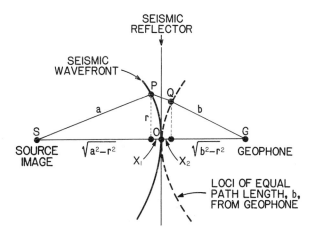

Figure 7-15   Raypath SPG is longer than raypath SOG by an amount PQ. If r is the radius of the first Fresnel zone, then PQ = $\lambda/2$. If r is small compared to a and b, then a good approximation is PQ = $X_1 + X_2$.

in Fresnel zone calculations is that the distances a and b are much larger than λ. This requirement leads to the equivalent condition that the Fresnel zone radius, r, shown in Figure 7-15, is small compared to distances a and b (Jenkins and White, 1957, p. 355; Halliday and Resnick, 1960, p. 1088). In such a case

$$PQ = X_1 + X_2, \tag{13}$$

and the distance, X_1, can be expressed as

$$X_1 = a - \sqrt{a^2 - r^2} \approx \frac{r^2}{2a} . \tag{14}$$

Thus

$$PQ = \frac{r^2}{2a} + \frac{r^2}{2b} = \frac{a+b}{2ab} r^2 . \tag{15}$$

The radii, r_m, of successive Fresnel zones then have values such that

$$m\frac{\lambda}{2} = \frac{a+b}{2ab} r_m^2 \tag{16}$$

Therefore, the radius of the first order Fresnel zone is related to the source-to-reflector distance, a, and to the reflector-to-geophone distance, b, by the equation

$$r^2 = \lambda \frac{ab}{a+b} . \tag{17}$$

If the geophone is at the earth's surface, then b=a, and

$$r^2 = (\lambda a)/2. \tag{18}$$

This latter expression is probably the most common form used to describe the Fresnel zones involved in surface seismic recordings. A more rigorous mathematical development of the size of the reflecting area involved in reflection seismology is given by Zavalishin (1975). His analysis allows the source to be offset from the observation point, rather than being coincident with, or vertically above, the geophone as assumed in this model. When the source and receiving point are laterally offset from each other, the Fresnel zone is elliptical rather than circular. Some investigators change Equation 18 to one involving frequency, f, velocity, V, and two-way time, T, by using the relationships $\lambda = V/f$ and $a = VT/2$.

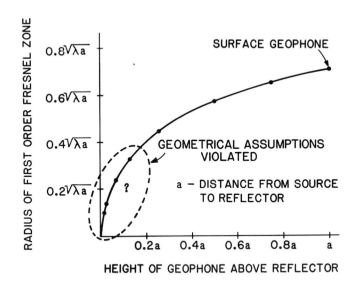

Figure 7-16   The radius of the first order Fresnel zone decreases as a VSP geophone
approaches a seismic reflector.   Consequently, the horizontal resolution
improves as the geophone is placed closer to the reflector.

For VSP measurements, $b \neq a$ and Equation 17 is the correct formulation to use to
define the size of first order VSP Fresnel zones as long as the assumption that a and b are
much larger than either λ or r is not violated. This equation is plotted in Figure 7-16 for
geophone-to-reflector distances restricted to the range $0 \leq b \leq a$. This curve shows that
the Fresnel zone involved in VSP measurements is indeed smaller than the Fresnel zone
associated with surface-recorded data. Thus, VSP data must necessarily provide a better
horizontal resolution of subsurface anomalies than do surface seismic reflection data.
The decrease in Fresnel zone size shown in Figure 7-16 is perhaps not as dramatic as some
VSP enthusiasts are initially inclined to claim. For example, a VSP geophone has to
approach to a distance a/7 from the reflector before the Fresnel zone radius is halved
relative to the radius that exists at the surface. On the other hand, any increase in
horizontal resolution is significant and encourages one to use VSP data in exploration
work. As b decreases to values smaller than a/7, the assumption that $\lambda \ll r$ is questionable,
and a mathematical model other than this geometrical optics approach must be used to
determine horizontal resolution.   The conditions under which one can specify horizontal

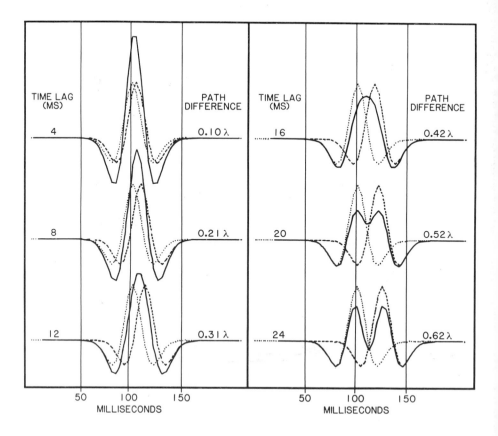

Figure 7-17   Each wavelet plotted as a solid line is the sum of the two 20 Hz Ricker wavelets plotted as dotted and dashed lines.   The dotted wavelet is stationary, and the dashed wavelet is delayed by the time lags labeled on the left.   These time lags correspond to the path differences indicated on the right, where λ is the predominant wavelength (Kallweit and Wood, 1982). The stationary wavelet is assumed to travel the fixed path SOG in Figure 7-15, whereas, the delayed wavelet travels the variable path SPG.

resolution when a receiver is very close to the surface that is being imaged do not appear to be stressed in geophysical literature.

Equations 17 and 18 are based on the criterion that the difference in travel path between the center and the edge of a first order Fresnel zone is $\lambda/2$. Certainly, wavelets arriving at a receiving point which have traveled paths differing by more than $\lambda/2$ will add destructively, but it is probably inappropriate to assume that path differences exceeding $\lambda/4$, let alone $\lambda/2$, should be included in a realistic Fresnel zone. A simple example that demonstrates how travel path differences affect wavelet shapes is shown in Figure 7-17.

This analysis shows the result of adding two identical 20 Hertz Ricker wavelets when one of them is delayed in time relative to the other. The wavelet shown as a dotted line remains stationary and represents a reflected wavelet traveling the fixed path SOG from source to receiver in Figure 7-15. The wavelet shown as a dashed line is delayed in time by varying amounts and represents a reflected wavelet traveling path SPG as point P moves away from the center of the Fresnel zone located at point O. The path difference associated with each travel time delay is expressed in terms of the wavelet's predominant wavelength, λ, which is defined as $\lambda = VT$, where V is the interval velocity at point O and T is the trough-to-trough wavelet breadth. The composite trace differs significantly from the original waveshape when the path difference exceeds 0.43λ, which is Ricker's resolution criterion (Kallweit and Wood, 1982). However, a significant time delay, as well as some distortion, exists even for a path difference of 0.3λ.

This calculation shows only the effect of a pure time lag on the compositing of two identical noise free signals. When noise is added, the distortions would be greater than those shown. Also, the lagging signal from the edge of a Fresnel zone, because it experiences a larger reflection angle, would have a smaller amplitude than would the signal reflected from the center of the Fresnel zone. When the effects of noise and reduced amplitude of the lagging signal are considered, it is reasonable to propose that for practical purposes, the radius of a first order Fresnel zone should be determined by defining the path difference between the center and the edge of the zone to be $\lambda/4$, and perhaps as small as $\lambda/8$, rather than $\lambda/2$.

A specific calculation may help illustrate the difference between the Fresnel zone sizes involved in vertical seismic profiling and surface reflection recording. Assume that the depth, a, to a flat horizontal reflector is 8000 feet, that a VSP geophone is positioned at a height, b, 2000 feet above this reflector, and that the seismic wavelength of interest is 400 feet long. The Fresnel zone radii calculated from Equations 17 and 18 are 800 and 1265 feet, respectively. These Fresnel zones are shown as the dashed circles in Figure 7-18, which is a plan view of the reflector. The dot at the center of the circle represents a vertical VSP wellbore. If we assume that the effective Fresnel zone size occurs when the path difference is $\lambda/8$ rather than $\lambda/2$, then the radii reduce to 400 and 632 feet, respectively. These Fresnel zones are shown as the solid circles, and they may more likely define the true imaged area of the reflector.

All of these circles are drawn to scale in order to emphasize that VSP measurements, because of their smaller Fresnel zones, provide an improved lateral resolution of subsurface geological conditions as compared to surface measurements. Also, note that typical well log data, such as a sonic log or a density log, can investigate only a few inches beyond the wall of a borehole. Thus, well log data can define reflection properties only within the dot at the center of each Fresnel zone.

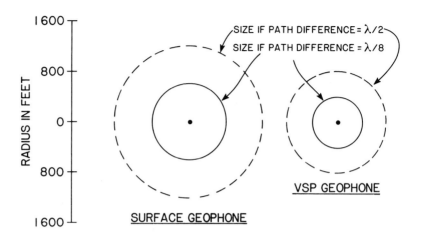

Figure 7-18   A plan view, drawn to scale, comparing the first order Fresnel zones
              involved in VSP and surface recording geometries.   Assumptions are: (1)
              depth to reflector = 8000 feet, (2) height of VSP geophone above reflector =
              2000 feet, (3) wavelength = 400 feet.

Seismic Amplitude Studies

With the advent of high quality digital field recorders which can preserve seismic
reflection amplitudes with great fidelity over a wide dynamic range, the petroleum
industry has placed increased emphasis on the geological significance of seismic reflection
amplitudes. For example, the bright spot technology, whereby subsurface lithology and
fluid content are interpreted from seismic reflection amplitudes, is now widely used. The
concepts of this technology are simple, but several assumptions and approximations have
to be made in the data processing. Specifically, there is some question as to what gain
function one should use to remove spherical divergence effects, and also what polarity
should be assigned to the final data displays.

The calculation of synthetic impedance logs from stacked seismic traces is a second
development which leads to a geological interpretation of seismic reflection amplitudes.
The calculation procedure assumes that the final processed versions of the reflection
amplitudes are reliable band-limited estimates of subsurface reflection coefficients. This
technology is becoming just as widely used in stratigraphic trap exploration as is the
bright spot technique. There is, therefore, a need for a direct physical measurement of
the behavior of seismic reflection amplitudes which will permit the accuracy of the final
stacked surface-recorded seismic amplitudes to be verified. A vertical seismic profile is

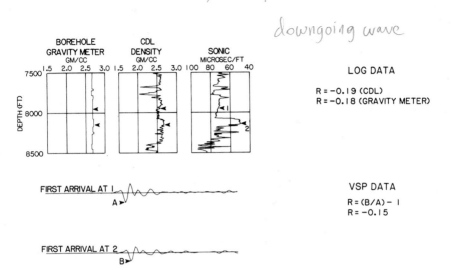

shown.

downgoing wave

LOG DATA

R = −0.19 (CDL)
R = −0.18 (GRAVITY METER)

FIRST ARRIVAL AT 1

A ➤

VSP DATA

R = (B/A) − 1
R = −0.15

FIRST ARRIVAL AT 2

B ➤

Figure 7-19   The arrowheads beside each log curve indicate two positions "1" and "2", located above and below a prominant shale-limestone interface.   The magnitude of the reflection coefficient, R, associated with this interface is traditionally estimated from log data. VSP data offer an alternative way to estimate reflection coefficients and calibrate reflection amplitudes. The VSP estimate should be more reliable since it results from a real geophone response to an actual seismic wavefront.   The value of R here is negative since the VSP geophone reacts to a downgoing particle velocity wave.

invaluable in this regard because it allows one to measure seismic wavelet amplitude behavior in situ, and thereby assign precise seismic reflection amplitude and polarity behavior to subsurface geological conditions.

An example showing how VSP data can be used to interpret reflection amplitudes is shown in Figure 7-19.   These data show the incident wavelet (at depth 1) and the transmitted wavelet (at depth 2) above, and below, a shale-limestone interface in an experimental well.   The amplitudes of both wavelets are adjusted by a factor of AT^n to account for spherical divergence (Equation 12, Chapter 5).   These wavelets allow an interpreter to calculate both the polarity and the magnitude of the reflection coefficient at the top of the limestone unit.   This calculation is shown in the lower right of the illustration.   The sonic and density logs recorded in the well are also shown in the figure, and the reflection coefficients calculated from the log data (shown at the upper right) can be compared with that obtained from an analysis of the VSP wavelet amplitudes.   The amplitudes of the incident and transmitted seismic wavelets are arbitrarily defined as the magnitude of the first trough of these wavelets.   Other measures of the wavelet amplitudes could be used.

The log data shown are unedited, and there has been no attempt to correct them for borehole effects, such as rugosity or mud invasion. In this example, the density information obtained via the borehole gravity meter measurements yields a reflection coefficient more like that obtained with the seismic amplitude calculations than do the standard borehole compensated density data. This result is probably true because the borehole gravity meter averages rock bulk density over a wide lateral expanse away from the borehole, just as does a seismic wavefront.

If good quality reflections can be seen in VSP data throughout a drilled section, then this technique can be used to calibrate surface reflection amplitudes for all reflecting interfaces penetrated by a VSP well. Superior stratigraphic and lithological interpretations of surface-recorded data and improved synthetic velocity log calculations should result from this application of VSP data.

Another example showing how the amplitudes of compressional first arrivals measured in a vertical seismic profile are affected by subsurface acoustic impedance changes is given in Figure 7-20. The amplitude decay caused by spherical divergence is removed from these data by multiplying all sample points of the wavelets by a factor, AT^n. T is the recording time associated with a sample point, and A and n are determined from an analysis of first break times (See Equation 12, Chapter 5). No density log is shown, so the sonic log provides the only estimate of acoustic impedance behavior through this stratigraphic section. Note that when the sonic log shows a velocity decrease at a depth of 9060 feet, then the amplitudes of the first arrivals increase below that depth. When the velocity increases at a depth of 9160 feet, the amplitudes decrease. Thus, there is an inverse relationship between the magnitude of the acoustic impedance in a stratigraphic unit and the amplitude of the seismic response measured by a velocity sensitive geophone in that unit. If the downhole transducer is pressure-sensitive, rather than velocity-sensitive, the wavelet amplitudes would behave opposite to those shown in Figure 7-20, i.e., low amplitude direct arrivals would be recorded in low impedance sections, and high amplitudes would occur in high impedance intervals. These wavelet behaviors can be explained by the transmission coefficient definitions given in Equations 11 and 12. According to Equation 11, the amplitude of a downgoing transmitted particle velocity wavelet is proportional to the impedance in the layer <u>above</u> the transmitting interface. Thus, a large transmission coefficient exists for a downgoing particle velocity wave arriving at the interface at 9060 feet, but a small transmission coefficient occurs at 9160 feet. For an upgoing particle velocity disturbance, the opposite situation exists; i.e., the transmission coefficient is proportional to the impedance <u>below</u> these interfaces (Equation 12). Thus, a higher particle velocity amplitude occurs in the low impedance interval between 9060 and 9160 feet, regardless of whether a particle velocity pulse is downgoing or upgoing.

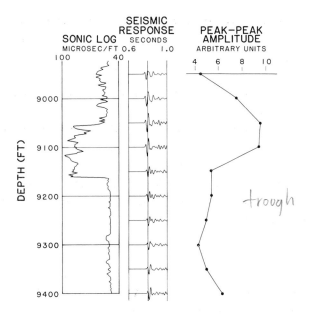

Figure 7-20   For a velocity sensitive VSP geophone, there is an inverse relationship between the acoustic impedance of a stratigraphic unit and the amplitude of a particle velocity pulse propagating through the unit as shown here; i.e., low particle velocity amplitude occurs within high acoustic impedance units. (For a pressure sensitive transducer, high pressure amplitudes would occur in intervals of high acoustic impedance.) The sonic log shown here is assumed to approximate the real acoustic impedance.

Since the amplitude behavior of surface-recorded seismic data plays such a vital role in the precise interpretation of subsurface stratigraphy and lithology, it is advisable to rely on real, physical measurements of in situ amplitude behavior in addition to synthetic calculations of seismic reflection data. VSP measurements like those described in Figure 7-20 are thus becoming more important in seismic exploration.

Determining the Physical Properties of Rocks

Considerable research has been done to determine if various parameters of full-waveform sonic logs can be correlated with the physical properties of the formation around a borehole in which these data are recorded (Cheng and Toksoz, 1981a; Pickett, 1963; White and Tongtaow, 1981; White and Zechman, 1968; Leslie and Mons, 1982). From field measurements, it has been established that some physical properties of the rocks penetrated by a borehole can be predicted by acoustic waveform characteristics under the following conditions:

1. The frequency content of the source wavelet extends from 5 to 25 kilohertz.

2. The source and receiver are located within the same borehole.

3. The source-receiver spacing is of the order of 3 to 15 feet.

4. Only transmitted wavelets are analyzed.

5. The waveform analysis includes a numerical characterization of the compressional, shear, and tube wave modes propagating along the borehole wall.

In particular, Leslie and Mons (1982) report that the amplitudes of the compressional and shear modes of sonic waveforms respond to texture changes in the rocks surrounding a borehole. They show that the amplitude of the shear wave mode is inversely proportional to the clay content of a lithological unit, and that shear wave amplitude is low in mud supported limestone, but high in grain supported limestone. In addition, they confirm Pickett's (1963) finding, that the shear wave interval transit time changes more dramatically than does the compressional wave interval transit time, when the porosity in a given rock type varies. One conclusion that can be made from this latter observation is that porosity prediction can be more reliable if shear wave velocity is used in the estimation process. Cheng and Toksoz (1981a) show that the amplitude of events occurring between the compressional and shear wave arrivals changes when the Poisson's ratio of the formation changes, and that the amplitudes of all guided waves diminish more than do the amplitudes of compressional and shear body waves as the attenuation of the formation increases. Thus, several physical properties of rock layers penetrated by a well can be inferred from amplitude and velocity information contained in full waveform sonic logs. A data example showing how the character of acoustical waveforms, recorded under the conditions listed above, varies for different lithologies (i.e. different physical rock properties) is given in Figure 7-21.

The data shown in this display were recorded with a fixed gain, and are displayed at a vertical depth increment of one foot with a constant plotter gain. Thus, differences in wavelet amplitudes, as well as differences in interval transit times, are geologically significant. The waveforms show three distinct energy modes. The onset of the compressional first arrival is indicated by the line labeled "C", the shear wave arrival is marked as "S", and the tube wave onset is located along line "T". The onset times of these wave modes are an interpretation and may need to be slightly altered at some depths. It is particularly difficult to distinguish between the onset of the shear body wave and the guided pseudo-Rayleigh modes (See Figure 3-14). The stratigraphic section in which these data were recorded is a cyclothem with repeated limestone, sand, and shale units, such as is found in the Mid-Continent of the United States.

An important feature of these data is that wavelets recorded in one type of lithological unit (e.g. limestone) do not look like wavelets recorded in other types of lithological units - a fact already documented by Leslie and Mons (1982) and Cheng and Toksoz (1981a). For example, every limestone unit shows a waveform with a compressional event, a strong shear wave, and a pervasive, high amplitude tube wave.

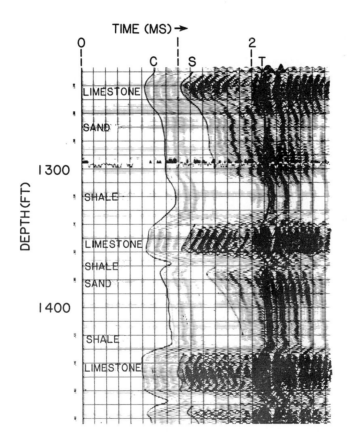

Figure 7-21   Sonic log waveform data recorded with a digital sonic logging tool. The onset of the compressional, shear, and tube wave events are marked by the lines labeled C, S, and T. Note the distinct waveshapes that occur in each lithological unit. These measurements suggest that some physical properties of rocks can be predicted from acoustical waveform character.

298

Every sand waveform contains a compressional arrival, a shear wave with amplitude less than that in limestone, and a tube wave which extends over less time and has less amplitude than does a limestone tube wave. The shale units in this stratigraphic section do not support a recognizable shear wave. The sand starting at a depth of 1375 feet becomes shalier with increasing depth. The shear wave in this sand occurs later in time, and weakens in amplitude, as the shale content increases, until it eventually blends into the tube wave when the lithology becomes pure shale at a depth of 1425 feet.

The fact that distinct acoustic waveshapes are recorded in rocks having different mineralogy, elastic constants, porosity, and grain size suggests that seismic waveshapes may also be sensitive to these rock properties. VSP measurements of propagating seismic wavelets are one way to investigate this possibility. An example of one such experiment, where both sonic waveform data and VSP data were recorded in the same borehole, is shown in Figure 7-22. The sonic waveforms, shown at the left, are displayed at vertical depth increments of one foot; the VSP data at the right were recorded every 50 feet through this stratigraphic section. A compressional vibrator sweeping from 8 to 64 Hz was used as the VSP energy source. The VSP data were recorded by a vertically oriented, moving-coil geophone in a vertical borehole. The lithologies in this interval are limestone and shale, as shown by the lithology strip chart.

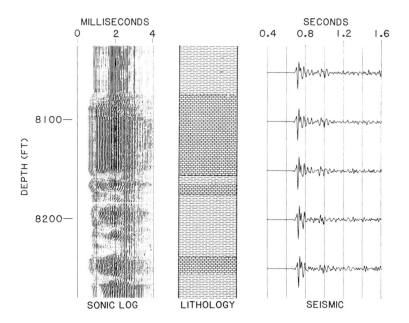

Figure 7-22   These data show the contrast between the sensitivity to rock properties expressed by sonic log waveforms, on the left, which contain kilohertz frequencies, and the apparent lack of sensitivity expressed by compressional VSP waveforms, on the right, which have frequencies of only a few tens of Hertz.

Although difficult to see details at this plotting scale, the sonic waveforms recorded in limestone are considerably different from those recorded in shale. The differences are much like those shown in Figure 7-21. However, it is difficult to see any appreciable or consistent distinction between the shale and limestone VSP wavelets. If information about the physical properties of the rocks is contained in the VSP compressional first arrivals, it is quite subtle. Thus, it remains to be shown that waveform character can define lithological content or physical properties of rocks in an interval when:

1. The frequency content of the propagating wavelet is restricted to less than 100 Hertz.
2. The source is at the earth's surface, and the receiver is in a borehole.
3. The distance from source to receiver is several thousand feet.
4. Both reflected and transmitted wavelets are recorded.
5. The waveform analysis is limited to a single energy mode; e.g., the compressional arrival portion of the total seismic response.
6. The waveform analysis is limited to a single component of particle motion; e.g., only the vertical component.

These six conditions describe circumstances commonly encountered in vertical seismic profiling. Note that, if conditions 2 and 4 are changed, so that the receiver is on the earth's surface and only reflected wavelets are recorded, we have a description of standard surface seismic recording.

A vertical seismic profile can provide measurements by which the lithological effects on transmitted and reflected seismic waveshapes can be detected, if such effects exist. The measurement concept is shown in a simplified form in Figure 7-23. Again, the data shown here represent only the vertical component of particle motion. At the present time, this application is only a possible benefit that can be obtained from VSP data, but the problem must continue to be rigorously investigated. The possibility that VSP waveform character can be used, together with surface-recorded reflection waveform character, to predict subsurface distributions of rock facies over a prospective area is too valuable to hydrocarbon and mineral exploration to be ignored. Many VSP researchers take the position that VSP data can predict rock types if three-component particle motions are recorded, so that all P-wave and S-wave modes can be included in a numerical analysis. Evidence to support this view will be illustrated and discussed in the last section of this chapter, which deals with combined compressional and shear wave interpretations.

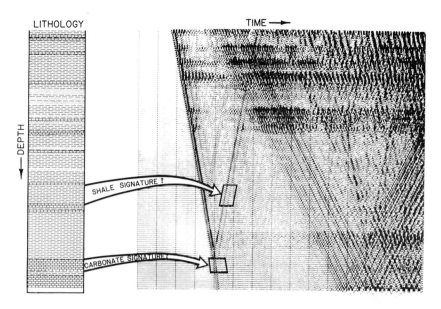

LITHOLOGY TIME ⟶

DEPTH

SHALE SIGNATURE ?

CARBONATE SIGNATURE ?

Figure 7-23   VSP data allow one to investigate the possibility that the physical properties of rocks can be inferred from numerical analyses of seismic wavelets in a manner similar to the way that such information is provided by the numerical analyses of sonic log waveshapes.

Seismic Wave Attenuation

Seismic waves diminish in amplitude as they propagate through the earth as the result of several factors. Some of the mechanisms that affect seismic wavelet amplitudes are discussed in the Amplitude Processing discussion of Chapter 5. Only the effects of transmission losses and spherical wavefront divergence were analyzed in that section, but the effect of seismic wave attenuation should also be considered because of its potential value as a diagnostic parameter that is sensitive to subsurface facies changes. The term "seismic wave attenuation" will be used to describe any irreversible energy losses, other than spherical divergence, transmission losses, and energy mode conversions, which a seismic wave experiences as it propagates through a stratigraphic section. This definition is a broader concept than is the mechanism of dissipation, which will be used to describe energy lost due to friction between moving rock particles, or due to fluid motion within rock pores. For instance, scattering losses and frequency effects introduced by intrabed multiples would also be included in this definition of seismic wave attenuation in addition to dissipation losses. Attenuation measurements are important because attenuation losses affect seismic wave amplitudes, and also because attenuation values describe the lithology, fluid saturation, overpressuring, and the general physical state of subsurface rocks. Several papers on the topic of seismic wave attenuation exist in geophysical literature. The geophysical reprint series prepared by Toksoz and Johnson (1981) provides an excellent summary and overview of this literature.

Vertical seismic profiling affords a unique and valuable data base for seismic wave attenuation studies since the measurement allows a propagating seismic wavefront to be analyzed in real earth conditions as it travels through a stratigraphic section. One useful analysis technique for retrieving attenuation information from VSP data is the spectral ratio method (Hauge, 1981; Kan, et al., 1981; Tullos and Reid, 1969). The concept is summarized in Figure 7-24. In panel A, first arrival wavelets, $g_1(t)$ and $g_2(t)$, are shown

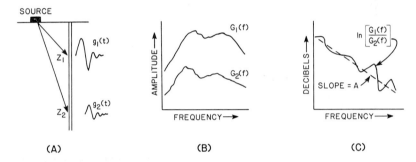

(A)   (B)   (C)

Figure 7-24   Determining cumulative seismic wave attenuation, A, by the spectral ratio method. The interval attenuation between depths Z_1 and Z_2 is defined as $A/(Z_2-Z_1)$.

recorded at depths Z_1 and Z_2. The amplitude spectra, $G_1(f)$ and $G_2(f)$, of these two geophone responses are plotted in panel B as a function of frequency.

Field and laboratory measurements suggest that seismic wave attenuation is a linear function of frequency (Toksoz and Johnson, 1981). In such a case

$$G_2(f) = K\,G_1(f)\,e^{-Af}, \tag{19}$$

where f is frequency and K is a frequency independent factor that accounts for amplitude effects such as the magnitude of the rock impedance at each recording depth (See Figure 7-20), spherical divergence, variations in recording gain, and changes in source and receiver couplings. The exponent, A, is the cumulative seismic wave attenuation between depths Z_1 and Z_2, and it is also independent of frequency. This equation can be rewritten as

$$\ln\left[\frac{G_2(f)}{G_1(f)}\right] = -Af + \ln(K).$$

(20)

The left side of this equation, plotted in panel C, is the spectral ratio of the two VSP responses recorded at Z_1 and Z_2. The cumulative attenuation value, A, is determined by the slope of the best straight line fit to this spectral ratio trend. A second attenuation estimate, which is more important for geological purposes than is the cumulative attenuation, is the interval attenuation, a_z, defined as

$$a_z = \frac{A}{(Z_1-Z_2)}$$

(21)

Tullos and Reid (1969) and Hauge (1981) report that distinct values of the interval attenuation of compressional waves are observed in different types of sedimentary rocks. Consequently, seismically derived interval attenuation values have the potential of mapping subsurface distributions of source and reservoir rock facies.

Most spectral ratio plots exhibit an oscillatory behavior like that shown in part C of Figure 7-24. Spencer et al. (1982) point out that these oscillations are most likely caused by upgoing primary reflections and downgoing multiples that are generated within a travel time distance of T/2 from the depth where the recording geophone is located, where T is the time duration of the seismic pulse. These authors call this behavior the stratigraphic component of attenuation and recommend field procedures to reduce its effect since it does not indicate actual energy dissipation in rocks. They show that when the distance between the receiver depths, Z_1 and Z_2, in Figure 7-24 is small, the spectral ratio of the recorded data is more influenced by the local stratigraphy at the geophone positions than it is by the value of seismic wave attenuation between the two depths. Thus, accurate rock attenuation values cannot be determined in small depth intervals. Besides receiver separation, other factors that affect the spectral ratio calculation are the thicknesses of the sedimentary layers between Z_1 and Z_2, the frequency content of the seismic wavelet, variations in shot wavelets created by the energy source, signal-to-noise conditions, and variations in geophone coupling at Z_1 and Z_2. If upgoing reflection events with good signal-to-noise properties are recorded, an analysis technique which multiplies the spectra of downgoing and upgoing wavelets can be used to cancel frequency effects introduced by variations in source excitation and geophone coupling (Spencer, et al., 1982, p. 18). One mathematical modeling study in this same reference used realistic values for layer thicknesses, reflection coefficients, seismic Q values, and wavelet bandwidth to show that the separation between Z_1 and Z_2 should be at least 300 feet. One important conclusion suggested by this modeling study is that more accurate attenuation estimates should be obtained in finely stratified intervals.

More seismic energy is dissipated in unconsolidated near-surface layers than in any other part of a stratigraphic section. In some areas, it is not uncommon for the amount of energy dissipation and high frequency loss in the top few hundred feet of section immediately below the surface to be larger than that in the total remainder of a stratigraphic column down to the deepest reflector. Thus, seismic wave attenuation values determined between two deeply placed VSP geophones are not representative of near-surface attenuation values. As a result, the Q values determined by vertical seismic profiling are sometimes difficult to confirm from surface-recorded reflection data which involve a two-way travel path through a high loss near-surface layer. In order to use relationships between attenuation values and geological conditions determined from VSP data in the interpretation of surface-recorded reflection data, it is important that VSP attenuation studies also concentrate on the seismic loss mechanisms that occur in the near-surface.

Several shallow VSP experiments that analyzed three-component particle motion in the top 200 meters of the earth have been reported by Newman and Worthington (1982). Both P-wave and S-wave surface energy sources were used in these studies. Some conclusions resulting from these experiments are: (1) seismic wave attenuation values are very sensitive to lithology changes in porous, near-surface strata, which suggests that attenuation measurements may be a valid parameter for mapping subsurface distributions of some rock facies; (2) compressional wave attenuation is usually more sensitive to changes in lithology than is shear wave attenuation; (3) compressional wave attenuation is more sensitive to reduced water saturation and to rock fissures than is shear wave attenuation, at least when attenuation is measured in the vertical direction; and (4) seismic wave attenuation in homogeneous, completely water-saturated rocks is primarily a shear type mechanism, but becomes more of a bulk-type of mechanism if fissures are present or if the rock is only partially saturated. More near-surface attenuation experiments of this nature need to be performed in order to distinguish near-surface loss mechanisms from those occurring deep in the earth where reservoir and source rocks exist, and where facies mapping needs to be done.

If VSP data are not recorded in the shallow near-surface, a spectral ratio technique can still be used to determine near-surface attenuation values. However, instead of working with direct arrivals as shown in Figure 7-24, the analysis focuses on comparing the spectral properties of an upgoing reflection event, measured at a depth Z just below the near-surface layer, with the spectral properties of the downgoing multiple (also measured at depth Z) which this upgoing reflection creates at the surface. The upgoing reflection event represents a seismic wave that has passed through the near-surface only once before it is recorded; whereas, its downgoing multiple has traversed the near-surface layer three times. The spectral ratio of these two events represents the effect of a two-way path through the near-surface, just as occurs for surface-recorded seismic reflection data. One example of a VSP study investigating attenuation losses in the near-surface has been reported by Michon and Omnes (1978). If sufficient frequency differences exist between the spectra of upgoing primary reflections and downgoing surface multiples, then digital filtering may help attenuate multiples so that reflections can be more obvious.

Exploration Applications of Tube Waves

The principal view concerning tube waves presented in this text, and shared by many people engaged in vertical seismic profiling, is that tube waves are a coherent noise that should be avoided during VSP experiments by any available, practical means. In general, this attitude toward tube waves is proper. However, tube waves do provide useful data describing some physical properties of the formation around a borehole, and some explorationists deliberately create tube waves in order to obtain this information.

Two physical rock properties that tube waves can provide are (1) estimates of the shear wave velocity in formations around a VSP well, and (2) indications of highly permeable zones in a drilled stratigraphic section. To some extent, tube wave characteristics also imply the permeability values to assign to permeable zones. Since estimates of shear wave velocities and the determination of permeable zones are critically important in exploration work, the attitude that tube waves are undesirable noise is not always a proper one.

The identification of permeable zones with tube waves is presently confirmed only in shallow investigations, and so is of most benefit in mineral exploration, evaluation of shallow acquifers, and near-surface environmental studies. With proper recording instrumentation and optimum noise conditions, the procedure may work at greater depths and become a factor in hydrocarbon exploration.

The technique described here was developed for the Canadian Atomic Energy Commission as a means of locating permeable zones penetrated by a wellbore in which nuclear waste were intended to be disposed (Huang and Hunter, 1981). The field setup is not greatly different from that of any vertical seismic profile in which a surface source

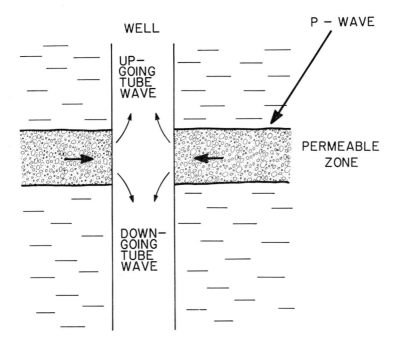

WELL

P – WAVE

UP-
GOING
TUBE
WAVE

PERMEABLE
ZONE

DOWN-
GOING
TUBE
WAVE

Figure 7-25   A compressional body wave is shown impinging on a subsurface permeable
zone, which is penetrated by an uncased, fluid-filled borehole.   The fluid
motion which the compressional pulse creates in the permeable zone is
transmitted to the fluid column in the wellbore. The result is that two tube
waves are created traveling in opposite directions.

near a VSP well creates a compressional wavefront that propagates down through a
stratigraphic section.   In this case, small dynamite charges were exploded in holes 2
meters deep, which were offset 6 to 29 meters from the wellhead. However, the borehole
data were recorded with pressure-sensitive crystals embedded in a free hanging cable,
rather than with a wall-locked geophone. Thus, the receiver system was selected so as to
emphasize tube waves, rather than discriminate against them.

Suppose that an uncased, fluid-filled wellbore penetrates a highly permeable zone,
as shown in Figure 7-25. When a downgoing compressional event reaches this zone, it will
cause the fluid in the pore (or fracture) system to move. When this fluid movement
reaches the fluid column in the well, two tube waves are created. One tube wave travels
upward, and the second tube wave travels down the borehole. Thus, the tube waves of

306

interest are created not at the earth's surface, as illustrated in Figure 3-22, but in situ at depths corresponding to highly permeable zones. The effect observed in VSP data recorded in such a well is shown in simple form in Figure 7-26. A tentative conclusion reached by investigators who have recorded and analyzed such VSP data is that the amplitude of these two tube waves is directly proportional to the permeability of the zone where they are created, and that the formation permeability is sampled out to a distance $\lambda/2$ from the wellbore, where λ is the dominant wavelength in the incident compressional wave (Huang and Hunter, 1981).

These same investigators describe the propagation velocity, V_t, of a tube wave in a fluid-filled, uncased borehole as:

$$V_t = \frac{V_s}{\sqrt{\frac{\rho_w}{\rho} + \left(\frac{V_s}{V_w}\right)^2}} \qquad (22)$$

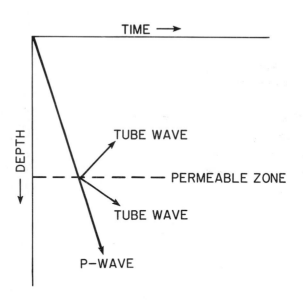

Figure 7-26 In real VSP data, the physical process described in Figure 7-25 would appear like this simplified picture. Both upgoing and downgoing tube wave events would emerge from the compressional first arrival at the depth of the permeable zone.

In this expression, ρ and V_s are the bulk density and the shear wave velocity in the rock formation surrounding the borehole, and ρ_w and V_w are the density and the compressional wave velocity in the fluid in the borehole. V_w and ρ_w are measurable constants, and ρ can be provided by a reliable formation density logging tool. If tube waves are created in a borehole, then V_t can be measured in selected stratigraphic intervals. The only remaining undetermined parameter is the formation shear wave velocity, V_s, which can then be calculated. Tube waves thus allow shear wave velocities in rocks around a VSP well to be determined. Similar mathematical expressions for tube wave velocity are given by White (1965) and Cheng and Toksoz (1981b, 1982a).

The VSP Polarization Method for Locating Reflectors

Surface-recorded seismic reflection data are used primarily to image subsurface reflectors in order to locate drilling sites. Accurate imaging of subsurface reflectors should thus also be a principal objective of vertical seismic profiling. Most of the VSP data processing techniques discussed in previous sections, and many of the exploration applications of VSP data presented in this chapter, assume that the reflecting interfaces in the vicinity of a VSP well are flat and horizontal. This assumption is often violated by the actual structural and stratigraphic situations at exploratory well sites. Consequently, every opportunity must be taken to broaden the concepts involved in VSP data processing so that dipping interfaces and faulted reflectors can be properly imaged. In addition, changes in field recording procedures which will enhance the detection of reflectors having arbitrary geometrical configurations need to be considered.

One data recording option which increases the imaging capability of VSP data in areas of complex geology, and which also introduces several new exploration possibilities for VSP data, is the recording of three-component particle motion. Three-component VSP data allow a more complete description of a seismic wavefront than does just the vertical component of motion, which comprises the majority of data examples shown in this text. Two portions of a VSP data trace are particularly amenable to three-component analysis, these being

1. the downgoing first arrivals, and
2. upgoing reflections.

This section will focus on data processing and interpretational techniques that are applicable for upgoing compressional reflections. Three-component analyses of upgoing P-wave events are particularly important since they allow one to image the interfaces that created these events in three-dimensional space. Three-component analyses of downgoing compressional and shear wave first arrivals, as well as unified interpretations of both upgoing compressional and shear wave reflections, are also important, but these topics will not be considered until the last section of this chapter. All of these analysis techniques are based on determining the nature of the linearly polarized particle motion created by propagating body modes. Hodogram analyses made from three-component data are particularly dependent on accurately knowing the relative gains of each geophone element in a triaxial geophone system, so that the voltage outputs of all three geophones can be converted into true particle motion.

Good quality three-component VSP measurements can be used to determine the direction from which reflected compressional wavefronts arrive at a subsurface recording position. The interpretation is based on the fact that a compressional disturbance causes rock particles to oscillate in a polarized direction normal to the compressional wavefront.

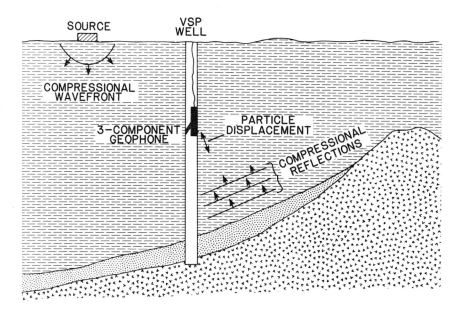

Figure 7-27   Determining reflector orientation by determining the particle displacement polarization created by reflected compressional wavefields.

Thus, particle displacements created by compressional waves "point" in the direction from which a compressional wavefront arrives (or in the direction it is going). In general, this direction will be the raypath's arrival angle at the borehole, and this vector direction will be normal to the interface that created the reflection, if the VSP geophone is reasonably close to the reflector so that appreciable raypath refraction has not occurred. Thus, from three-component analyses of compressional reflection events, one can, in concept, determine the orientation of a reflecting interface in three-dimensional space. An illustration of a field measurement employing this concept is shown in Figure 7-27.

VSP polarization methods have been widely used for structural and stratigraphic investigations in the Soviet Union for several years (Gal'perin, 1977). Non-Soviet geophysicists are just beginning to use these VSP methods and are in the process of constructing appropriate downhole instrumentation. Some valuable work in this technology has been reported by Aki, et al. (1982) and Albright and Pearson (1980). Their investigations concentrated not on upgoing reflection events, but on VSP measurements of direct arrivals created by induced microseisms in geothermal reservoirs. Their data analyses, however, are essentially the same as those which could be used to analyze VSP reflection data.

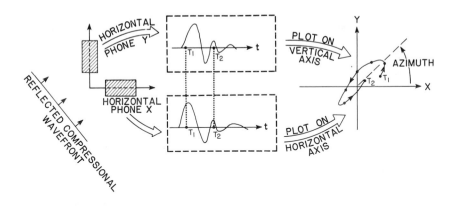

Figure 7-28   Determining the azimuthal direction of travel of upgoing VSP compressional wavefronts by particle velocity hodograms. The axis of the hodogram defines the azimuth direction in which the P-wave event is traveling relative to a chosen geophone axis. The direction that this reference geophone axis is pointing when a VSP tool is locked in a wellbore must be determined by independent inclinometer or magnetometer data.

One way to determine the direction of particle displacement created by a linearly polarized compressional wavelet is to construct hodograms from the responses recorded by pairs of geophone elements in a triaxial VSP geophone package (White, 1964; DiSiena, et al., 1981). In this discussion, we will assume that the geophone elements are oriented along Cartesian XYZ axes, with the vertical Z axis being along the longitudinal axis of the downhole VSP tool. If the geophone tool is vertical, then a hodogram constructed from the responses created when an upgoing compressional wavelet activates the two horizontal geophone elements will define the azimuthal direction of arrival of that event, as shown in Figure 7-28. A hodogram made from data recorded by the vertical geophone, and either of the horizontal geophones, will indicate the angle by which the upgoing wavefront is inclined away from the plane of the horizontal geophones, as shown in Figure 7-29. In these figures, the azimuth angle is arbitrarily measured relative to the direction in which the axis of horizontal geophone X points, and the reflector dip angle is measured relative to the direction in which the axis of the vertical geophone points. Consequently, it is essential to know the spatial orientation of these geophones. This requirement means that the downhole geophone package should contain magnetometers, which define the horizontal orientation of geophone X relative to north, and inclinometers, which measure how much the axis of geophone Z is rotated in the XZ plane relative to true vertical. Since magnetometers cannot function properly inside casing, this type of downhole tool orientation instrumentation can be used only in uncased intervals.

Seavey (1982) has described a VSP geophone system which determines the in situ orientation of its geophone elements by a compass unit when operating in an uncased hole, and by a gyroscope unit when working in a cased well. However, in actual field work, this

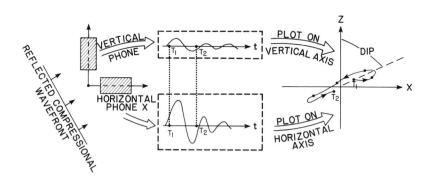

Figure 7-29   Determining the dip direction of travel of upgoing VSP compressional wavefronts by particle velocity hodograms. The axis of the hodogram defines the dip angle relative to the axis of the vertical geophone.

Sandia group found that a more reliable technique for determining geophone orientation was to detonate three or four explosive charges at surveyed locations around the well, and then construct hodograms of the direct arrival from each shot. The technique worked satisfactorily in their tests down to recording depths of 7000 feet, as long as the offset distances to the shot holes were as large as the recording depth. Because this procedure is time consuming and expensive, it will likely be a viable orientation technique only in those vertical seismic profiles where data are recorded at only a few depth levels. Other procedures should be considered when data must be recorded at a large number of depths, when the borehole geophone must be moved often, or when the source-to-geophone raypaths are highly contorted due to complex geology.

DiSiena, et al. (1981) show how three-component VSP data can be analyzed by polarization techniques to mathematically orient a triaxial VSP geophone system so that the vertical geophone and one horizontal geophone lie in the plane containing the particle displacement vector created by a compressional first arrival traveling downward from a surface source. This mathematical rotation allows the downgoing VSP wavefield to be separated into three components; i.e., a component containing all particle displacements polarized in the direction that the compressional first arrival is traveling, and two components containing displacements perpendicular to this direction. These latter two components define SV and SH motions if the SV and SH modes arrive along the same ray-path as the P-wave does.

A similar procedure can be used to analyze upgoing compressional reflections if these events exhibit good signal-to-noise behavior. The analysis technique basically consists of constructing hodograms so that the azimuthal angle of the compressional arrival can be determined, as shown in Figure 7-28. Once this angle is known, data recorded by the X and Y geophones shown in this illustration can be converted to data equivalent to that which would be recorded by geophones positioned along new orthogonal axes, X' and Y', where X' lies in the azimuthal direction of compressional energy arrival. Thus, the new X' axis would be the dashed line passing along the axis of the hodogram. These new geophone data are given by the expressions

$$X'(t) = X(t)\cos(\phi) + Y(t)\sin(\phi) \tag{23}$$

and

$$Y'(t) = -X(t)\sin(\phi) + Y(t)\cos(\phi) \tag{24}$$

where ϕ is the azimuth angle, and $X(t)$ and $Y(t)$ are the responses of the X and Y geophones. If the horizontal axis in Figure 7-29 is now defined to be this new X' axis, then a second mathematical rotation similar to that defined in Equations 23 and 24 can orient the Z geophone to a new position, Z', so that it points in the dip direction of compressional energy arrival. In performing this rotation, the dip angle along which the compressional wave travels is used, rather than the azimuth angle. The Z' signal then defines particle motion polarized in the direction of compressional displacement. The orthogonal X' and Y' signals are now isolated from each other and from the linearly polarized P-wave arrival, Z'. If it is appropriate to assume that SV and SH modes are traveling in the same direction as the P-wave when they arrive at the geophone position, then Y' defines the SH response. The X' data represent the SV mode, if one is analyzing only first arrivals, but X' can be a combination of both P and SV modes for events that follow the first arrival.

It has been demonstrated that this type of mathematical rotation of geophone axes, when applied to downgoing compressional first arrivals generated by a surface source, points the Z' axis at the source location with acceptable accuracy if the source offset distance is small, and if the stratigraphy consists of flat, horizontal layers that do not create highly refracted ray paths (DiSiena, et al., 1981). In non-horizontal layering, or when large source offsets are involved, the azimuth of the final rotated angle is simply an empirically determined parameter specified by the data. The calculation does not provide an azimuth value measured relative to true North. In situations where the source location is not known, as would be the case when analyzing upgoing reflections, or when interpreting downgoing wavefields in complicated geology, true azimuth can be determined only if instrumentation is included in the geophone package which measures the directions in which the geophones are pointing at the time the VSP data are recorded.

Hodogram results are strongly affected by noise contamination, thus the time window, T_1 to T_2, used to generate the hodogram displays in Figures 7-28 and 7-29 must be carefully chosen. Data processing steps which attenuate downgoing events, random and coherent noise, and upgoing events traveling at non-compressional velocities may have to precede the selection of an upgoing reflection analysis window. The choice of this window is largely a matter of interpretation. A highly elliptical hodogram is preferred, and several analysis windows may have to be tested before a reliable hodogram is obtained.

Multichannel data processing techniques are not necessarily linear mathematical processes when applied to three-component data, since such processes may alter the relative phase relationships among data components. For example, velocity filtering (a multichannel process), followed by a mathematical rotation of triaxial geophone data via Equations 23 and 24, will usually not yield the same result as does a mathematical rotation followed by multichannel velocity filtering. Thus, the order of the processing sequence applied to three-component data is important if a multichannel technique must be used in order to produce a satisfactory hodogram.

The ellipticity of a hodogram is accentuated, and thus its orientation angle is easier to define, if each sample point is squared before it is plotted, rather than just plotting the magnitudes of the sample points, as is done in Figures 7-28 and 7-29. The squared values, $(X^2(t), Y^2(t))$, must be plotted along the radial line pointing in the angular direction, $\tan^{-1}(Y(t)/X(t))$, and the squared values, $(X^2(t), Z^2(t))$, must be plotted along the radius vector defined by $\tan^{-1}(Z(t)/X(t))$. This technique creates an energy hodogram rather than an amplitude hodogram.

In hodogram studies that focus on events which follow first arrivals, it is essential to know the relationship between the direction that a geophone case moves and the polarity of the geophone output. For example, if a horizontally oriented geophone moves toward the right, is the signal output positive or negative? Changing the polarity of the horizontal geophone signal in Figure 7-29 will rotate the hodogram, and thus the direction of arrival for an event, by 90 degrees. Changing the polarity of both geophone signals will rotate the hodogram 180 degrees. Geophone tap tests are one way to assign signal polarity conventions to all directions of geophone movement and avoid these types of directional uncertainties. These tests, or shake table tests of the complete geophone assembly, also establish relationships between the magnitudes of the voltage outputs of

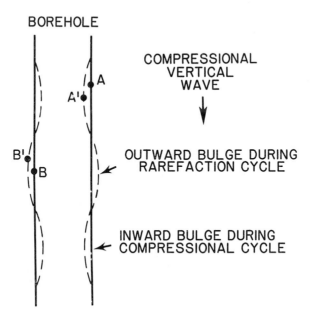

Figure 7-30   Model showing that a vertically traveling compressional wave creates both vertical and horizontal displacement of the wall of a wellbore (point A moves to A', point B to B', etc.)

314

triaxial geophone elements. For example, if the amplitude of the signal from a geophone oriented along the X axis is 30 percent larger than the signal amplitude of the Y geophone when both geophones are equally impulsed, then this fact must be known in order to properly adjust the amplitudes of the recorded data before constructing hodograms.

When making hodograms of compressional wave particle motion from data recorded by a wall-locked, three-component geophone, one caution should be kept in mind. The geophone is clamped to a cylindrical rock interface that is free to expand and contract radially. Since a non-rigid medium (drill mud) fills a VSP borehole, a rock particle on the fluid-solid interface at the borehole wall moves with greater freedom and amplitude in the radial direction than does a rock particle far removed from the interface. This excessive radial motion causes a hodogram to be slightly biased along the graphical axis where the radial component of motion is plotted. Consider for example the situation sketched in Figure 7-30, where a compressional wave is traveling vertical'y along a vertical borehole. Any rock particle located far away from the borehole will be displaced only in a vertical direction, normal to the wavefront. However, since the fluid-filled borehole is less rigid than the formation, it will contract or expand radially, depending on whether it is subjected to a compression or rarefaction cycle of the passing compressional wave. Points such as A and B do not move vertically. Instead, they move along curved, oblique trajectories to positions A' and B'. A three-component geophone locked at point A would not create a P-wave hodogram with a true vertical axis normal to the direction of P-wave travel.

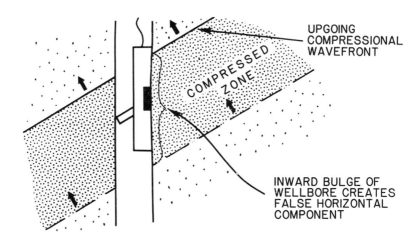

Figure 7-31   Schematic showing that the particle velocity measured by a multi-axis wellbore geophone, which is locked to a borehole wall,   does not point exactly along the normal to a compressional wavefront.

This same argument applies to a situation where a compressional wave arrives at a three-component, wall-locked geophone at an arbitrary angle of incidence, such as shown in Figure 7-31. The radial contraction of the borehole in the depth interval spanned by a compressional cycle of the incident wave creates an additional radial component of motion at the borehole wall, thus causing the three-component geophone to move obliquely to the direction of propagation of the compressional wavefront.

The measurement and interpretation of subsurface particle trajectories is discussed in detail by Gal'perin (1974 - Chapter 5; 1977). Soviet researchers refer to this type of vertical seismic profiling as CDR (Controllable Directional Reception). The term CDR is also used to describe migration techniques applied to surface-recorded data (Zavalishin, 1982).

Thin Bed Stratigraphy

In many stratigraphic sections, reflecting interfaces are so closely spaced that their individual reflection and transmission coefficients cannot be determined by analyzing incident, reflected, and transmitted seismic wavelets. Instead, it is necessary to think in terms of the reflectivity and transmitivity of several of these interfaces taken as a collective group. In thin bed stratigraphy, one must work with reflectivity functions and transmitivity functions rather than with individual reflection and transmission coefficients.

This type of reflection and transmission process is illustrated in Figure 7-32 where an incident VSP wavelet, $f(t)$, is shown approaching a series of thin, reflecting beds indicated by the hatchured zone. A reflected wavelet, $g(t)$, and a transmitted wavelet, $h(t)$, are recorded above and below this zone. These wavelets are numerically related to the incident wavelet by the expressions

$$g(t) = R(t) * f(t) \qquad (25)$$

and

$$h(t) = T(t) * f(t), \qquad (26)$$

where $R(t)$ is the reflectivity function of the thin bed sequence, and $T(t)$ is the transmitivity function of the thin bed zone. The symbol * implies convolution. Since $f(t)$, $g(t)$, and $h(t)$ are measured VSP quantities, $R(t)$ and $T(t)$ can be calculated.

316

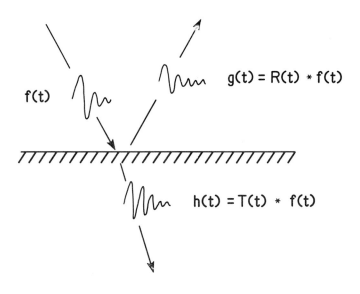

$$g(t) = R(t) * f(t)$$

$$f(t)$$

$$h(t) = T(t) * f(t)$$

Figure 7-32   The reflectivity, R(t), and transmitivity, T(t), of a thin-bed stratigraphic sequence can be numerically estimated if the incident wavelet, f(t), reflected wavelet, g(t), and transmitted wavelet, h(t), can be isolated in VSP data.

Both reflectivity and transmitivity are valuable diagnostic parameters in seismic stratigraphy studies. An example showing how the physical properties of a stratigraphic interval can be inferred from reflectivity calculations is given in Figure 7-33. The VSP data used in these calculations were recorded as part of a project to determine parameters by which one could map water aquifers in the Madison Group - Red River interval of the Powder River Basin from surface-recorded seismic reflection data (Balch, et al., 1980a). The calculations show distinct differences in the reflectivity functions determined for tight, barren zones and for porous aquifers. The data panel on the right side of Figure 7-33 shows how these two types of aquifer conditions would appear in surface-recorded data. In these calculations, the maximum amplitude of the incident VSP wavelet, f(t), was always scaled to unity. Therefore, the amplitudes of the reflectivity functions, R(t), can be compared to each other. In this stratigraphic interval, porous zones exhibit significantly larger compressional reflection functions than do non-porous zones. Also, more reflection events are generated in the porous intervals than in the tight zones. These two observations establish which characteristics of surface-recorded compressional reflection data should be used to map subsurface distributions of water bearing Madison aquifers near this test site.

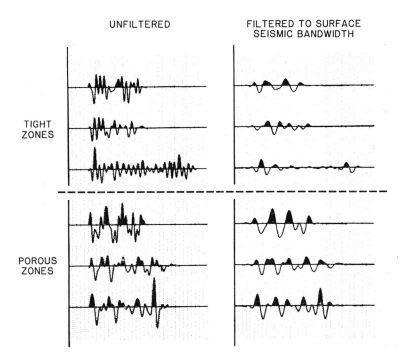

UNFILTERED   FILTERED TO SURFACE
                    SEISMIC BANDWIDTH

TIGHT
ZONES

POROUS
ZONES

Figure 7-33   Reflectivity functions calculated from VSP data recorded in the Madison
               Group - Red River interval, eastern Powder River basin, Wyoming-Montana.
               (After Balch, et al., 1980a, Copyright SPE-AIME).

Transmitivity functions also contain valuable stratigraphic information, particularly
information describing the seismic attenuation and dispersion occurring in thin bed
sequences. These functions are often displayed in the frequency domain rather than in the
time domain. The Fourier transform of Equation 26 is

$$H(\omega) = T(\omega) \cdot F(\omega) \tag{27}$$

where $H(\omega)$, $T(\omega)$, and $F(\omega)$ are the Fourier frequency spectra of $h(t)$, $T(t)$, and $f(t)$. Thus,
the Fourier transform, $T(\omega)$, is similar to the spectral ratio function described in Equation
20.

Velocity Anisotropy Measurements

The velocity with which a seismic wavefront propagates through a layered stratigraphic section depends upon its direction of travel. Generally, a seismic disturbance is observed to travel faster in a horizontal direction than it does in a vertical direction (Uhrig and Van Melle, 1955). As a consequence, seismic velocity has to be viewed as an anisotropic function of subsurface coordinates. There are several factors that contribute to this anisotropic behavior, the principal ones being sedimentary grain shape, sedimentary layering, fracturing, and tectonic stresses. Each of these factors causes the elastic properties of the earth in a horizontal direction to differ from its elastic properties in a vertical direction. There is no intent to discuss here the theoretical basis of seismic velocity anisotropy. Interested readers are referred to Berryman, 1979; Brekhovskikh, 1960; Dunoyer de Segonzac and Leherrere, 1959; Kraut, 1963; Keith and Crampin, 1977a, 1977b, 1977c; Levin, 1978, 1979, 1980; Postma, 1955; Uhrig and Van Melle, 1955; and White, 1982.

Seismic velocity anisotropy creates several problems in seismic data processing. For instance, the stacking velocity used to composite surface-recorded seismic reflection data is a measure of propagation velocity in the horizontal direction in which geophones are deployed, but velocities in the vertical direction are needed in order to convert reflection times to reflector depths. Empirically derived adjustment functions, which alter stacking velocities by variable percentages, are commonly used to predict reflector depths. It is particularly important to understand the principles of velocity anisotropy in the newly developing technology of 3-D seismology where surface geophones may be deployed in several horizontal directions. Stacking velocities and migration velocities have to be known as a function of azimuth in order to process 3-D data so that subsurface features are properly imaged.

Velocity anisotropy results in the wavefront geometry shown in Figure 7-34 (Vander Stoep, 1966). At an instant in time, the wavefront created by a surface, or near-surface, seismic energy source is shown by the solid line that approximates an ellipse. The velocity at any non-vertical angle, $\emptyset$, is $V(\emptyset)$. The vertical velocity, $V(0)$, is labeled V_z, and the horizontal velocity, $V(\pi/2)$, is V_x. In an anisotropic section of the earth, V_x is greater than V_z by as much as 20 percent (Uhrig and Van Melle, 1955). Compressional waves are usually less anisotropic than are shear waves (Keith and Crampin, 1977a,b). For values of $\emptyset$ less than $\emptyset_M$, the wavefront can be represented as an ellipse with its major and minor axes having a ratio, A, such that

$$V_x(z) = A(z) \, V_z(z) \qquad (A(z) > 1.0) \qquad (28)$$

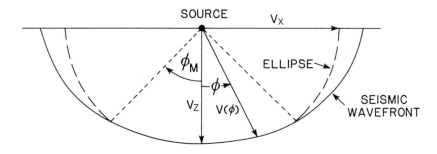

Figure 7-34   Seismic wavefront geometry in a layered anisotropic medium.

The ratio, A, must be defined as a function of depth since the velocities V_X and V_Z also change with depth and rock type. The angle ϕ_M is typically 45 degrees, thus, an elliptical approximation of the wavefront is satisfactory for surface recordings, if the geophones spread lengths do not exceed the depth to the deepest reflector to be imaged.

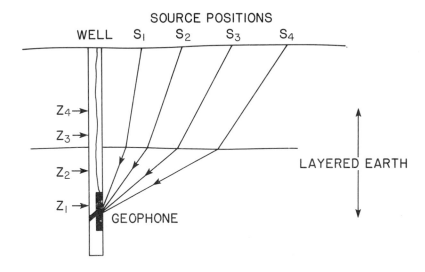

Figure 7-35   Field geometry for velocity anisotropy measurements.

A general VSP field technique which is used in velocity anisotropy studies is shown in Figure 7-35. The subsurface wavefront arrival time is recorded at a sequence of geophone depths, Z_1, Z_2, Z_3, . . ., for a variety of source offset distances, S_1, S_2, S_3, ... This recording geometry is identical to that used in other applications described in Chapter 8, with the exception that first arrivals, not reflections, are the critical data to be captured. Numerical techniques for reducing these arrival time measurements to elliptical ratio values, A, are described by Vander Stoep (1966).

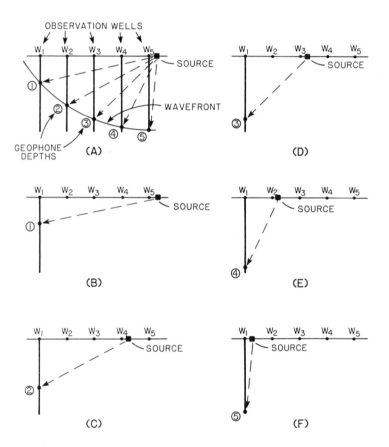

Figure 7-36   (A) An ideal field situation for measuring seismic wavefront shapes would be to have several observation wells W_1, W_2, W_3, ... with geophones in each well. (B) - (F) Approximating the ideal situation in (A) with only one well at W_1, a single geophone, and an ambulatory surface source.

An ideal field geometry would be to have several borehole geophones simultaneously suspended in several adjacent observation wells, so that the propagating wavefront created by a single seismic shot could be measured as a function of horizontal distance, vertical depth, and travel time. Such a field experiment is shown in Part A of Figure 7-36. In the real world, however, one is ordinarily restricted to making these measurements in a single well and with a geophone system that records data at only one depth. Under these circumstances, the measurement can proceed as diagrammed in Parts B through F. The source is positioned at several horizontal distances from the one existing well, W_1, where the wavefront is to be sampled, and downhole data are recorded at appropriate vertical increments throughout the entire drilled layered sequence. These data then have to be analyzed so that the wavefront position is known as a function of depth, travel time, and horizontal separation between W_1 and the source. Experiments of this type are described by Garotta (1978) and Robertson, et al. (1979). Both studies show that shear waves propagating in layered media exhibit more anisotropy than do compressional waves. In the stratigraphic interval studied by Garotta, shear waves exhibited an anisotropy factor of 1.20.

This field procedure assumes that the wellbore is vertical, and that appropriate travel paths for velocity anisotropy investigations are created by positioning a VSP energy source at various offset distances from the wellhead. If the well is not vertical, then travel paths that allow velocity anisotropy effects to be measured can often be achieved with a stationary source. This type of VSP shooting geometry is becoming more common in offshore vertical seismic profiling, where severely deviated wells are often drilled. Such a well is shown in Figure 7-37. Source 1 remains stationary at the drilling platform as the VSP geophone moves uphole. The resulting travel paths, such as L_1 and L_2, describe seismic propagation velocity behavior in both the vertical and horizontal directions, just as do the travel paths shown in Figure 7-36. By using a second source (Source 2 in Figure 7-37) that is continually repositioned vertically above the downhole geophone, standard VSP data can be recorded in the deviated well while velocity anisotropy data are recorded using Source 1.

There is increasing evidence that velocity anisotropy is a parameter that is sensitive to lithology, and that surface anisotropy measurements may allow subsurface mapping of lithofacies. In particular, velocity anisotropy has been shown to increase in fractured carbonates, and when the shale content in a clastic interval increases (Garotta, 1978). Vertical seismic profiling is probably the most rigorous way by which seismic velocity anisotropy effects can be correlated with subsurface geology.

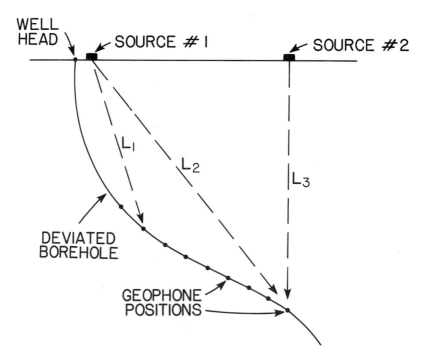

Figure 7-37 Velocity anisotropy measurements can be made in a deviated well by keeping a source fixed near the wellhead (Source #1). Conventional VSP data can be acquired by keeping an ambulatory source vertically above the geophone (Source #2).

Combined Compressional and Shear Wave Interpretations

Historically, seismic shear waves have not been widely used in hydrocarbon and mineral exploration. This fact is easily confirmed by tabulating the number of kilometers of surface seismic data recorded worldwide in which compressional sources and vertically oriented geophones were used, and comparing this amount of data with the number of kilometers recorded when using shear wave sources and horizontal (or three-component) geophones. Even though some excellent field studies are reported in which shear waves reveal important subsurface properties, shear wave seismology is still viewed by many explorationists as an unproven exploration technology. Thus, seismic exploration is presently a discipline dominated by the generation, recording, and analysis of compressional wave modes. This almost complete reliance on compressional wave seismology is to be expected in marine exploration, but it is not a necessary condition that needs to be imposed in onshore exploration.

There are several reasons for the slow development of surface shear wave reflection seismology. Some principal factors being:

1.   Energy sources - Sources built for the explicit purpose of generating seismic shear waves usually have two major short comings; viz, they either produce weak shear wave signals that do not allow deep targets to be imaged, or if they do produce strong shear wavefields, they create so much damage to the earth's surface that they cannot be used in some areas.

2.   Field procedures - It requires more time to plant the horizontally oriented geophones used in shear wave recording than it does the vertically oriented geophones used in recording compressional waves. Horizontal geophones usually have a viewing port on the top of the geophone case in which a bubble floating in liquid can be adjusted so that the geophone coil is horizontal within an allowable tolerance. Likewise, these geophones must be rotated so that they have the same polarity output for geophone motions directed horizontally to the right and left. The time required to carefully orient each horizontal geophone is considered by some explorationists to result in an inefficient use of field recording resources when compared with the speed at which compressional geophones are deployed. Gimbal mounted horizontal geophone coils are used by some crews in order to reduce the time required to vertically position geophones cases.

3.   Processing Deficiencies - Most computing software that processes seismic data was structured with the intent of being used only on compressional data. Consequently, velocity analysis, static adjustments, and other programs have to be rewritten to accommodate shear wave data.

4.   Exploration Objectives - Hydrocarbon exploration has been basically a search for subsurface structure, which compressional wavefronts image very well. Thus, there has been no urgent need to develop an additional

seismic imaging technique. Because of the increasing emphasis on finding non-structural traps, there is now a need to use all seismic measurements that are sensitive to lithology, porosity, fracturing, and pore fluids.

5. Shear Wave Velocity Measurements - In those cases where surface shear wave reflections have been recorded, it is often difficult to know exactly what subsurface anomalies created the shear reflection events. Shear wave reflections usually do not look exactly like compressional wave reflections recorded along the same profile, and until recently, shear wave sonic logs which could be used to convert reflector depths to reflection times, or to make synthetic shear wave seismograms, have not been available. It is in this area of interpreting subsurface geology in terms of both compressional and shear wave responses that vertical seismic profiling offers one of its greatest potentials.

A rather comprehensive overview of the industry attitude toward shear wave seismology, which discusses limitations of shear wave measurements and emphasizes potential benefits of shear wave data as an exploration tool, has been published by Helbig and Mesdag (1982).

In a unified compressional and shear wave interpretation of a stratigraphic section, the principal question to be answered is, "Which compressional and shear reflections measured at the surface are created at a specific subsurface impedance interface?" An example of P-wave and S-wave surface reflection data recorded along the same profile is shown in Figure 7-38 (Garotta, 1980). If an interpreter wishes to use both sets of data to interpret the subsurface beneath these seismic lines, then each depth interval of interest must be accurately defined on both reflection profiles. If subsurface structure is present, then shear reflections can usually be correlated with their equivalent compressional reflections, as exemplified by the structure created by the dolomite pod in these data. However, when structure is not present, it is more difficult to establish which shear reflections should be equated with selected compressional reflections. For example, which shear reflections define the top and bottom of intervals A, B, C, and D marked on the compressional section? Are the indicated dashed correlation lines correct? Some P-wave and S-wave data windows from these two seismic sections, which are interpreted by Omnes (1978b) as spanning equivalent stratigraphic units, are listed to the right of the curves plotted in Figure 7-39. Garotta (1980) shows other examples of unified interpretations of spatially coincident compressional and shear wave profiles where equivalent compressional and shear wave reflections are identified.

shown

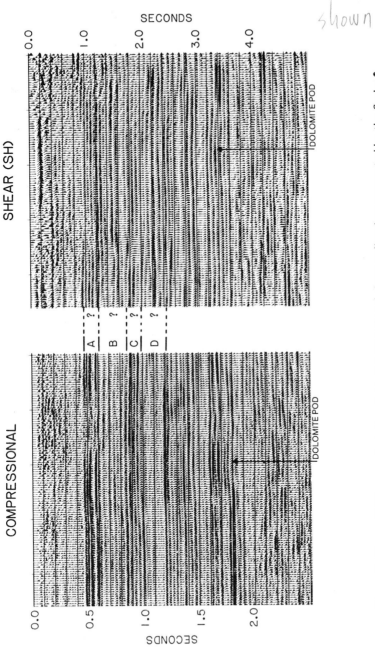

Figure 7-38   2400 percent stacks of compressional and shear reflections generated by the Syslap®
technique. A dolomite pod is shown at 1.75 seconds on the compressional wave
section and at 3.2 seconds on the shear wave section. (After Garotta, 1980)

® - CCG trademark

A vertical seismic profile in which good quality three-component data are recorded can indicate how one should correlate compressional and shear wave reflections generated in a stratigraphic section in the vicinity of a VSP study well. As shown in Figure 7-40, VSP measurements can show whether both compressional and shear reflections are generated at depths Z_1 and Z_2, and if so, what the one-way travel time, reflection amplitude, and polarity of each reflection is. These reflection characteristics must be known in order to correctly identify these events in surface-recorded P-wave and S-wave data near the well. In addition, other important P-wave and S-wave parameters, such as interval velocity, velocity anisotropy, and attenuation, can be extracted from the VSP data.

One valuable diagnostic parameter that can be obtained from combined compressional and shear wave surface-recorded reflection data is the ratio of the interval shear wave velocity, V_s, to the interval compressional wave velocity, V_p. This ratio is defined as

$$\gamma_t = V_s/V_p = T_p/T_s \tag{29}$$

where T_p is the two-way interval reflection time between two selected reflectors measured from compressional data, and T_s is the interval reflection time between shear wave events originating at these same two interfaces. This ratio appears to be sensitive to lithology, as is demonstrated by the values plotted in Figure 7-39 (Omnes, 1978b). These ratios were calculated from the stacked compressional and shear wave sections in Figure 7-38. Since there is essentially no structural complications in these particular seismic sections, these velocity estimates should contain negligible structural influences, and should be primarily sensitive to subsurface elastic parameters. Note particularly how the interval velocity ratio diminishes as shale content increases toward the left, and how it increases when carbonate content increases. Tatham (1976, 1982) shows that the ratio, V_p/V_s, is usually more sensitive to crack and pore geometry than it is to mineral content. Similarly, Moos and Zoback (1983) use VSP data and sonic logs to show that a high density of macrofractures causes V_p and V_s to decrease, but the ratio V_p/V_s to increase. Thus, some caution should be used before establishing direct relationships between these velocity ratios and lithology. It is because this ratio is sensitive to several rock properties that it is so potentially valuable in stratigraphic trap exploration. However, in order to use the correct values of interval travel times, T_p and T_s, in the ratio calculation, it is essential that equivalent reflection events be identified on corresponding compressional and shear wave cross-sections. A rigorous way to identify these events is by means of vertical seismic profiling.

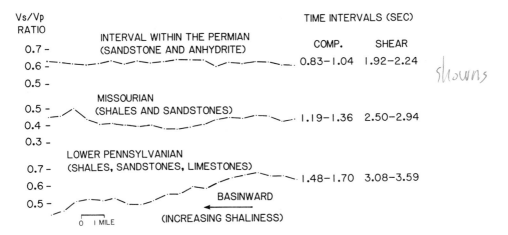

Vs/Vp RATIO

INTERVAL WITHIN THE PERMIAN
(SANDSTONE AND ANHYDRITE)

MISSOURIAN
(SHALES AND SANDSTONES)

LOWER PENNSYLVANIAN
(SHALES, SANDSTONES, LIMESTONES)

BASINWARD
(INCREASING SHALINESS)

0   1 MILE

showns

TIME INTERVALS (SEC)	
COMP.	SHEAR
0.83–1.04	1.92–2.24
1.19–1.36	2.50–2.94
1.48–1.70	3.08–3.59

Figure 7-39   Relationship between lithology and Vs/Vp = Tp/Ts ratios calculated from the compressional and shear data shown in Figure 7-38 (after Omnes, 1978b). The time windows listed at the right are interpreted as imaging equivalent depth intervals.

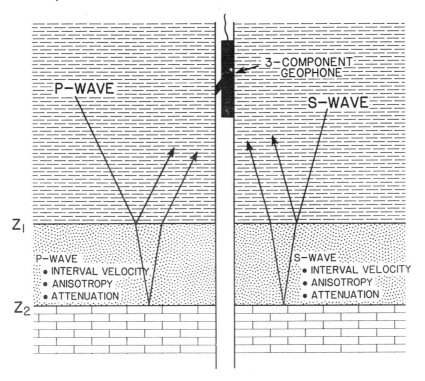

Figure 7-40   The use of vertical seismic profiling in unifying compressional and shear wave interpretations of the subsurface.

An example in which three-component VSP data provide the basis for a unified interpretation of spatially coincident, surface-recorded P-wave and S-wave seismic reflection data is shown in Figures 7-41 to 7-43 (G. Omnes and D. Michon, CGG, private communication). Several surface energy sources were used to generate the VSP signals and the surface reflection data in this experiment, but the responses created by surface vibrators will be emphasized. Two types of downhole VSP geophones were employed. The data recorded by one downhole geophone system is labeled 'H' in the illustrations. This probe has two locking arms which are driven by a hydraulic system that is energized by an electrical motor. The geophone elements inside this probe are arranged in an orthogonal XYZ configuration, and are located at the top of the tool above both locking arms. The response of the second downhole geophone system is labeled 'M'. This probe has only one locking arm which is driven by a mechanical spring. The spring release and retraction is controlled by an electrical motor. This probe also has three-component geophone elements arranged in an orthogonal XYZ geometry. Both probes have nearly the same length (~ 3 meters), diameter (~ 9 centimeters), and mass (~ 100 kilograms).

Shown in Figure 7-41 are raw, unfiltered VSP data recorded with a surface P-wave vibrator (Part A) and a surface S-wave vibrator (Part B). These sources were offset approximately 40 meters from the wellhead. The only processing performed on these data is a trace amplitude normalization for plotting purposes. The P-wave vibrator sweep extended from 10 to 130 Hertz, and the S-wave vibrator sweep ranged from 10 to 70 Hertz. In each display, data are plotted at depth decrements of 10 meters, starting at a borehole depth of 820 meters. This particular hole was cased and well-cemented when these data were recorded. A prominent tube wave is created by the P-wave vibrator, but no tube waves are created by the S-wave vibrator.

Upgoing compressional and shear wave reflections can be seen, even though the only processing performed on the data is trace normalization. Numerical velocity filtering of these data should thus yield robust estimates of upgoing compressional and shear reflections. After velocity filtering to attenuate downgoing events, upgoing P-wave and S-wave primary reflections are obtained by statically time shifting the velocity filtered traces to vertically align upgoing reflections from flat, horizontal interfaces, deconvolving the data, and then vertically summing the traces to create a single composite trace. The resulting estimates of upgoing primary compressional and shear wave reflections created by this processing sequence for various combinations of surface energy sources and downhole probes are shown in Figure 7-42. A non-linear depth scale is included in order to identify equivalent depth positions in the compressional and shear wave responses, and all data are plotted as a function of two-way travel time. Reflections originating from below the deepest geophone depth (820 meters) are recorded

Shown

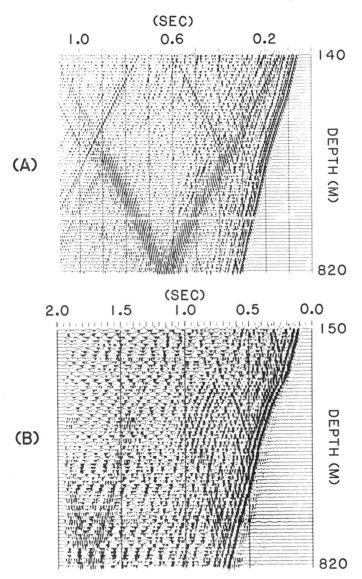

(SEC)

1.0   0.6   0.2

140

(A)

DEPTH (M)

820

(SEC)

2.0   1.5   1.0   0.5   0.0

150

(B)

DEPTH (M)

820

Figure 7-41 (A)   Raw P-wave VSP data recorded in well "X". P-wave vibrator sweep = 10-130 Hz. Trace amplitudes equalized before plotting.

         (B)   Raw S-wave VSP data recorded in well "X". S-wave vibrator sweep = 10-70 Hz. Trace amplitudes equalized before plotting. (Courtesy of CGG)

330

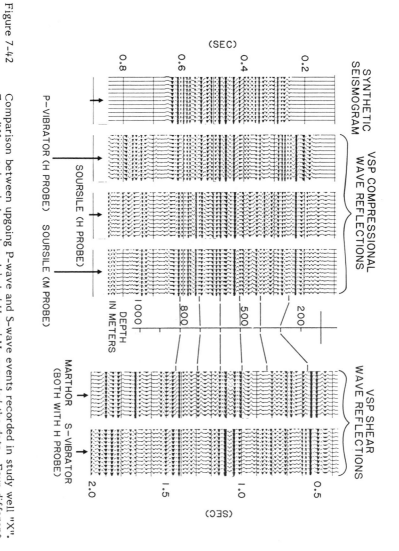

Figure 7-42   Comparison between upgoing P-wave and S-wave events recorded in study well "X". Two different downhole probes, labeled H and M, recorded the data. Four different energy sources, labeled P-vibrator, S-vibrator, Marthor®, and Soursile®, were used. (Courtesy of CGG)

® – Marthor is a registered trademark of Institut Francais du Petrole.

® – Soursile is a registered trademark of CGG.

since both the P-wave and S-wave responses extend to two-way times greater than those corresponding to a depth of 820 meters. A synthetic seismogram constructed from a sonic log recorded in this study well is also shown in order to provide an alternate estimate of the upgoing primary compressional reflections.

Having now established via these VSP data the depths at which each P-wave and S-wave reflection is created, and the two-way time at which each of these reflections should arrive at the surface, the identification of equivalent P-wave and S-wave events in surface-recorded data near this well site is straight forward. Both compressional and shear wave surface reflection data were recorded across the VSP well using field procedures designed to yield high subsurface resolution. These procedures included utilizing a single compressional (or shear) vibrator, rather than an array of vibrators, employing small geophone groups (e.g. 3 phones) separated by only 10 meters, burying all geophones a few inches below the surface in order to reduce noise, creating 48 fold stacks, and recording at a sample rate of one millisecond. The resulting CDP stack of the surface-recorded P-wave data is shown in Part A of Figure 7-43, and the S-wave stack is plotted in Part B. Inserted at the well location in each cross-section is the VSP determination of the upgoing reflections. Using time-depth relationships derived from the VSP data, equivalent depth intervals and equivalent compressional and shear reflection events can now be identified in the P-wave and S-wave seismic cross-sections. Thus, compressional-to-shear velocity ratios, γ_t values, and Poisson's ratio can now be calculated in the same depth intervals across the entirety of the surface coverage.

Several seismic parameters that react to lithological changes, or to physical changes in rock properties such as fracturing, can be calculated once compressional and shear VSP data are obtained in the same study well. One vertical seismic profiling investigation of this type from the Paris Basin is summarized in Figure 7-44 (Omnes, 1980). Wavefields created by compressional and shear wave surface sources were recorded by a three-component borehole geophone locked at increments of 20 meters between well depths of 500 and 1420 meters. The VSP data, which are not shown, are similar to the data plotted in Figure 7-41 in that the P-wave and S-wave first break arrival times, and hence the P-wave and S-wave interval velocities, can be measured at each recording depth. The lithological column penetrated by the study well is shown in panel B. The velocity ratio, γ_t, determined from compressional and shear first break times is shown in panel A. The ratio varies with lithology, and in general, is greater than 0.5 for competent rocks and less than 0.5 for shales and clays. The interval velocity data, shown in panel C, were determined every 20 vertical meters. Note that the velocity scales are logarithmic, and that the shear wave interval velocity, V_s, reacts more strongly to lithological changes

332

shown

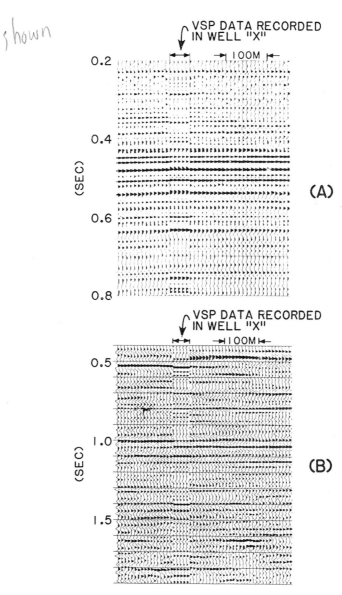

Figure 7-43 (A) Comparison between upgoing P-wave events determined from VSP data and a P-wave cross-section constructed from surface-recorded P-wave reflections.

(B) Comparison between upgoing S-wave events determined from VSP data and an S-wave cross-section constructed from surface-recorded S-wave reflections. (Courtesy of CGG)

Shawn

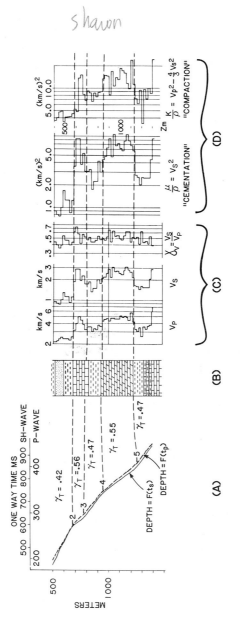

Figure 7-44   An example of lithological and petrophysical properties calculated from compressional and shear wave data recorded in a VSP study well (Omnes, 1980, Copyright SPE-AIME). γ_t and γ_v are measures of velocity anisotropy, μ is an indication of cementation, and K is sensitive to compaction.

than does the compressional interval velocity, V_p. As a result, the ratio

$$\gamma_v = V_s/V_p \tag{30}$$

is also quite sensitive to lithological variations, fracturing, and any velocity anisotropy effects.

Two elastic properties logs which can be constructed from the interval velocity values, V_s and V_p, are shown in the display at the far right in panel D. These log curves are defined by the elastic velocity equations

$$V_s^2 = \mu/\rho \tag{31}$$

and

$$V_p^2 - \frac{4}{3} V_s^2 = K/\rho \tag{32}$$

where ρ is the rock bulk density, μ is the shear modulus of the rock, and K is the rock bulk modulus. Since V_s exhibits a strong reaction to lithological changes, then so does the μ/ρ curve. The shear modulus, μ, is a measure of how well rock grains are cemented together, and so the μ/ρ curve is labeled "cementation". The bulk modulus, K, indicates the incompressibility of a material, and thus implies compaction history. No detailed petrophysical data are available from this study well to confirm the cementation and compaction estimates provided by these data.

Another example of vertical seismic profiling establishing a relationship between compressional and shear velocities and subsurface facies distributions is a field experiment reported by Audet and Garotta (1982). This study demonstrated that a porous oolitic limestone reservoir could be distinguished from a tight, low porosity lagoonal limestone by an analysis of the interval velocities, Vs and Vp. The VSP data showed that the Vs/Vp ratio was 0.6 in the non-commercial carbonate, but reduced to 0.48 in the oolitic reservoir facies.

More VSP experiments of this nature are needed in order to firmly establish how much lithological detail, and which physical properties of subsurface rocks, can be inferred from combined P-wave and S-wave seismic velocities. The geological interpretations provided by these VSP velocity measurements can then be extended to P-wave and S-wave velocities determined from surface-recorded reflection data so that subsurface facies distributions can be mapped over extensive distances. It should be emphasized that relationships established between subsurface rock properties and V_s/V_p velocity ratios, or T_p/T_s travel-time ratios, which are measured in VSP experiments where the raypaths from the source to the borehole geophone are vertical, may not directly transfer to V_s/V_p and T_p/T_s ratios measured from surface-recorded reflection data which involve highly oblique travel paths and very few vertical travel paths. Some investigators do not consider the velocity ratio and travel-time ratio in Equation 29 to be an equality if the velocities and travel times are evaluated from surface-recorded reflection data. Instead, the quantity, Vs/Vp, is viewed as a ratio of horizontal velocities, since Vs and Vp are determined from moveout corrections that depend upon horizontal velocity behavior, and the quantity, Tp/Ts, is viewed as a ratio of vertical velocities because travel times are measured from stacked, zero offset data traces. In such a case, these two ratios should then be written as

$$\gamma_v = (V_s/V_p)_{hor} \tag{33}$$

and

$$\gamma_t = T_p/T_s = (V_s/V_p)_{vert} \tag{34}$$

where the subscripts "hor" and "vert" are added to designate horizontal and vertical velocities. Thus the quantity

$$\gamma = \frac{\gamma_v}{\gamma_t} = \left[\frac{V_s}{V_p}\right]_{hor} \left[\frac{V_p}{V_s}\right]_{vert} \tag{35}$$

is the ratio of shear wave velocity anisotropy to compressional wave velocity anisotropy. This parameter, γ, should be a valuable diagnostic measurement, since shear wave anisotropy is particularly sensitive to fracturing in carbonate units and to increasing shale content in clastic units. Although no confirming field data yet exist, it is possible that γ is also diagnostic of the type of pore fluid in rocks and can indicate overpressured zones.

VSP measurements thus need to be performed in order to confirm the P-wave and S-wave velocity relationships in Equations 33 through 35. In particular, VSP experiments employing variable source offsets and triaxial geophones should be designed so that the recorded raypaths closely approximate the vertical and oblique travel paths occurring in P-wave and S-wave CDP recording geometry. Experiments also need to investigate those situations where upgoing SV modes are created by downgoing compressional waves. An example of marine CDP data processed to reveal both compressional and converted SV reflections is described by Tatham and Stoffa (1976). A ray model study illustrating how to recognize and process converted SV data has been reported by Graves (1979).

CHAPTER 8

PRODUCTION AND DRILLING APPLICATIONS
OF VERTICAL SEISMIC PROFILING

Information obtained from vertical seismic profiling can be used in drilling, well completion, and secondary recovery efforts because VSP data provide answers to such questions as:

1. How far below current drilling depth is it to the next seismic reflector?
2. What type of geological conditions are most likely to occur below the drill bit?
3. How far laterally does a borehole need to be deviated to reach an objective?
4. How can the subsurface be more accurately imaged near a well?
5. How is a given secondary recovery process propagating through a reservoir?

Some petroleum companies are beginning to record VSP data at critical decision points in their drilling programs in order to "look below the bit" and decide whether to drill ahead, or to "look sideways" from the bit to decide if they need to deviate a hole. Only a few people are using vertical seismic profiling to monitor secondary recovery processes in reservoirs, although the technology should also be helpful in this area of production engineering.

This chapter describes some of the valuable information that vertical seismic profiling can provide for drilling and production engineers. Some applications cannot be as clearly described as they should be because at this time only limited amounts of VSP data have been recorded that address them, and these data are still held proprietary by those companies who have performed the experiments. In spite of the absence of abundant, confirming field data, it is hoped that the information that is presented will help engineers decide what role vertical seismic profiling should play in future drilling and production programs.

The applications of vertical seismic profiling that have the greatest potential in drilling and production activities can be grouped into three general areas; viz,

1. Those which involve looking ahead of a drill bit.
2. Those which involve looking laterally away from a borehole.
3. Those which take advantage of the recording geometry offered by a deviated well.

The material presented in this chapter is organized so that it approaches drilling and production applications of vertical seismic profiling from each of these points of view.

Predicting Depths to Seismic Reflectors

Predicting drilling depths to key seismic reflectors is a common activity in oil and gas exploration, and many geophysicists are quite adept at making accurate depth estimates via surface determinations of seismic propagation velocity. However, these estimates become less accurate when a well is in a virgin area where there is little drilling history, or in an area where surface-recorded seismic data are poor quality. Consequently, using VSP data to predict reflector depths will have the greatest benefit in those places where surface-recorded seismic data cannot be used to make reliable depth estimates. Certainly, wells have been, and will continue to be, drilled in areas where surface seismic data are poor quality, and where depth interpretations made from these data are speculative. A factor that works in favor of vertical seismic profiling in such areas is that deep boreholes are usually a seismically quiet environment, and high quality seismic data can often be recorded below the ground surface when only poor quality seismic data can be recorded at the surface. A specific, but hypothetical, example of a geological situation where subsurface seismic measurements can yield better depth estimates than can surface measured data is shown in Figure 8-1. This example assumes that near-surface scatterers create a significant variation in the travel times of reflected wavefronts just before they reach the surface geophone arrays, and that there are no deep scatterers. This problem of near-surface scattering exists in such places as the karst areas of Florida and West Texas and some volcanic-covered regions of Idaho and Utah. Several types of severe lateral changes in near-surface weathered zones occur worldwide, and they all, in general, make it difficult to record good quality surface seismic data. Other geological situations could lead to the decision to rely on subsurface VSP data, rather than surface-recorded data, for making depth estimates below drilling TD. The main point to be emphasized is that, when surface seismic data are poor quality for any reason, serious consideration should be given to making subsurface interpretations via vertical seismic profiling.

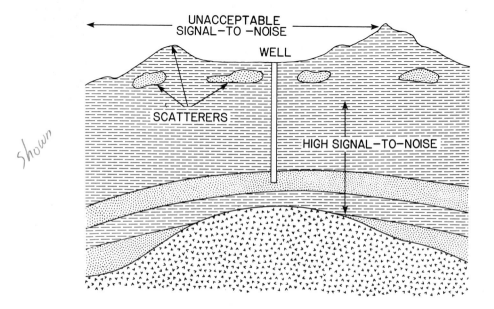

Figure 8-1   Combating surface "no record" areas with VSP data.

The data shown in Figure 8-2 will be used to illustrate the ability of vertical seismic profiling to predict reflector depth below the drill bit. These data are also illustrated and discussed in Figure 7-3. The vertical distance between each recording depth in this example is 50 feet. This well had a single $5\frac{1}{2}$ inch casing in this interval, which was cemented from TD up to about 8000 feet below KB. The drilled depth was 10,750 feet, but a frac ring set in the casing at 9750 feet prohibited passage of the geophone tool, so effective TD for recording VSP data is 9750 feet. These data are relatively good quality since several seismic events can be seen reflecting upward from the downgoing first arrival.

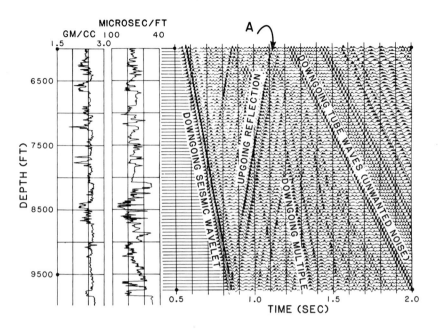

Figure 8-2   VSP data showing a strong reflector, "A", originating at a depth of 9850 feet.

The seismic reflection event labeled "A" marks the top of a prominent limestone unit. The concept that will be illustrated from these data is that one can look ahead of the bit, when the drilling depth is several hundred feet above the "A" limestone, and estimate the depth to this unit.

A way that these VSP data could have been used to predict the distance from the drill bit to the "A" formation is illustrated in Figure 8-3. The same data which are plotted in Figure 8-2 are shown here, except it is assumed that the well has been drilled to only 8000 feet, and that VSP data are recorded from TD (8000 feet) upward far enough so that deep reflection events can be seen and interpreted. This requirement usually means that data should be recorded from the bottom of the borehole to about 2000 feet vertically above the bottom of the well. The question to be considered is, "How far below current drilling depth is it to reflector A"? As a first approximation, it can be assumed that the downgoing first arrival wavelet continues below 8000 feet with the same time-

340

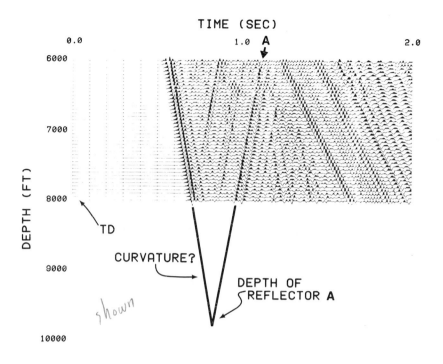

Figure 8-3   Looking ahead of the drill bit with VSP data.

depth curvature that it has in the recorded data interval (6000 to 8000 feet), and that reflection event "A" also propagates upward from the "A" unit to the bottom of the well at 8000 feet with the same curvature that it has in the data recorded above TD. Because the curvature of the first break times below TD is assumed, it is indicated as a questionable data behavior in Figure 8-3. The intersection of these two extrapolated travel paths estimates the predicted depth at which the reflection occurs. In this example, this extrapolation estimate is accurate to within $\pm$ 50 feet. One flaw in this technique is that one must assume the magnitude of the seismic propagation velocity below current drilling depth. Likewise, it may be difficult to decide whether event A in Figure 8-3 is an upgoing multiple or an upgoing primary reflection. It is often helpful to compare the time delays by which downgoing multiples follow the first arrival with the time intervals between upgoing events in order to identify which upward traveling events are multiples and which are primaries. A more rigorous technique for determining whether or not event A is an upgoing primary reflection would be to (a) separate these data into downgoing and upgoing wavefields by velocity filtering, (b) design deconvolution operators that attenuate multiples in the downgoing wavefield, (c) apply these deconvolution operators to the upgoing wavefield to reduce upgoing multiples, and (d) recombine these deconvolved versions of downgoing and upgoing wavefield into a single display like Figure 8-3. This new display would be less contaminated by multiples. In spite of the difficulty of distinguishing between primary and multiple reflections, and of having to make the velocity assumptions for the stratigraphic section below TD, there will be situations where depth estimates made using this technique will be sufficiently accurate.

It should also be emphasized that a more fundamental question than, "How far downward is it to the next reflector?", can be answered by these VSP data, and that is the question, "Is there a reflection of any type below TD?" Again, there are instances where wells are drilled in areas where surface seismic data quality may not allow one to identify reflections below a certain depth. In these cases, a properly executed VSP survey can often offer definitive answers to the questions, "Are there seismic reflectors below TD?" and "Should we drill ahead"?

An example of vertical seismic profiling being used to resolve a problem of this nature has been reported by Balch, et al. (1980b, 1982a). The objective of this VSP experiment was to confirm and identify seismic reflections from the Paleozoic basement beneath Yucca Flat, Nevada. This area is a potential site for underground testing of nuclear weapons, and it is essential that the depth to basement, together with the local configuration of the basement, be known at each proposed test site. Initial efforts to map the Paleozoic contact via surface seismic reflection methods "met with limited success probably due to near surface effects" (Balch, et al., 1980b). Consequently, a vertical seismic profiling investigation was done to confirm whether or not a Paleozoic reflection exists, and to determine what type of energy source should be used to image the Paleozoic surface.

An interpretation of the subsurface by Byers and Quinlivian (USGS, unpublished), based on well control and geophysical data at the experimental site, is shown in the top panel of Figure 8-4. VSP data were recorded in well UE10bd, which is a large diameter borehole (diameter greater than 64 inches). This well barely penetrates the Paleozoic rocks, as shown in this cross-section. A second well, U10bd, which is 100 feet away from UE10bd, was used to record VSP data from downhole airguns. Surface airguns and vibrators were also used as energy sources.

An example of VSP data recorded in this well is shown in the panel at the bottom of Figure 8-4. Data were not recorded to the total depth of the well, the deepest VSP trace being 1470 feet below ground level. These data have been processed with a maximum coherency velocity filtering technique (Chapter 5), which attenuates downgoing events and accentuates upgoing events. A small remnant of the large amplitude downgoing

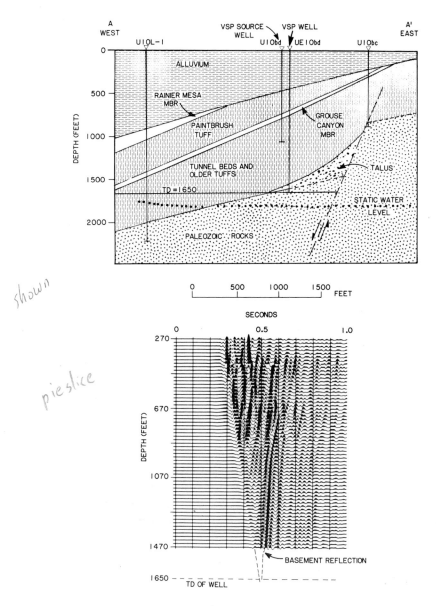

shown

pieslice

Figure 8-4   An example of VSP data being used to identify subsurface reflectors in a surface "no record" area (After Balch, et al., 1980b).

direct arrival can still be seen in the data. The energy source in this case was on the surface and offset from the well. An upgoing basement reflection is observed in the bottom half of the data set. Thus, the possibility that a downgoing shot wavelet is completely attenuated by the loose alluvium section at the surface cannot be true, as some investigators originally proposed. Neither can a second proposal, that the Paleozoic surface is so severely eroded that it cannot create a coherent reflection, be true.

The issue to be resolved is to determine why the upgoing basement reflection disappears at about 700 feet below ground level, where several upgoing multiples begin to appear. Balch, et al. (1982a) tentatively assume that the Paleozoic rocks are block faulted close to the wellbore in a way that prevents raypaths from the basement from reaching the upper half of the well. No structural cross-section or raypath calculations are presented to clarify this assumption. Nonetheless, this experiment is particularly valuable as an example of vertical seismic profiling being used in drilling decisions because it demonstrates the ability of VSP data to look below the deepest recording depth in order to identify deeper reflections and reveal reflection events that cannot be identified in surface reflection measurements.

Predicting Rock Conditions Ahead of the Bit

Some rock conditions can be predicted ahead of the bit by means of a calculation procedure which converts upgoing VSP reflection amplitudes into estimates of the reflection coefficients at the interfaces where the reflections were generated. These reflection coefficient estimates can, in turn, be converted into compressional seismic velocities which allow such parameters as rock type, porosity, and pore pressure to be approximated below current drilling depth. These approximations are, of course, limited by the assumptions involved in the calculation and by the fact that compressional seismic velocity does not define unique values for these rock properties. The computational procedure is similar to the Seislog*, G-log**, and other techniques which are often used to convert amplitude processed, surface-recorded compressional reflection data into synthetic compressional velocity logs (Lindseth, 1979; Becquey, et al., 1979).

---

*Trademark of Teknica
**Trademark of Geophysical Services Inc.

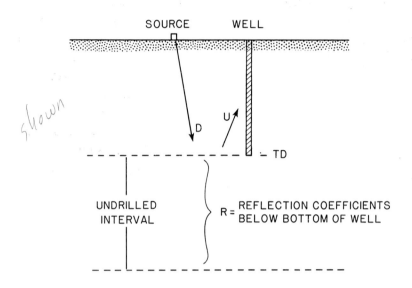

Figure 8-5   A primary objective in looking ahead of the bit is to define the reflection coefficients, R, below current drilling depth. If the downgoing wave, D, entering an undrilled interval, and the upgoing response, U, exiting from that interval are known, then the reflection coefficient series, R, within the interval can be recovered from the relationship, U(t) = D(t) * R(t).

The basis of the concept is illustrated in Figure 8-5. Here it is assumed that a well is drilled to a depth, TD, and a vertical seismic profile is recorded in the well in order to predict rock conditions below TD. Suppose the downgoing wavefield, $D(z,t)$, and the upgoing wavefield, $U(z,t)$, can be separated out of the total VSP measurement. Then the seismic input entering the undrilled interval below TD is $D(TD,t)$, and the seismic response returned from that interval is $U(TD,t)$. By definition,

$$U(TD,t) = D(TD,t) * R(t), \qquad (1)$$

where $R(t)$ represents the unknown reflection coefficient series occurring below TD.

Since U(TD,t) and D(TD,t) are known functions, R(t) can be calculated from the relationship

$$R(t) = D^{-1}(TD,t) * U(TD,t). \tag{2}$$

Once these reflection coefficients are known, the reflection coefficient equation (Equation 1, Chapter 7) can be inverted to obtain the acoustic impedance in the undrilled section. This inversion equation can be written as

$$(\rho V)_n = \frac{1 - R_n}{1 + R_n} \ (\rho V)_{n-1} \tag{3}$$

where ρ is the bulk density and V is the compressional velocity in the rock layers creating the reflections. The subscript n denotes the depth of these rock layers as a function of seismic recording time (see Figure 7-1). If the impedance value is known at depth TD, then the impedances at deeper depths can be recursively calculated from Equation 3. These calculated values of ρV allow rock type, rock porosity, pore pressure, and other acoustically sensitive rock properties to be estimated so that drilling decisions can be formulated.

In this calculation, the function U(TD,t) should contain no multiples; therefore, considerable care must be taken when performing the numerical processes that retrieve U(TD,t) from the data. All multiples generated above TD will be contained in both D(TD,t) and U(TD,t), and theoretically these multiples can be removed from U(TD,t) by appropriate deconvolution operators designed from D(TD,t). However, one problem is that multiples generated below TD may not be adequately removed from U(TD,t) since they are not in the downgoing function D(TD,t) used to calculate the deconvolution operators.

An illustration of the calculation process is shown in Figure 8-6. Shown at the left is a set of VSP data which have been rotated 90 degrees relative to the data shown in Figure 8-2 and 8-3; i.e., the depth coordinate is now horizontal, and the recording time axis is vertical. The horizontal dashed line labeled 'TD' represents the bottom of the well in terms of seismic recording time. All upgoing events below this line represent either primary or multiple reflections originating from beneath the deepest penetration of the bit. The data are time shifted so that upgoing events created by flat, horizontal reflectors occur at the same two-way time at all depths. No other processing has been done other than for plotting purpose, a numerical gain function has been applied to bring all events to the same amplitude level.

The object of the data processing is to extract all upgoing primary compressional reflections from the data set shown at the left so that their amplitudes and polarities

346

shown

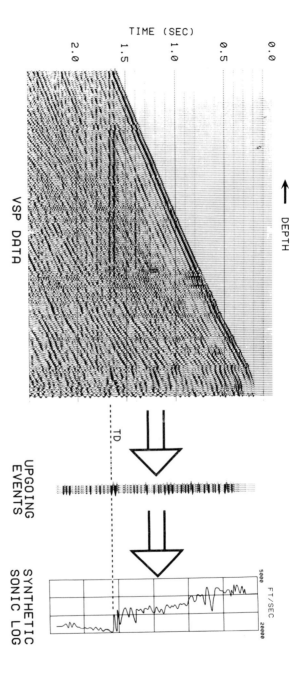

TIME (SEC)

DEPTH

VSP DATA

UPGOING
EVENTS

TD

SYNTHETIC
SONIC LOG

FT/SEC

Figure 8-6   Predicting geological conditions ahead of the bit with VSP data. The traces labeled "Upgoing Events" can be viewed as a bandlimited estimate of the reflection coefficients.

relative to each other are correct. A possible processing procedure might include the following sequence of operations:

1. Remove the amplitude decay caused by spherical divergence with a gain function of the form expressed in Equation 12 of Chapter 5. The behavior of this function below the bottom of the well must be extrapolated from its behavior measured in the drilled stratigraphic section. A properly designed amplitude recovery process is essential so that deconvolution operators can be calculated which will correctly attenuate multiples.

2. Segregate the data into upgoing and downgoing wavefields by appropriate velocity filters. Spatial mixing of data should be minimized in this step if robust deconvolution operators are to be calculated.

3. Attenuate multiples in the upgoing wavefield data by deconvolution operators designed from the downgoing wavefield data. This procedure is discussed in Chapter 5.

4. Sum several of these deconvolved relative amplitude VSP traces to improve the signal-to-noise of the upgoing events. Since the number of traces summed usually increases with two-way time, each data point of the composite trace should be divided by N, where N is the number of traces summed at that two-way time.

A seismic trace which could be produced by this processing sequence is shown repeated several times immediately to the right of the original VSP data set in Figure 8-6. It is assumed that all events in this trace, but particularly those that occur below the dashed line, are primary reflections whose amplitudes are directly proportional to the magnitude and polarity of the impedance changes where they were created. In other words, this trace is a bandlimited estimate of the reflection coefficients occurring between 0.3 and 2.3 sec two-way travel time.

The synthetic sonic log inversion procedure uses Equation 3 and the known compressional velocity behavior down to the total depth of the well, together with the bandlimited estimates of the reflection coefficients, to generate the synthetic sonic log shown at the right. The low frequency character of the sonic log should be obtained from seismic interval velocities, which are known from VSP first break times down to TD. These interval velocities may or may not be known below the bottom of the well. Consequently, even though the high frequency character of the synthetic sonic log below

TD is usually reliable if the VSP data are properly processed, the low frequency trend is speculative. Stone (1982) shows examples of the velocity detail that can be achieved by applying this technique to VSP data. Sometimes a satisfactory estimate of the impedance sequence in the rocks below current drilling depth can be constructed by just manually drawing a synthetic sonic log curve so that its magnitude increases and decreases in phase with the amplitude peaks and troughs of the final processed VSP trace (Kennett, Ireson, and Conn, 1980). This same reference states that these interpretations have been successful to distances in excess of 1500 meters ahead of the bit. If a sonic log has been recorded in the well, the synthetic sonic log can be compared with it to establish confidence in the process. The important point is that a synthetic sonic log can be generated which extends below current drilling depth. Therefore, rock properties below TD that affect seismic velocity (e.g., lithology, pore pressure, and matrix porosity) can be inferred from the synthesized velocity behavior. If good quality three-component VSP data are recorded, both compressional and shear velocity behavior can be approximated below TD. Combined P-wave and S-wave velocity analyses allow better estimates of rock types and physical conditions to be made.

Conceivably, this technique can also be used to predict the impedance sequence above the shallowest level where VSP data are recorded. An image of the shallow impedance layering at a VSP well is contained in the downgoing wavefield, which contains the multiple reflection of the direct arrival created by each reflecting interface above the topmost VSP trace. Thus, one analyzes the amplitude and polarity of these downgoing multiples in order to estimate the impedance sequence back toward the surface in the same way the amplitude and polarity of upgoing primary reflections are used to predict impedance behavior below the bottom of a well.

Looking Laterally Away From a Borehole

There are situations where it is important to record a vertical seismic profile in such a way that the data can be used to construct a seismic image of a reflector extending laterally away from a borehole. For instance, an explorationist may want to determine where a fault plane is located relative to a wellbore, or what kind of lateral facies changes occur around a well. Some common exploration questions that can be answered by the ability to look laterally away from a well with VSP data are depicted in Figure 8-7.

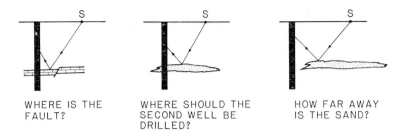

Figure 8-7   Subsurface questions that can be answered by side-looking VSP data.

There are two basic VSP field procedures by which the subsurface around a borehole can be imaged. One method keeps the energy source at a fixed offset distance and moves the geophone over an extended vertical depth interval. This procedure is currently the most common field recording technique used in vertical seismic profiling. The second method keeps the geophone at a fixed depth and moves the energy source to various offset distances. This technique is called a walkaway VSP. Both recording geometries are shown in Figure 8-8. In each technique, the lateral extent of the recorded subsurface image depends on the magnitude of the source offset distance and on the dip of the reflecting interface. Applications using each approach are presented in the following sections.

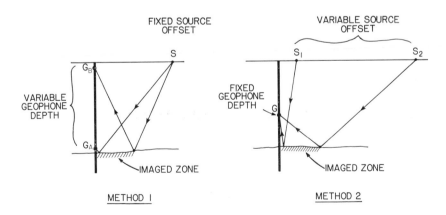

Figure 8-8   Two VSP field procedures by which one can look laterally away from a borehole.

Locating Faults

The hypothetical well shown in Figure 8-9 represents an attempt to drill "close" to a fault zone. A real example of this situation could be the Austin Chalk trend in Texas, where production occurs in fracture systems commonly confined to within 30 or 40 meters of faults extending through the chalk. A second example could be the fault fractured carbonate reservoirs in the Williston and Michigan Basins. Even when these faults can be seen in surface-recorded seismic data, it is still difficult to terminate a well within a narrow fracture zone, 30 to 40 meters wide, at a depth of a few thousand meters. Thus, operators in these areas are beginning to run VSP surveys in wells that miss a fault in order to determine, "How far, and in which direction, is the fault from the borehole"?

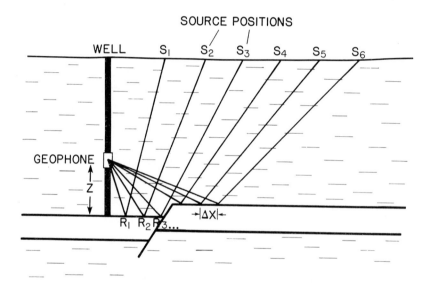

Figure 8-9   Shooting offset source (walkaway) VSP data in order to detect faults near a borehole.

The general appearance of data recorded in an offset-source, or walkaway, VSP survey is illustrated in Figure 8-10. Each trace represents the response recorded when the source is at one of the positions, S_1, S_2, ..., shown in Figure 8-9. The trace data are

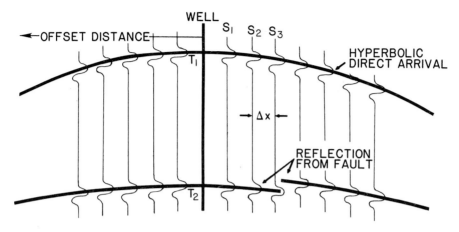

Figure 8-10   Expected VSP response for a walkaway source shooting into a geophone kept at a fixed depth. Fault diffractions and downgoing multiples are ignored in this sketch.

drawn symmetrically about the well position, as if the source were offset both to the left and to the right of the wellhead. If the seismic velocity above the reflector is known or can be reasonably approximated, then the source positions, S_1, S_2, ..., can be calculated via ray trace modeling before starting the field experiment so that reflection points R_1, R_2, ... in Figure 8-9 will occur at regularly spaced increments Δx away from the wellbore. Examples of this type of ray trace modeling are shown in Figures 6-15 and 6-16. Although the distance Δx is used as the trace spacing in Figure 8-10, it should be emphasized that Δx is the horizontal separation between reflection events only along the faulted reflector. A different horizontal separation, and thus a different trace spacing, exists between successive reflection points that occur at any other depth, unless the source and/or receiver positions shown in Figure 8-9 are adjusted to keep the same Δx value at that depth.

In lieu of computer ray trace modeling, simple one-layer models, which allow the source, geophone, and reflection point positions to be calculated by hand, can be used. These models are useful to those people who have to quickly develop an impromptu recording geometry in the field without the benefit of analyzing the situation with a computer ray tracing program. One possible model is shown in Figure 8-11. The distance from the source to the wellhead is L, the depth to the geophone is D, and the depth to the interface to be imaged is Z. Some obvious physical assumptions involved in the model are

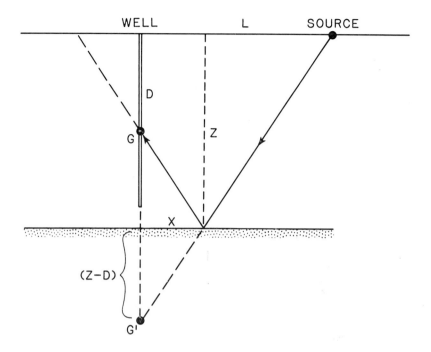

Figure 8-11   A simple, straight raypath model which allows the source location, L, and geophone depth, D, to be calculated so that a reflection point along a horizontal interface at depth, Z, will occur at a specified distance, X, from a vertical borehole.

a vertical borehole, a flat, horizontal reflector, and a single constant velocity layer between the surface and the reflector. Even with these limitations, the model is still useful for imaging any feature, such as a fault, on low dip interfaces. Because the velocity is constant, raypaths are straight, and the following relationship exists among the geometrical distances

$$\frac{X}{Z-D} = \frac{L}{2Z-D} \qquad (4)$$

Consequently, the position, X, of the reflection point can be hand calculated from the known distances D, L, and Z. This simple equation affords a good rule of thumb relationship that is quite helpful in positioning a surface source and a borehole geophone so that a desired spot on an interface is imaged. In multi-fold walkaway experiments, data are recorded at several different geophone depths. In such a case, this equation allows one to calculate a new surface source position which will keep the reflection point at X, even though the geophone depth, D, has changed. Anstey's patent (1980) discusses some of these concepts in greater detail.

The curvature of the reflection event created by the faulted unit in Figure 8-10 can be controlled by changing the distance between the geophone and the reflector. The farther the geophone is above the reflector, the less will be the curvature. If the fault is seismically visible, it should appear as a discontinuity in the reflection event, as shown in Figure 8-10. To estimate how far the fault is away from the borehole, one multiplies the number of traces between the well location and the reflector discontinuity by the calculated value of Δx. In the example in Figure 8-10, the fault is a distance, $3 \Delta x$, to the right of the wellbore. The sources S_1, S_2, S_3, ... can be located so that the separation, Δx, between reflection points is set to whatever resolution is needed for a particular problem, but typical values of Δx are 10 to 30 meters. The throw of the fault cannot be directly estimated by measuring the one-way time displacement of the reflection between traces S_3 and S_4. An example of the errors that can result if the oblique nature of the travel paths is not considered when calculating fault throw can be found in Chapter 6 in the discussion of the ray models shown in Figure 6-15 and 6-16.

The simple VSP response illustrated in Figure 8-10 is usually not realized in actual field recordings because shallow reverberations cause many downgoing multiples to follow the direct arrival. One or more of these multiples can interfere with the upgoing primary reflection from the faulted zone and not allow detailed fault interpretations to be made. These downgoing multiples are particularly bothersome since they often have higher amplitudes than do upgoing reflections. Also, diffraction tails created at reflector terminations, or at sharp curvatures associated with the fault, make it difficult to decide exactly how many trace spacings there are between the well and the reflection break.

A walkaway vertical seismic profile reported by Kennett and Ireson (1977) is shown in Figure 8-12. The strong first arrival is clipped in this plot and appears as the curved dashed line marking the first break of each trace between 0.7 and 0.8 second. One primary reflection is labeled P_1. The moveout of this primary event is not symmetrical about the well position, which suggests that the reflector is dipping, as demonstrated by the walkaway models shown in Chapter 6. However, this same moveout asymmetry is also apparent in the direct arrival, and so it must be an effect created by the source geometry, or by the stratigraphic section above the geophone, and is not an indication of reflector dip.

Interpretational problems exist when one has to decide if events such as A, B, and C are upgoing reflections, downgoing multiples or perhaps upgoing multiples. Obviously, great care must be used to separate primary and multiple events in this data set on the basis of differences in their moveout behavior. The moveout of the downgoing direct arrival is noticeably greater than the moveout of upgoing primary event P_1, and since the moveout of event A is similar to that of P_1, then A is likely a primary reflection also.

354

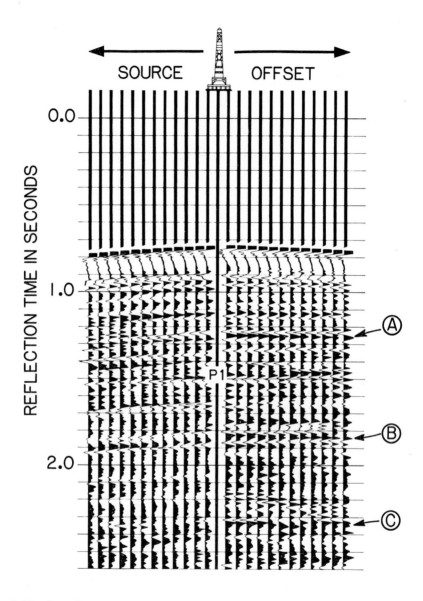

Figure 8-12 A walkaway source vertical seismic profile illustrating how downgoing
multiples and upgoing reflections overlap. One primary reflection is labeled
P_1. Other events such as A, B, and C could be upgoing reflections or
downgoing multiples. (After Kennett and Ireson, 1977).

Considerable phase changes occur in event B across the recording spread, which implies that this event is either an upgoing composite of several closely spaced reflections or an interference between an upgoing reflection and a downgoing multiple. No reflector in this data set appears to be faulted. These same data are discussed by Anstey (1982) and Kennett (1979).

In spite of the problem of distinguishing primary reflections from multiples, there are advantages in a walkaway type of recording geometry. In particular, a more favorable primary-to-multiple amplitude ratio exists than does for surface-recorded data because the geophone is closer to the reflector and farther from the reverberating interfaces in the near-surface which create most multiples. The reverse of this situation exists for a surface geophone. It is closest to the near-surface reverberating system and farthest from the subsurface reflector. In addition, a walkaway geometry allows a detailed analysis of the moveout behavior of the direct arrival so that the seismic propagation velocity, including any velocity anisotropy effects, can be determined for the entire stratigraphic section above the geophone.

One VSP walkaway source technique that some production people have used to detect faults and facies changes near a drilled well is shown in Figure 8-13. The field procedure consists of executing several walkaway source measurements with a downhole geophone located at a different depth for each walkaway. Only two geophone positions are shown in this example, but three or more geophone levels are recorded in most experiments. The objective is to record several VSP responses from the same subsurface

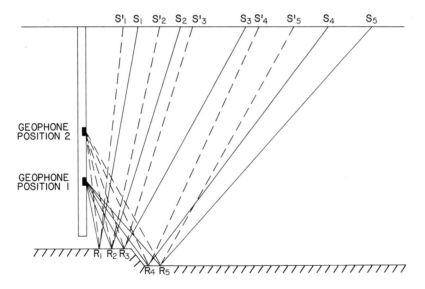

Figure 8-13   High resolution, multi-fold common reflection point imaging around a VSP well.

reflection points so that the responses can be summed to improve the signal-to-noise character of the data and also to attenuate multiples. A critical requirement of the measurement is that, for each geophone recording level, the source must be positioned at successively increasing offset distances so that the first order Fresnel zones created at the stratigraphic unit to be investigated are always centered at the same reflecting points, R_1, R_2, R_3, ... Because of this requirement, this procedure is usually restricted to reservoir development and production environments, where previous drilling has allowed seismic propagation velocities through the stratigraphic section down to the target unit to be determined. This known velocity profile then permits surface source positions to be calculated by ray trace modeling, so that for any geophone recording depth, the recorded raypaths reflect from the same subsurface points, R_1, R_2, R_3, ... For instance, source positions S_1, S_2, S_3 ... are calculated for geophone position 1, and different positions, S_1', S_2', S_3', ..., for geophone position 2. The relationship defined in Equation 4 can also be used to approximate these source and geophone positions. Obviously, considerable pre-survey planning and preparation must be done in order for this field experiment to be properly designed, and in addition, the source and receiver must be carefully positioned during the field work.

The general appearance of VSP data recorded at these two geophone positions is shown in Part A of Figure 8-14. The numbers used to define the reflection points are the same numbers used as subscripts for the reflection points in Figure 8-13. The one-way times required for the first arrivals to reach geophone position 1 are greater than the one-way times required to reach geophone position 2, as shown by the dashed line comparing direct arrival times T_1 and T_3. However, the reflection from the faulted unit arrives at

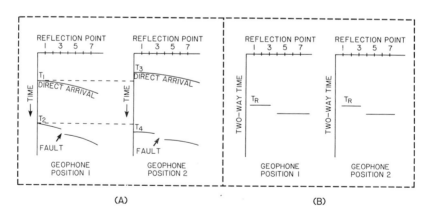

(A)   (B)

Figure 8-14   (A) - The shape of the VSP responses measured at geophone positions 1 and 2, shown in Figure 8-13. (B) - These same VSP responses adjusted so that the upgoing fault reflection is positioned at its surface measured two-way travel time, T_R.

geophone position 1 before it arrives at geophone position 2, thus, reflection time T_2 is less than reflection time T_4.

This observation about arrival times is quite important, because it means that if downgoing multiples interfere with the upgoing fault reflection at geophone position 1, they likely will not interfere with the fault reflection at geophone position 2 since the reflection does not occur at the same time at these two recording depths. In Part B, the data are adjusted in time by a normal moveout correction so that upgoing reflections from every reflection point are positioned at their proper two-way surface arrival times. These two time-adjusted geophone responses can now be summed to improve the signal-to-noise ratio of the data since the fault reflection occurs at the same time, T_R, in both data sets. More importantly, however, the timing relationship between the upgoing fault reflection and the downgoing multiple wavefield is different at the two recording depths, so that the summation process amplifies the fault reflection but attenuates downgoing multiples. If data are recorded at three or more closely spaced depth levels, then multichannel velocity filters can also be designed to discriminate against downgoing multiples. Again, it must be emphasized that only one reflector depth has been imaged. A completely new set of source and receiver positions must be determined in order to position reflection points at the same horizontal spacing at other reflector depths. Soviet geophysicists use the term RRO (reverse reflection observations) to describe a similar type of VSP source-receiver geometry (Karus, et al., 1975). This same reference contains two impressive examples of RRO data which are stacked so that they image the subsurface in an extensive depth interval that extends horizontally for several kilometers (see their Figures 7 and 8).

The VSP horizontal stacking technique described in Chapter 5 is particularly valuable for detecting faults near a wellbore. The field geometry used when recording the data is shown as Method 1 in Figure 8-8. An example of a VSP horizontal stack constructed by Wyatt and Wyatt (1981b, 1982b) is shown in Figure 8-15. These authors use the term VSPCDP to describe their data processing technique. A surface-recorded reflection profile crossing a prospective area is shown on the left. Small faults, which can determine a reservoir's lateral distribution and vertical extent, are likely in the zone from 1.25 to 1.40 seconds. The trace spacing between the surface-recorded data is 165 feet, so three traces span a lateral distance of 500 feet. A VSP source was offset 1000 feet from the well location in order to examine the reservoir unit with better resolution than provided by the surface-recorded data. Shown next to the surface-recorded data is a horizontal stack constructed from these VSP data. This stack is constructed so that the horizontal distance between each VSP derived trace is 25 feet at all reflector depths. Thus, the horizontal resolution is improved by a factor of almost seven compared to that achieved with the surface reflection data. The enlarged data window on the right shows the detailed subsurface imaging that results from the VSP horizontal stack. A small fault

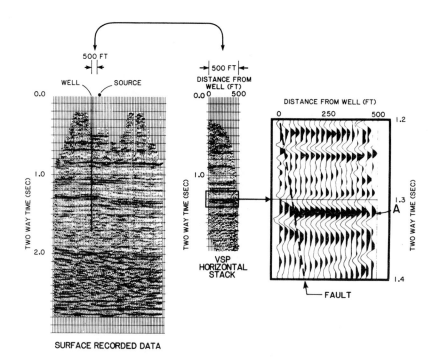

Figure 8-15   A comparison of the horizontal resolution obtained with surface-recorded seismic reflection data and a horizontal stack made from VSP data. Different time datums were used when processing the surface-recorded data and the VSP data.   As a consequence, the VSP data should be moved approximately 40 ms earlier in time in order to tie the surface data. (After Wyatt and Wyatt, 1981b, 1982b)

interpreted by these authors is shown which is impossible to detect in the surface-recorded data.

The fact that the VSPCDP imaging technique creates stacked reflection events that have uniform horizontal spacing, regardless of reflector depth, is a major improvement over some commonly used VSP data processing procedures which create reflection points that have a depth-dependent horizontal spacing.   The VSPCDP procedure also has the following additional advantages that make it an attractive imaging technique.

1.   The subsurface velocity profile does not need to be known before commencing a VSP experiment in order to construct ray trace models to

properly position the sources and receivers. The velocity layering only needs to be known after the data are recorded so that NMO corrections and adjustments for refracted raypaths can be made. This velocity information can be determined from the VSP first break times, or by iteratively stacking the VSP data with a suite of velocity functions similar to what is done when processing surface-recorded data in order to determine proper stacking velocities.

2.   It is not necessary to position the sources and receivers at exact, predefined positions; it is necessary only to know the spatial coordinates of the sources and receivers so that the proper geometrical corrections can be applied to the data for stacking purposes.

3.   A minimum amount of data, and therefore a minimum amount of field acquisition time, is needed to image an extensive vertical slice of the earth near the wellbore with a uniform horizontal sampling. A larger amount of data is needed to accomplish uniform horizontal sampling for a walkaway technique, or any other technique which creates a horizontal spacing between reflection points that varies with depth.

Although examples of VSPCDP processing are available (Wyatt and Wyatt, 1981b, 1982a, 1982b), the mathematical steps needed to construct this imagery have not been published. A U.S. patent application has been filed by Phillips Petroleum Company which relates to the VSPCDP concept. This patent, if issued, will describe the mathematical algorithms involved in the technique and will contain a computer code that performs the data manipulations.

Monitoring Secondary Recovery Processes

Most secondary recovery processes involve the introduction of a foreign fluid into the pore system of a reservoir. If this injected fluid causes the bulk density of the reservoir rock, or the seismic velocity through the reservoir unit, to change, then there is a possibility that the movement of the injected fluid in the reservoir can be monitored by measuring both compressional and shear wave reflection behavior across the reservoir. A schematic representation of the concept is shown in Figure 8-16.

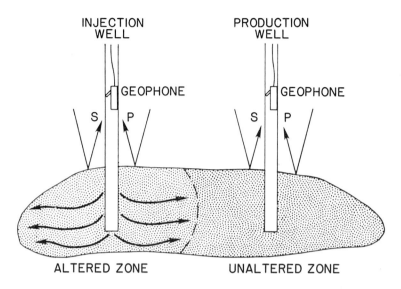

Figure 8-16   Monitoring secondary recovery processes by recording "before" and "after" compressional (P) and shear (S) reflection behavior in VSP wells.

Both compressional and shear reflection data need to be measured in several reservoir wells before the secondary recovery process begins. Both compressional and shear wave seismic energy sources should be used, and a downhole geophone tool having three mutually orthogonal geophone elements that can detect compressional and shear wave events should be positioned above the reservoir, and even into and below the reservoir if possible. The experiment can then be repeated at specified time intervals during the secondary recovery project, duplicating the field recording geometry, energy sources, and recording instrumentation as closely as possible each time. At each well, borehole data should be recorded for several offset positions of the energy source. The movement of the secondary recovery process through the reservoir can be mapped if the ratio of compressional to shear reflected energy in the altered zone of the reservoir differs from the ratio measured in the unaltered zone, or if some property of the reflected compressional and shear waves varies between the flushed and unflushed zones. A measurable difference should allow a production engineer to monitor the expansion of the flushed zone. This application of VSP data will be limited to onshore reservoirs because shear waves cannot be reliably generated in a marine environment. A basic assumption in this application is that compressional waves are more affected by changes in pore fluid than are shear waves.

Detection of Manmade Fractures

The deliberate creation of fractured zones within some reservoirs is critical for proper reservoir development. For instance, tight, low permeability sands and carbonates often must be fractured before initial oil and gas production can be established; and after a period of production, marginally commercial reservoirs can sometimes be revived by fracturing selected productive intervals. Likewise, some secondary recovery processes in depleted reservoirs can proceed only if preceded by reservoir fracturing.

In recent years, several groups have stressed the development of alternate energy sources other than standard oil and gas reserves, and some of these alternate energy efforts also involve deliberate subsurface fracturing. As an example, in situ combustion processes in oil shale require that extensive fracturing exist in the oil shale in order for the combustion to sustain itself. Often oil shales must be converted into true subsurface rubble zones, not simply fractured zones, before in situ combustion can be initiated and maintained. A second example is geothermal reservoir development. Since geothermal wells are almost exclusively located in hard rock country, where granular porosity is essentially non-existent, they are efficient energy producers only if considerable fracturing exists around the wells so that high volumes of steam and/or hot water can enter production boreholes. Geothermal exploration focuses on searching for naturally occurring fracture zones, but in some instances, mechanically induced fracturing may be necessary in order to establish geothermal production. Efforts have been made to locate the positions of these induced fracture zones by recording three-component VSP data in wellbores near fracturing experiments. (Aki, et al., 1982; Albright and Pearson, 1980).

Most of these fracturing requirements can be satisfied by hydrofracing, or by detonating explosive charges implanted in the reservoir interval. However, a common problem in any formation fracturing effort is that one seldom knows how large a volume of the earth has actually been fractured or rubblized. One purpose of this section is to summarize some of the parameters that can be extracted from VSP data which allow volume estimates of subsurface fracture zones to be made. The measurement concept involves a "before and after" seismic picture of the subsurface, just as is described in the preceding section discussing secondary recovery processes. Consequently, the analyses described here can address only manmade fracture zones where one knows exactly when the fracturing will occur, rather than natural fracture zones where there is no opportunity to record VSP data before the fractures existed. The field measurement concept is shown in Figure 8-17. As implied in this figure, the data analysis concentrates on the transmitted direct arrivals measured in a VSP borehole near the fractured zone. The dashed raypath in the "after" situation, which is recorded at depth Z_0, represents backscattered energy from the fracture zone rather than reflected energy. A massive

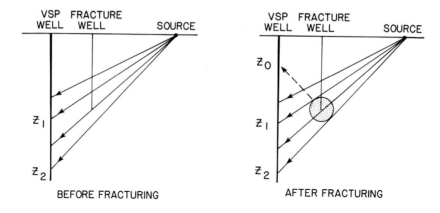

Figure 8-17   The detection and measurement of manmade fracture zones by vertical seismic profiling.

fracture zone could conceivably generate a seismic reflection, but in real earth situations, manmade fracture zones are usually small enough so that they behave as scatterers to a seismic wavefront.

Since the interpretation depends on measuring seismic energy that has traveled through the fractured region, the VSP well must be reasonably close to, and extend below, the fractured space. This requirement means that the technique would, most likely, benefit shallow reservoir studies where numerous wellbores can be economically drilled. A specific example could be the measurement of rubble zones in shallow oil shales.

The following four physical characteristics of seismic data are particularly significant when measured and compared before and after fracturing a subsurface interval:

1.   Propagation Velocity - The propagation velocity of a seismic body wave which passes through a fractured zone is reduced by the fractures. This phenomenon results whether the propagating wave is a P-wave or an S-wave mode.

2.   Energy Content - More seismic energy is dissipated in a fractured zone than in a non-fractured interval.

3.   Wavefield Scattering - Some energy will be scattered out of a propagating wavefront by a fractured zone and will travel in directions different from that in which the wavefront travels.

4. <u>S-Wave Polarization</u> - The polarization of a shear wave propagating
    through a fractured interval can change.

These parameters were measured and documented by Toksoz, et al., (1980) from VSP data
recorded in fractured Antrim oil shale in Michigan. Some aspects of this study are also
found in other reports by Turpening, et al. (1980) and Stewart, et al. (1981a).

The field geometry used in their VSP experiment is the same as that indicated in
Figure 8-17. VSP data were recorded at closely spaced intervals throughout a vertical
VSP borehole before and after explosive charges created a fracture zone of unknown size
in a nearby well. A three-component VSP geophone was used to record both
compressional and shear body waves. The VSP well depths, Z_1 and Z_2, shown in Figure
8-17, define the upper and lower boundaries of the recording interval where energy passes
through the fractured zone and arrives at the VSP wellbore. These depths will be used in
several subsequent illustrations.

A novel energy source which generated highly repeatable source wavelets was used
to create the "before" and "after" P-wave and S-wave modes observed in this experiment.
This source consisted of a military mortar, cemented to the ground, which fired plastic
bags of water rather than hard projectiles. By keeping the elevation and azimuth of the
barrel constant and firing the same volume of water each time, essentially identical
wavelets were propagated into the earth in both the "before" and "after" experiments.
Interestingly, a water cannon was one of the earliest sources used to generate shear waves
(Jolly, 1956).

The "before and after" VSP data behavior observed in this experiment is summarized
in Figure 8-18. Both the compressional and shear body waves recorded between depths Z_1
and Z_2 exhibited velocity slowdowns after fracturing, as shown in panels A and B. In
addition, the amplitude of the compressional wave that traveled through the fractured
interval was significantly reduced, when compared to the amplitude of the compressional
wave recorded before fracturing, as shown in panel C. No doubt the amplitude of the
shear wave mode is also affected by the fracture system, but an analysis of the shear
wave attenuation observed in this experiment has not yet been reported. The
compressional wave attenuation created by the fracture system can be calculated by
making spectral ratio comparisons of "before" and "after" compressional direct arrivals.
Stewart, et al. (1981a) report a Q value of 2.0 in this instance, which implies that a highly
fractured rock volume was created. Also, an amplitude anomaly was observed at depth Z_0
above the fractured zone. This anomaly is shown in panel C as an amplitude reduction.
Toksoz, et al. (1980) interpreted this effect as a destructive interference between the
compressional direct arrival and scattered energy from the fractured zone.

364

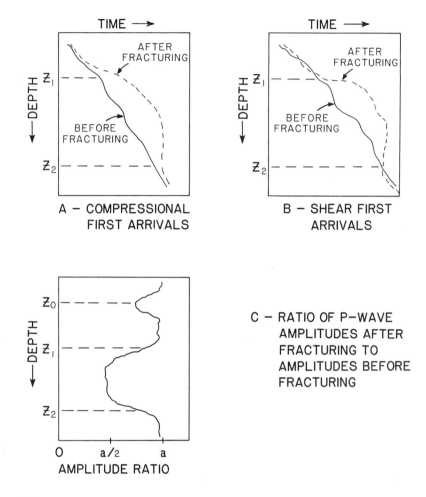

Figure 8-18   Three diagnostic parameters which "before and after" VSP data can provide to identify a fractured zone are reduced compressional and shear seismic wave velocities (A and B); increased P-wave attenuation (C); and evidence of backscattered mode (depth Z_0 in C).

A scattering model that can explain the amplitude effect at Z_0 is shown in Figure 8-19. If it can be assumed that the fractured zone is spherically shaped, and that the wavefronts arriving at the fractured zones are planar, then Yamakawa (1962) shows that the compressional and shear wavefields radiated by the sphere are

$$a(r,\theta,t) = Am^2R^3(C_1 - C_2\cos\theta - C_3(3\cos2\theta + 1)) \, (1/r) \, \exp(i(mr - \omega t)) \qquad (5)$$

and

$$b(r,\theta,t) = Ak^2R^3(C_4\sin\theta + 0.75(k/m)\sin2\theta) \, (1/r) \, \exp(i(kr - \omega t)) \qquad (6)$$

where

    $a(r,\theta,t)$ is the scattered compressional wave amplitude,

    $b(r,\theta,t)$ is the scattered SH shear wave amplitude,

    A is the amplitude of the incident plane wave,

    m is the compressional wavenumber,

    k is the shear (SH) wavenumber,

    R is the radius of the scattering anomaly,

    ω is frequency, and

    the C_i's are constants.

These equations predict backscatter lobes at angles of 120 degrees relative to the propagation direction of the incident plane wave, as shown in Figure 8-19. Therefore, if one of these backscatter lobes can be identified in the VSP data (e.g., the interference at depth Z_0), then the spatial location of the center of the scatterer can be estimated. In the case considered here, the center of the scatterer is already known since the location of the subsurface explosive charge is known. Also, the direction of travel of the incident plane wave is known. By using these facts, coupled with the required scattering angle of 120 degrees, one can predict the depth, Z_0, where the backscatter lobe should appear at the VSP wellbore, and carefully examine the VSP data's amplitude behavior at that depth. If the amplitude decrease at depth Z_0 can be assumed to be the effect of backscattering, the amplitudes $a(r,\theta,t)$ and $b(r,\theta,t)$ can be determined. Knowing these amplitude values allows the radius, R, of the fractured volume to be estimated, which is the fundamental quantity needed to answer the question, "How large is the fractured zone"?

An alternative method for determining the size of the fractured volume would be to shoot across the fracture zone from several azimuths, so that the size of its shadow zone can be defined both vertically and horizontally. This procedure would require (1) a fixed source on one side of the fractured volume and laterally spaced wells on the opposite side in which VSP data could be recorded, or (2) a single receiver well and several sources located directly behind the fracture zone and extending laterally past both edges. Either approach could create a fan of raypaths crossing the complete height and width of the fracture zone. In such a technique, as well as in the situation shown in Figures 8-18 to 8-20, the distance between the fractured region and the VSP well must be small enough so that the wavefront does not heal itself before it reaches the geophone. The seismic travel path from the source to the geophone should be a straight line path passing through the fracture system, and not a curved path passing around the fractured region.

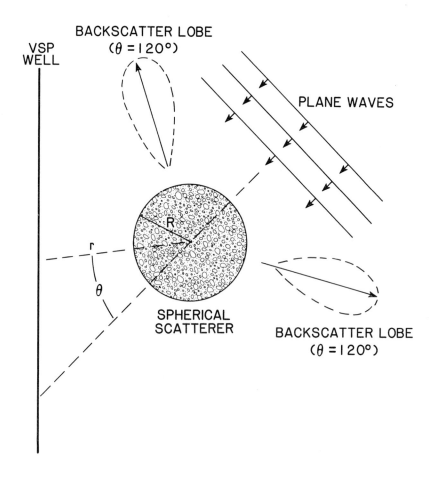

Figure 8-19   Yamakawa's (1962) plane wave scattering model applied to VSP measurements by Tokoz, et al. (1980)

Some investigators refer to this wavefront healing on the backside of an anomaly as "wash around." As an example of the necessity for recording seismic transmission effects close to the fracture system, Toksoz, et al. (1980) report that, when the receiver was 1000 feet from the fractured zone which they studied, they observed no travel time differences and no energy attenuation anomalies when comparing "before" and "after" direct arrivals.

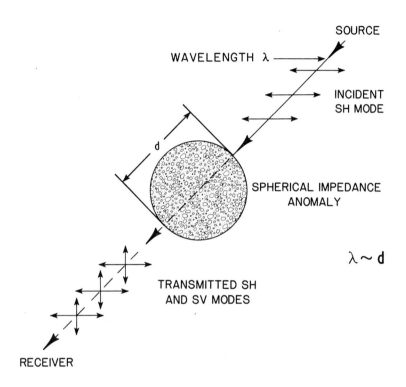

Figure 8-20   Surface shear wave seismic sources generate primarily SH type energy. A large spherically shaped impedance anomaly can act as an acoustic lens to rotate some of this SH energy into the SV mode. In this model, the diameter of the scatterer is of the same order as the incident wavelength.

In some induced fracture studies, such as a carefully monitored hydrofrac, the physical process of fracturing creates both P-wave and S-wave seismic modes which can be identified in three-component VSP data. The distance, D, from a monitoring geophone to an induced fracture can be calculated if the P-wave velocity, V_p, and S-wave velocity, V_s, in the intervening medium are known. If the P-wave event arrives at the VSP geophone at time T_p, and the S-wave event arrives at time T_s, then

$$\frac{D}{V_p} = T_p - T_o \tag{7}$$

and

$$\frac{D}{V_s} = T_s - T_o \tag{8}$$

where T_o is the time when the P-wave and S-wave modes were simultaneously generated by the fracture. T_o is usually unknown, so it can be eliminated by subtracting Equations 7 and 8 to obtain

$$D = \frac{V_s V_p (T_s - T_p)}{(V_p - V_s)} \tag{9}$$

This approach to fracture location is recommended by Seavey (1982), in conjunction with a hodogram analysis of the compressional arrival, in order to determine the position of a hydro-fracture relative to the position of a monitoring geophone.

Another important physical effect that helps identify the existence of fractures results when a seismic shear wave interacts with an extensive fracture system. A large fracture zone, particularly if somewhat spherical in shape, can act as an acoustic lens and rotate the polarization vector of a shear mode. If an incident shear wave contains predominantly SH energy, which is the most common mode created by surface shear wave sources built for exploration purposes, this rotation means that the shear energy transmitted through the fracture zone is comprised of both SH and SV modes, as illustrated in Figure 8-20. By examining the vertical component of a triaxial geophone monitoring seismic transmission through a zone "before" and "after" the zone was fractured, Toksoz, et al. (1980) and Stewart, et al. (1981a) found that the shear wave energy recorded by this geophone behaved as shown in Figure 8-21. The SH mode generated by the surface source used in this experiment creates only a small vertical component of particle displacement between recording depths Z_1 and Z_2 before the fractures exist. After the zone is fractured, the amount of SV energy observed between Z_1 and Z_2 increases dramatically due to the rotation of the shear wave polarization vector away from the horizontal plane.

These observations about shear wave behavior have lead some explorationists to prospect for naturally occurring fracture zones with combined surface-recorded compressional and shear wave reflection data. The concept involved in the measurement is illustrated in Figure 8-22. If the carbonate unit AB is naturally fractured between X_1 and X_2, then the preceding VSP results demonstrate that below depth A, compressional

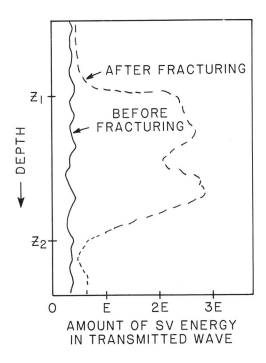

Figure 8-21 If a VSP shear wave source creates predominantly SH mode energy, the amount of SV energy recorded by vertically oriented VSP geophones depends upon how large a fracture zone the shear wave must cross before it reaches the geophones.

and SH body waves will be attenuated more between X_1 and X_2 than they will to the left of X_1, or to the right of X_2. Consequently, reflections B, C, D, and reflections below D, will dim between X_1 and X_2. If both compressional and SH modes are generated and recorded at the surface, then present thought is that surface-recorded SH modes will dim more than compressional modes do because the SH mode is subjected not only to attenuation and scattering, as is the compressional mode, but also to a rotation of its polarization vector. If SV reflection modes can be identified in the surface responses, then this rotation of the SH polarization vector will result in an increase in the SV reflection amplitudes marking interfaces B, C, and D. Three important surface reflection amplitude criteria that imply subsurface fracturing between X_1 and X_2 are listed at the top of the figure.

370

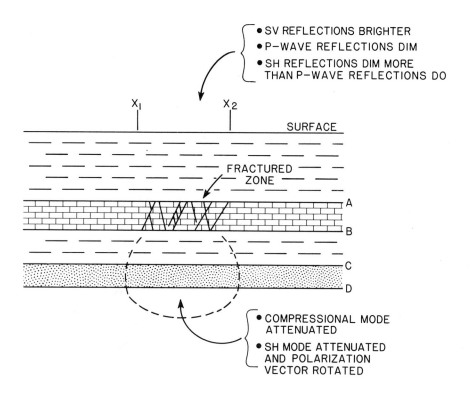

Figure 8-22   Detecting naturally fractured zones by surface recorded compressional and
shear wave reflections.

Seismic Tomography and Reservoir Evaluation

A point emphasized in earlier sections of this text is that the seismic resolution of a
subsurface anomaly is increased when a geophone is positioned deep in the subsurface near
the anomaly rather than at the surface. The next logical step in the quest for improved
seismic resolution is to place both source and geophone deep in the subsurface. In some
cases, the raypath geometry describing the propagation of seismic wavefronts from a
subsurface source to subsurface receivers is analogous to the physical arrangement of
radiation source, patient, and receiver used in medical tomography; consequently, the

term seismic tomography seems appropriate for this type of field geometry. By definition, tomography means, "section (tomos) drawing (graphy)", and sectional viewing of the subsurface is the ultimate aim of detailed reservoir evaluation.

A simplified picture of a typical source-patient-receiver geometry involved in medical tomography is shown in Figure 8-23. This illustration shows that there are some

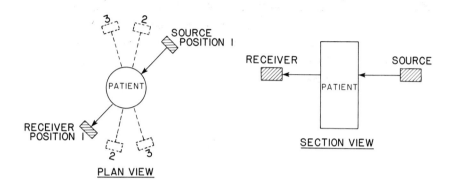

Figure 8-23   A simplified picture of medical tomography.

important aspects of medical tomography that cannot be achieved with seismic sources and geophones positioned in wellbores in the subsurface. The principal limitations are:

1. Narrow collomated beams of radiation are used in medical tomography. It is difficult, and in many cases impossible, to create focused beams of acoustical radiation with seismic sources. The radiation pattern of source arrays in deep boreholes has been theoretically studied (Heelan, 1953; White and Sengbush, 1963; Peterson, 1974; Lee and Balch, 1982), but little field data have been exhibited to substantiate these theoretical predictions.

2. Only one mode of radiation is created by the medical source. Commonly used sources emit x-rays or acoustic waves with carefully defined frequency ranges. Seismic sources on the other hand create both compressional and shear wave modes with frequencies spanning several octaves. These waves, in turn, create various converted waves. Consequently, seismic radiation patterns are more complicated than those used in medical tomography.

3.   The medical source and receiver can be freely moved (or the patient can be moved) so that a target area can be inspected from any direction. Subsurface seismic source and receiver positions are limited to available borehole locations, which may not always allow adequate imaging of a reservoir unit or a geological amomaly.

Even with these limitations, there are some potentially valuable advantages associated with a seismic tomography approach to reservoir evaluation. Possible ray-paths that seismic waves can follow as they propagate from a surface or subsurface source position to either surface or subsurface geophone locations are shown in Figure 8-24. Ray 1 is the only path which does not travel through the near-surface weathered layer, where high frequency components of a seismic wavelet are lost due to dissipation and scattering. Consequently, the greatest seismic bandwidth should be recorded, and the greatest resolution obtained, for the field geometry associated with path 1. Also, ray-path 1, since it is between a subsurface source and receiver, will normally be the shortest of the four possible types of source-receiver paths shown in the illustration. This fact is a second reason why higher frequencies, larger signal amplitudes, and greater resolving power occur along path 1 than along any of the other paths.

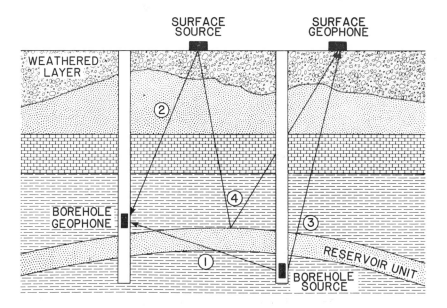

Figure 8-24   Raypaths involved in seismic wave propagation between surface and subsurface source and receiver positions.

Raypath 4 is the type involved in standard surface reflection measurements, and it differs from the other paths in three important ways:

1. It is the longest of the paths, and this fact limits its resolving capabilities.

2. Waves traveling along path 4 must pass through the near-surface weathered layer twice. Consequently, their high frequency components have the greatest likelihood of being attenuated.

3. Only seismic reflection information from the reservoir unit is measured by path 4; all other paths measure the reservoir's effect on seismic transmission, since either the source or the receiver is below the reservoir. A surface-to-reflector-to-surface raypath extending to a reflector below the reservoir unit would, of course, be affected by the seismic transmission properties of the reservoir.

Few public descriptions of seismic tomography field experiments exist. Bois, et al. (1971, 1972) describe how seismic velocities can be estimated along raypaths between wellbores and demonstrate their results with real field data. Soviet geophysicists have done considerable work in studying surface-recorded arrivals from seismic shots fired in deep wells. They call the technique "inverted seismic well shooting method" or "torpedoing" (Gal'perin, 1974). Weatherby (1936) was one of the early American geophysicists to propose this technique. Lytle and Dines (1980) compute models that illustrate some of the potential applications and practical limitations of seismic tomography. Their results are valuable even though the data are synthetic and not actual seismic measurements. Butler and Curro (1981) describe procedures for recording crosshole seismic data and illustrate some of the pitfalls that can occur if inappropriate field procedures are followed. Their conclusions are particularly valuable because they are based on real field data; however, they consider only shallow, high frequency data that can be used to predict the structural stability of the near-surface at construction sites. No doubt much of their work can also be applied to crosshole testing of deep reservoir units.

A field experiment involving a situation comparable to hydrocarbon reservoir evaluation by tomographic principles is described by Parrott (1980). The objective of this field work was to map thin marker beds of shale, anhydrite, and dolomite inside a salt structure being considered as a site for nuclear waste disposal. These beds are water bearing and thus are potential escape routes for radioactive material. The work was conducted by the United States Geological Survey, and papers by other USGS personnel describing additional results of this field test are pending. The subsurface structural problem, and the source-receiver geometry that was used, are shown in Figure 8-25. The source was an airgun positioned in an uncased part of a well which penetrated into the salt. A three-component VSP geophone system was positioned at closely spaced vertical positions 8 feet (2.4 meters) apart in a second well which also penetrated the salt structure. These small depth sampling intervals were required in order to avoid spatial aliasing, because data up to 400 hertz were generated and recorded.

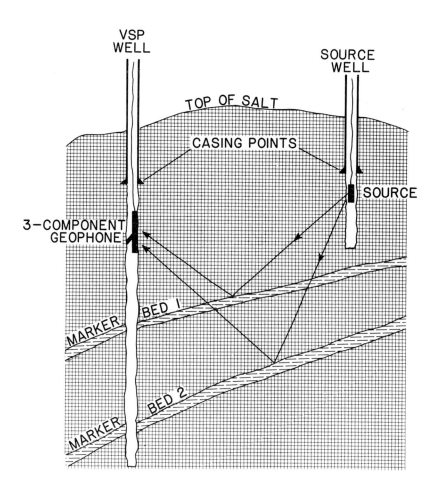

Figure 8-25   Cross sectional view of VSP field geometry employing a subsurface borehole source (After Parrott, 1980).

A seismic borehole source used for vertical seismic profiling must create a large number of identical seismic shot wavelets without physically damaging the borehole. There is essentially no experimental work describing how much physical damage airguns of low to moderate strength can do in a deep well. Two possible adverse results of seismic shooting in a deep borehole are shown in Figure 8-26. The main concerns are that repetitive shooting in an uncased hole may cause cave-ins that prevent the source from ever being retrieved from the hole, and that shooting many shots in a cased hole may create leaks in casing joints near the source. Neither of these borehole damages occurred with the 120 in^3 (1.97 liters) airgun operated at 3000 psi (20682 kPa) in the USGS experiment, but more field studies are warranted.

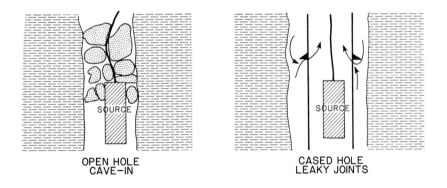

OPEN HOLE
CAVE-IN

CASED HOLE
LEAKY JOINTS

Figure 8-26   Possible problems associated with deep borehole airguns.

One serious problem encountered in a VSP application employing a borehole source, and confirmed by Parrott (1980), is that many secondary body waves are created by strong tube waves traveling up and down the borehole after the initial seismic impulse has propagated away from the wellbore. This same behavior has been noted by Gal'perin (1974, p. 21). The physical mechanisms creating seismic body waves in a borehole are shown in Figure 8-27. Both compressional and shear body waves are created by the primary seismic impulse (White and Sengbush, 1953). In addition, extremely strong tube waves are created which propagate up and down the borehole. Secondary compressional and shear body waves are created when these tube waves arrive at changes in borehole impedance. One such possible impedance contrast is shown as a change of borehole diameter in Figure 8-27. Other borehole impedance changes could result when a borehole penetrates abrupt changes in lithology. This wave propagation phenomenon is the reverse of the mechanism described in Chapter 3, where body waves interacting with strong borehole impedance changes create tube waves in the fluid column in a wellbore. These secondary body waves follow the primary body waves at various time delays and complicate the interpretation of the primary body wave signals.

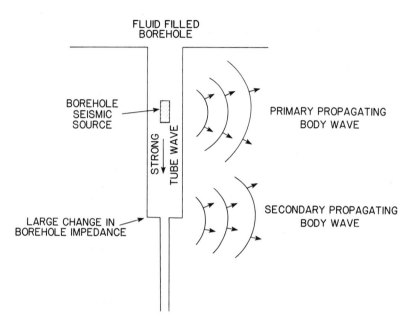

Figure 8-27   A problem created when a seismic energy source is located in a deep, fluid-filled borehole is that extremely strong tube waves are generated. These tube waves create secondary body waves at prominent borehole impedance contrasts which complicate the interpretation of signals received from the primary body wave.

    In the USGS experiment, the source well was 494 feet (150.6 meters) away from the VSP recording well, and both boreholes were vertical. The airgun was operated at a depth of 750 feet (228.5 meters) in the source well, and three-component data were recorded between depths of 1180 and 3100 feet (359.7 to 944.9 meters) in the VSP well. (See Figure 8-25.)   An example of the large number of different wave modes recorded in this experiment is shown in Figure 8-28.   These data represent the response of only one horizontal geophone as the geophone tool is positioned at a sequence of depths below the airgun source.   Often displays of other components of the particle velocity vector (particularly the vertical component) show these events better in some depth intervals in Parrott's paper.   The data are filtered to attenuate upgoing events. The labeled wave modes are generated in the following manner:

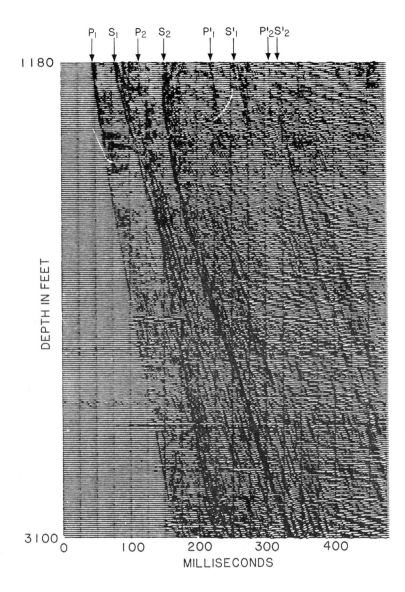

Figure 8-28   Downgoing body wave events identified from the horizontal Y-component of the recorded wavefield.   Source was an airgun positioned 750 feet (228.6 meters) deep in a well 494 feet (150.6) meters away from the VSP well (After Parrott, 1980).

P_1 and S_1 - primary compressional and shear waves generated by the initial airgun discharge.

P_2 and S_2 - secondary compressional and shear waves created when the tube wave arrives at the bottom of the source well the first time.

P'_1 and S'_1 - secondary compressional and shear waves created when the tube wave that is reflected from the bottom of the source well arrives at the airgun and its associated air bubble.

P'_2 and S'_2 - secondary compressional and shear waves created when the tube wave that created P'_1 and S'_1 reflects downward from the airgun and arrives at the bottom of the source well a second time.

If the source borehole has numerous large washouts, even more secondary wave modes would be generated. In this investigation, the shear wave velocity in the formation around the borehole was larger than the velocity of the tube wave. In situations where the shear wave velocity is less than the tube wave velocity, a tube wave's amplitude attenuates as it propagates along a fluid column, and the problem of secondary sources may not be as severe as that shown in Figure 8-28. Possibly, the strength of the secondary waves can also be controlled by using a heavy or viscous fluid in the source well so that tube waves are damped out rapidly. However, at depths where hydrocarbon reservoirs occur, the shear wave velocity will ordinarily exceed the tube wave velocity. Also, most wells will be filled with relatively non-attenuating fluid. Thus, this USGS experiment reveals the typical interpretational problems that must be overcome in order for seismic tomography to be a viable technology for imaging reservoirs. Any application of seismic tomography data involving a borehole source will have to deal with the interpretational complications introduced by numerous secondary wave modes. The interpretations are more difficult than those associated with standard VSP field geometries using surface sources; however, the potential benefits of imaging deep reservoir units with seismic waves containing frequencies up to 200 to 400 Hertz, as Parrott documented, warrant continued development of the technology.

Vertical Seismic Profiling in Deviated Wells

Many oil and gas wells, particularly offshore production wells, are deliberately deviated from vertical, and recording VSP data in these deviated wellbores can provide valuable reservoir evaluation data. Consider the cross-sectional view of such a deviated well in Figure 8-29. The dots represent stations where a borehole geophone is locked in order to record data. A surface energy source is assumed to be positioned vertically above each geophone station in the manner illustrated in Figures 4-20 and 4-21.

The geophone is moved both vertically and horizontally between these recording positions. The vertical movement is important because it allows upgoing and downgoing wavefields to be identified and separated by appropriate data processing procedures, such as f-k filtering, median filtering, and wavefield subtraction. Also, once upgoing and downgoing wavefields are separated, deconvolution operators can be determined from the downgoing wavefield which can effectively attenuate multiples in the upgoing wavefield. The horizontal geophone movement is important also, because it allows subsurface seismic responses to be recorded at small lateral intervals. In particular, upgoing reflections

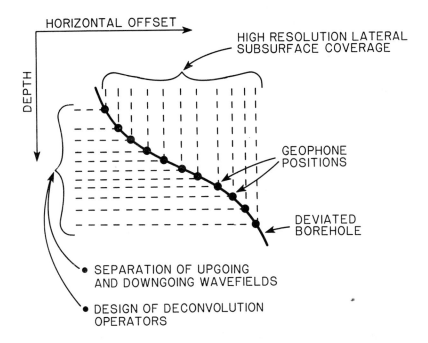

Figure 8-29 Basic diagnostic seismic properties that become available when VSP data are recorded in a deviated borehole.

originating from below the geophone positions are sampled at closely spaced points so that subsurface anomalies and reservoir conditions below a deviated wellbore are imaged with considerable lateral resolution.

The hypothetical geological situation shown in Part A of Figure 8-30 will be used to illustrate how VSP data recorded in a deviated production or water injection well can be used to map seismic reflectors defining reservoir facies. A flat, horizontal reflector R_1 is shown laying unconformably over a reservoir unit. Interfaces R_2 and R_3 define a faulted seismic reflector within the reservoir interval. A deviated borehole is shown penetrating the reservoir as a stepout well, and VSP data are recorded at the marked geophone positions in this wellbore. A surface energy source is assumed to be positioned at the locations labeled X_1, X_2, X_3, ..., which are vertically above each subsurface geophone station. In this example, the geophone locations are arbitrarily selected so that the distance (X_n-X_{n-1}) between successive source locations is constant. The depths corresponding to the geophone recording positions are labeled Z_1, Z_2, Z_3, ... Contrary to the way these data are depicted, it is usually preferable, for ease of processing, to record the data so that the traces are uniformly separated in the vertical direction, Z. This means that the horizontal increment between source locations will vary. However, either procedure for spatially sampling the seismic wavefield will suffice to illustrate the advantages of recording VSP data in deviated wells.

The princial features of VSP data recorded in this well are sketched in Part B. To simplify the concepts, the only downgoing event shown is the direct arrival, DA, and the only upgoing events shown are the primary reflections, R_1, R_2, and R_3. All downgoing and upgoing multiples are ignored. In real VSP data, multiples would have to be attenuated by deconvolution operators designed from the downgoing wavefields. Likewise, diffraction effects are ignored. At each recording depth, reflection wavelets are positioned in time so that they represent the shortest travel paths to their respective reflecting interfaces. Reflector R_1, since it is flat and horizontal, creates an upgoing event that has the same magnitude of time stepout along its trajectory as does the downgoing direct arrival. Because of their dip, interfaces R_2 and R_3 create events that exhibit time stepouts whose magnitudes differ from that of the downgoing direct arrival.

These same data are shown in Part A of Figure 8-31 after two processing procedures have been applied. First, all downgoing events have been attenuated and upgoing events have been emphasized by velocity filtering. Second, each trace is delayed in time by its first break time. The new location of the first break times is indicated by the inclined dashed line. All events at each recording depth are now defined in terms of two-way travel time and can be better compared with surface-recorded data crossing the well. This time shifting places reflection, R_1, from the flat, horizontal interface at the same two-way time at each recording depth. However, reflections R_2 and R_3, created by the dipping interfaces, still exhibit time stepout and have decreasing onset times as recording

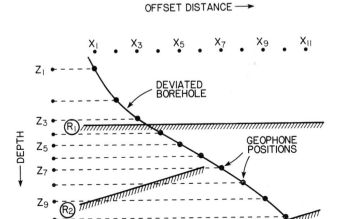

OFFSET DISTANCE ⟶

X_1 · X_3 · X_5 · X_7 · X_9 · X_{11}

Z_1

DEVIATED BOREHOLE

Z_3 (R₁)

Z_5

GEOPHONE POSITIONS

Z_7

Z_9 (R₂)

(R₃)

Z_{11}

DEPTH

(A)

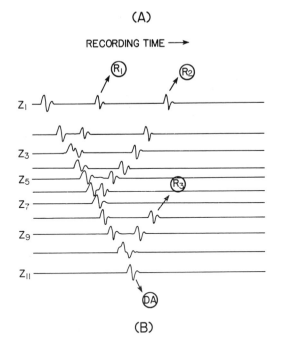

RECORDING TIME ⟶

(R₁)   (R₂)

Z_1

Z_3

Z_5 (R₃)

Z_7

Z_9

Z_{11}

(DA)

(B)

Figure 8-30 (A) - Deviated borehole penetrating a stratigraphic section containing reflectors R_1, R_2, and R_3. (B) - VSP data recorded through the stratigraphic section shown in A.

382

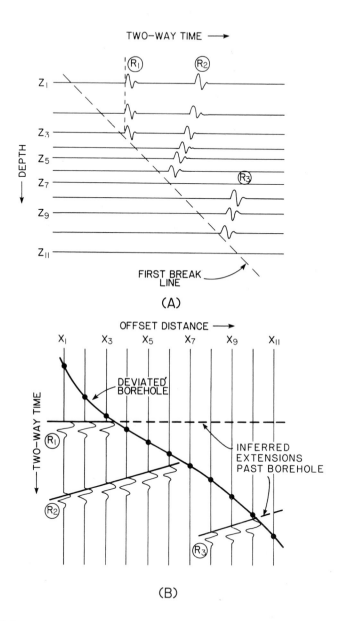

TWO-WAY TIME →

(R_1) (R_2)

Z_1

Z_3

Z_5

Z_7 (R_3)

Z_9

Z_{11}

DEPTH

FIRST BREAK
LINE

(A)

OFFSET DISTANCE →

X_1 X_3 X_5 X_7 X_9 X_{11}

DEVIATED
BOREHOLE

TWO-WAY TIME

(R_1)

INFERRED
EXTENSIONS
PAST BOREHOLE

(R_2)

(R_3)

(B)

Figure 8-31   (A) The data in Part B of Figure 8-30 shifted to align upgoing reflections
generated by flat, horizontal interfaces. (B) Same data as in Part A but
rotated 90 degrees and plotted horizontally as a function of offset recording
distance. The original reflecting interfaces from Part A of Figure 8-30 are
shown in their proper spatial positions by the solid lines R_1, R_2, and R_3.

depth increases. Note that the upgoing reflections R_2 and R_3 slope up to the right; whereas, upgoing events from the dipping reflectors modeled in Chapter 6 slope up to the left. The reason for the difference in slope is that all models in Chapter 6 were calculated for a vertical borehole, but in this example the recording geophones are not distributed vertically above each other. As a result of the horizontal movement of the geophone, these VSP data are, in effect, a horizontal seismic section comprised of single-fold, zero-offset traces, similar to what could be constructed from the inside trace of surface-recorded seismic shots that follow the track of the wellbore. However, an important distinction is that the VSP geophone is deep in the earth, near the feature that is being illuminated by the surface energy source. Consequently, the VSP data have a high signal-to-noise ratio, even though they are only single-fold.

Since the well is deviated, the geophone positions shown in Figure 8-29 are a function of both depth, Z, and horizontal offset distance, X. The VSP data in Part B of Figure 8-30 and in Part A of Figure 8-31 are plotted as a function of depth as VSP data typically are. However, the data can also be plotted as a function of the horizontal offset distance at which each geophone is located, as shown in Part B of Figure 8-31. The trace labeled X_1 is the same as trace Z_1 in Part A, trace X_3 corresponds to trace Z_3, etc. The true spatial positions of interfaces R_1, R_2, and R_3 are superimposed on the wiggle traces as solid lines. It is obvious that the VSP reflections closely approximate the true positions of the dipping interfaces, particularly in the immediate vicinity of the borehole. The VSP reflections fail to accurately define the position of the dipping interfaces at large distances from the borehole (e.g., the difference between the solid line position of interface R_2 and the first break time of the R_2 wavelet at offset distance X_1). A VSP migration procedure would have to be applied to the data if more precise reflector location is required at large distances from the borehole.

There are several positive aspects to this type of subsurface imaging; namely,

1. Reflectors below a borehole are seismically imaged with the best possible resolution since the geophone is much closer to them.

2. Reservoirs can be seismically sampled in the horizontal direction at very closely spaced intervals.

3. Upgoing reflections created at interfaces penetrated by a borehole are properly positioned in (X, Z) space and in reflection time at the borehole without even migrating the data.

One example of a vertical seismic profile recorded in a deviated well is shown in Figure 8-32. Actually, VSP data were recorded in both the vertical and the deviated wells

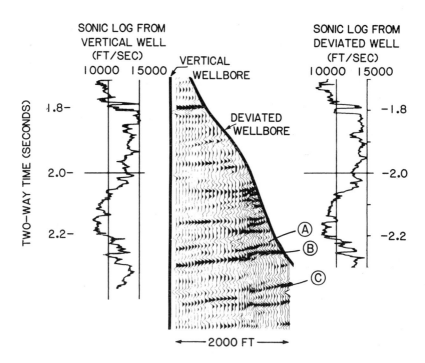

Figure 8-32 An example of VSP data recorded in a deviated well (After Kennett and Ireson, 1981).

shown in the illustration, but only the data recorded in the deviated well are displayed. The data recorded in the vertical well are shown in Chapter 5, Figure 5-45. The bottom of the deviated well is horizontally displaced 2800 feet from the vertical well. The lateral coverage of VSP data underneath the deviated borehole is 2000 feet, thus the vertical borehole is 800 feet away from the leftmost trace plotted in Figure 8-32. The lateral spacing between the VSP traces decreases as the wellbore trajectory approaches vertical, but the average horizontal trace spacing is about 50 feet. Sonic log data recorded in each well are plotted to the right and left of the VSP data to help correlate equivalent stratigraphic units across the 2800 feet separating the two wells.

The stratigraphic section below the deviated wellbore is imaged in considerable detail, and several faulted units can be seen in the VSP response. An interpretation of these data proposed by Kennett and Ireson (1981) is shown in Figure 8-33. The reflector terminations labeled 1, 2, 3 are interpreted from the data shown in Chapter 5, Figure 5-45, where the same labels 1, 2, 3 are used to identify the appropriate diffraction events and reflection terminations. Seismic units A, B, C shown in this interpreted cross-section are identified in Figure 8-32. In their paper, the authors also compare these VSP data with surface-recorded reflection data crossing the two wells.

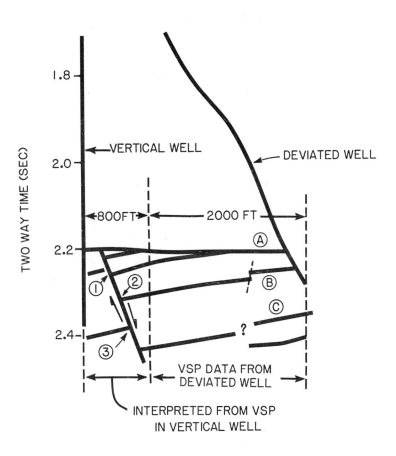

Figure 8-33   Stratigraphic interpretation of VSP data shown in Figures 5-45 and 8-32 (After Kennett and Ireson, 1981).

386

In Figure 8-33, the distances from the vertical wellbore to diffraction points 1, 2, and 3 are critical components of the subsurface structural picture near this VSP study well. These distances can be calculated from an analysis of the diffraction curves shown in Figure 5-45 if it can be assumed that straight raypaths adequately describe the propagation of seismic energy from the surface source to the subsurface diffractors.

A simplified picture of the response which should be recorded as a VSP geophone moves vertically past a point diffractor is shown in Figure 8-34. In this illustration, a diffractor D is assumed to be a distance G below the surface. The apex of the observed diffraction curve occurs when the travel time from this diffractor to a geophone in the VSP observation well is a minimum. For a vertical well, the minimum travel path results when the geophone is at the same depth as the diffractor, as can be shown by completing the raypath diagrams for points A and B in Figure 5-43. When the geophone is above or below depth G, the travel time from the diffractor to the geophone increases. This increased travel time creates the curved legs of the diffraction event above and below G. This diffraction event is a shear wave (SV) diffraction mode since the downgoing leg of the diffraction curve exhibits an apparent velocity much slower than that of the downgoing compressional first breaks.

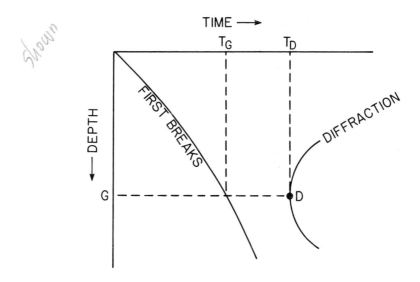

Figure 8-34   A simplified VSP response showing only downgoing first breaks and a diffraction event whose apex is positioned at depth G and time T_D.

In order to make an optimum interpretation of the subsurface, it is important to estimate the position of diffractor D relative to the wellbore where the VSP data are recorded. Ideally, one would like to predict the azimuth direction to the diffractor, but azimuth determinations cannot be made unless the geophone tool contains orthogonal three-component geophone elements, together with an orientation measuring device that defines the directions that these geophone elements point when data are recorded. In this analysis, we will assume that the data are recorded with vertically oriented geophone elements so that the azimuth direction from the wellbore to the diffractor will be indeterminate. Consequently, the only diffractor coordinate that can be calculated from the VSP data is the distance from the geophone to the diffractor. One computational scheme by which this distance can be estimated is illustrated by the data parameters shown in Figures 8-34 and 8-35.

The time required for the first arrival to travel from the surface source, S, to the borehole geophone at G is T_G. The distance from source S to geophone G is therefore VT_G, where V is the average velocity between the surface and depth G. When the geophone is at G, the time required for a shot wavelet to propagate from the source to the diffractor and then to the geophone is T_D. Both of these times, T_G and T_D, are labeled in the hypothetical VSP response shown in Figure 8-34. The critical time value needed to estimate the distance to the diffractor is the minimum time required for the wavelet to travel from the diffractor to the geophone. This time value will be called T_0,

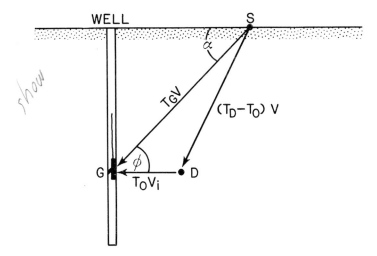

Figure 8-35   Raypaths describing seismic wave propagation from surface source S to diffractor D to geophone G. V is the average velocity to depth G, and V_i is the interval velocity at depth G.

and it, together with a good estimate of the interval velocity at depth G, allows the distance from the borehole to the diffractor to be calculated.

The time, T_0, cannot be directly measured from the VSP response; it must be calculated. If the interval velocity at depth G is V_i, then the distance from geophone position G to diffracting point D is $T_0 V_i$, as shown in Figure 8-35. The travel time from the source to the diffractor is the time difference, $(T_D - T_0)$, thus the distance from the source to the diffractor is $(T_D - T_0)V$.

Using a single velocity value, V, to determine the travel distances SG and SD allows the raypath geometry to be described in terms of straight lines. From the law of cosines, the raypath distances in Figure 8-35 are related as

$$((T_D - T_0)V)^2 = (T_0 V_i)^2 + (T_G V)^2 - 2 T_0 T_G V_i V \cos\emptyset \qquad (10)$$

or

$$(V_i^2 - V^2) T_0^2 - 2 (T_G V_i V \cos\emptyset - T_D V^2) T_0 - (T_D^2 - T_G^2) V^2 = 0 \qquad (11)$$

Except for T_0, all travel times and seismic velocity values in Equation 11 can be determined by analyzing either the VSP first breaks or the diffraction apex. The angles α and $\emptyset$ in Figure 8-35 can be calculated since the source offset distance and geophone depth are known. Thus, the time value, T_0, can be determined, and the distance, $T_0 V_i$, from the well to the diffractor can be estimated. If a diffraction curve is created by a converted shear wave, then the interval velocity, V_i, used in this calculation must be the interval velocity of a shear wave at depth G, not the interval compressional velocity determined from first break times spanning depth G.

A second example of VSP data recorded in a deviated well, described by Kennett and Ireson (1981), is summarized in Figure 8-36. VSP data spanning a lateral distance of 460 meters underneath a deviated wellbore are shown to the right of surface-recorded reflection data imaging the same vertical section of the earth. The distance between the surface-recorded data traces is approximately 25 meters; whereas, the VSP traces are separated by approximately 15 meters. These data illustrate one of the serious exploration problems encountered worldwide, which is the inability to record good quality seismic data below some unconformities. It is particularly difficult to image geological conditions immediately below the Jurassic Cimmerian unconformity in some areas of the North Sea. Consequently, this experiment is important because it compares the ability of surface-recorded data and VSP data to properly image reflecting interfaces occurring below part of that unconformity.

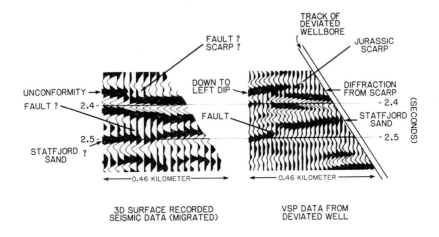

Figure 8-36   Comparison between surface-recorded reflection data and VSP data recorded in a deviated wellbore penetrating an unconformity (After Kennett and Ireson, 1981).

The unconformity is marked at 2.35 seconds on the surface-recorded data and appears as a continuous unit dipping down to the right. However, the vertical seismic profile shows that the unconformity is not continuous, but is, in fact, an irregular scarp surface which has sections that dip to the left. In the surface-recorded data, the scarp feature is barely recognizable, and there is no evidence of any leftward dip. The VSP data, as well as log data recorded in the borehole, show that the Statfjord sand intersects the wellbore at 2.45 seconds. In addition, the VSP data show that this sand reflector dips down to the left and is faulted. The Statfjord sand is not recognizable at 2.45 seconds in the surface-recorded data, and it is also very difficult to interpret a fault in that unit from the surface data. The left dip of the Statfjord sand measured from the surface-recorded data is much less than the dip shown by the VSP data.

One of the more thorough uses of VSP data in the development of a petroleum reservoir is the effort expended by Occidental, Getty, Thomson, and Union Texas in the Piper Field located in the UK sector of the North Sea (Johnson, Riches, and Ahmed, 1982). This paper is particularly important because it represents a deliberate attempt to demonstrate to engineers that vertical seismic profiling is a methodology that can solve many field development problems. A map of the Piper Field is shown in Figure 8-37. This reservoir, like so many others worldwide, is complicated due to the fact that several faults cross the hydrocarbon accumulation. Unless the fault pattern within the reservoir is accurately mapped, one does not know exactly where to position development wells, or how many wells should be drilled, in order to obtain optimum recovery from the field.

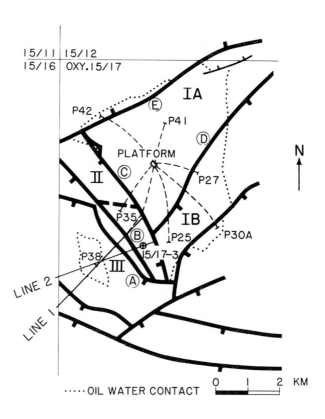

Figure 8-37   Location of VSP wells, major faults, and oil-water contact in the Piper field. VSP data were recorded in all seven of the deviated production wells shown by the dashed lines. No VSP data were recorded in exploratory well 15/17-3. Line 1 and Line 2 are two surface seismic lines that follow the track of deviated VSP well P38. The Roman numerals indicate some of the major fault blocks. A, B, C, D, and E are some of the bounding faults. The reserves shown by the dotted contour in Block III were added as a result of the VSP analyses done in well P38. (After Johnson, et al., 1982, Copyright SPE-AIME).

The locations of the faults crossing the Piper Field, as they are interpreted from surface seismic data and subsurface well data, are shown as the grid of heavy lines that dominate the map in Figure 8-37. Five faults occurring within the closed contour of the oil-water contact strongly affect the siting of development wells and are labeled A, B, C, D, E. VSP data were recorded in seven deviated production wells, shown radiating out from the platform at the center of the field, in an attempt to improve the structural interpretation across these fault blocks, and thereby allow subsequent production wells to be properly positioned. Only the VSP data recorded in wells P27, P35, and P38 will be shown in this discussion.

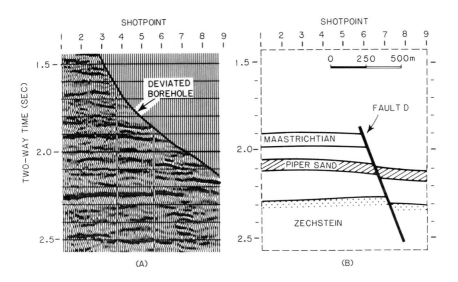

Figure 8-38 (A) Upgoing VSP wavefield recorded in deviated well P27 after applying deconvolution operators to remove multiple events. (B) A structural interpretation suggested by these VSP data. Note that the VSP coverage extends laterally for 1100 meters between shotpoints 1.0 and 8.8. Fault D is shown in Figure 8-37. (After Johnson et al., 1982, Copyright SPE-AIME).

Considerable field testing of energy sources and downhole geophones was done during the course of these VSP measurements. A shooting boat equipped with appropriate navigational positioning equipment was used to fire an airgun energy source directly above the borehole geophone at each recording depth. The diameter of the allowed shooting circles (Figures 4-20 and 4-21) was set at $2\frac{1}{2}$ percent of the depth to the geophone. Either a tuned twin-gun array, or a single airgun, was used as the energy source in different surveys.

During the field tests, it was noted that the most important requirements for a source were that it should create reservoir reflections with high signal-to-noise ratios and a consistent output wavelet throughout the entirety of a survey. The requirement that no bubble oscillations should occur was relaxed since bubble reverberations could be adequately removed during data processing. Consequently, a single large airgun had some advantages over an airgun array. These observations are in agreement with those noted by others, which are referenced in the discussion of airgun sources in Chapter 2 (page 27). The airguns were suspended approximately 30 feet below the shooting boat; however, some undesirable wavelet variations were recorded by the near-field hydrophone because the gun depth fluctuated due to large wave heights. A gimbaled geophone element was found to provide a sensitivity and a signal-to-noise ratio significantly better than a non-gimbaled detector.

VSP data recorded in well P27 are shown in Figure 8-38. These data represent the upgoing VSP wavefield after deconvolution operators have been applied in order to attenuate upgoing multiples. The bottom location of this well is in fault block IB, almost due east of the production platform. The location of fault D, which separates reservoir blocks IA and IB, and the change in the stratigraphic position of the Piper Sands created by this fault, must be known in order to efficiently develop the eastern portion of the field. A cross-sectional view of the track of the P27 wellbore is shown in Part A. The distance scale in Part B shows that the VSP data span a lateral distance of 1100 meters. The VSP trace data are, in effect, a high quality single-fold seismic cross-section, and since the average distance between traces is approximately 14 meters, the data provide a very detailed horizontal image of the subsurface below the wellbore.

A structural interpretation of the geological section below the deviated borehole made from these data is shown in Part B. Key objectives of this experiment were to locate the position of fault D more accurately, to define the Piper Sands on both the east and the west sides of the fault, and to detect any minor faulting that might exist in this part of the field. All of these objectives can be reasonably achieved with these data. However, any interpretation of the VSP data, including the one in Part B, is subjective and needs to be confirmed with as much supporting evidence as possible.

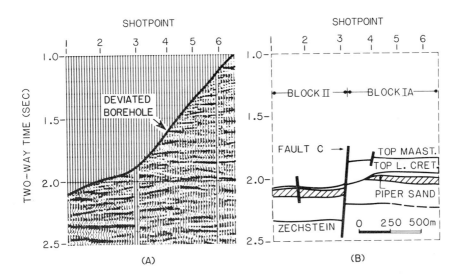

Figure 8-39   (A) Upgoing VSP wavefield recorded in deviated well P35 after applying
deconvolution operators to remove multiple events.   (B) A structural
interpretation suggested by these VSP data. The VSP coverage extends 1300
meters across the reservoir.   The location of fault C is shown in Figure
8-37.   Two data traces at shotpoint 3.0 were too noisy to be accepted as
valid data, so dead traces are inserted.   (After Johnson, et al., 1982,
Copyright SPE-AIME).

A similar VSP profile, recorded in well P35, is shown in Figure 8-39. This well
terminates southwest of the production platform in fault block II, which is downthrown
from block IA by fault C.   The VSP data in this well cover a lateral distance of
approximately 1300 meters with an average spatial separation of approximately 20 meters
between reflection points. A structural interpretation that can be made from these data
is shown in Part B.   The two dead traces at shotpoint 3.0 are included because the two
original traces were too noisy to be considered as valid data. The absence of data at this
shotpoint hinders the interpretation of fault C; however, the smaller faults shown in the
geological cross-section, as well as the geometry of the stratigraphic units, can be seen by
careful inspection of the trace data. The fault shown at shotpoint 1.7 corresponds to the

dashed line fault shown crossing block II in an east-west direction in Figure 8-37. Based on these VSP data, the well was redrilled so that it bottomed 1000 feet north of this east-west fault. Subsequent well data confirmed that this fault created some 45 feet of throw in the lower Piper Sands. A knowledge of the regional and local geology, coupled with confirming surface seismic data and subsurface drilling information, was used to make the specific interpretation shown in Part B. Without these facts, an interpreter could no doubt make a different structural picture of the subsurface from this VSP imagery.

A third VSP, recorded in well P38, is shown in Figure 8-40. The location of seismic Line #1 directly overlays the bottom half of the deviated track of this wellbore in Figure 8-37. Well P38 has the largest stepout distance of any of the seven VSP study wells. It extends almost 3 miles in a southwest direction from the platform, crosses three interpreted major faults (A, B, C), and eventually reaches a true vertical subsea depth of 9400 feet. The measured depth of the well is 18200 feet, thus on the average, the wellbore deviates from vertical by approximately 60 degrees. Interestingly, the geophone tool was lowered to total depth in this well without having to add weight to it, which is one advantage of recording VSP data in a cased well.

The average distance between the VSP traces in Figure 8-40 is approximately 48 meters, which is essentially the same as the trace spacing of typical surface-recorded seismic data. A structural interpretation of the VSP data is given in Part B. The evidence of fault A in these data is significant in that it confirms the existence of the postulated fault separating Blocks II and III. The throw of fault A is almost 1000 feet.

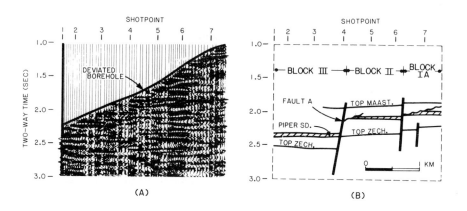

Figure 8-40 (A) Upgoing VSP wavefield recorded in deviated well P38 after applying deconvolution operators to remove multiple events. (B) A structural interpretation suggested by these VSP data. The VSP coverage extends laterally for 3 kilometers. Fault blocks IA, II, and III are shown in Figure 8-37. (After Johnson, et al., 1982, Copyright SPE-AIME).

Perhaps the most important contribution that VSP data made to the interpretation of this field is the improvement in the subsurface imaging that resulted from reprocessing some of the surface-recorded seismic data with VSP derived deconvolution operators. A technique for determining these operators is described in Chapter 5.

Referring to Figure 8-37, two surface seismic lines, Line #1 and Line #2, lie almost directly over the track of well P38. A state-of-the-art migrated version of each of these seismic lines is shown as panel A in Figures 8-41 and 8-42. The location of the bottom of well P38 is shown by the arrow in the left portion of each cross-section. The vertical exploratory well, 15/17-3, shown in Figure 8-37, is also spotted on Line #2.

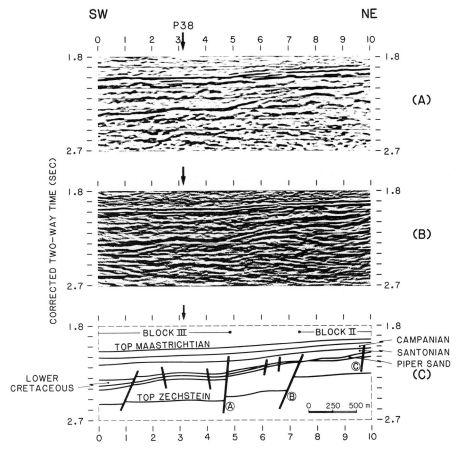

Figure 8-41 (A) Original version of surface seismic Line #1. (B) Surface seismic Line #1 after being reprocessed with deconvolution operators derived from the VSP data recorded in well P38. (C) Structural interpretation made from the reprocessed surface data. Note the distance scale in the lower right corner. The fault picks made from these VSP-reprocessed data do not correspond to those mapped from surface data in Figure 8-37. (After Johnson, et al., 1982, Copyright SPE-AIME).

Deconvolution operators were calculated from the VSP data recorded in the P38 well, via a methodology similar to that described in Chapter 5, and then used to reprocess these two seismic lines. The reprocessed data are displayed in panel B. These reprocessed lines are also migrated with a wave equation algorithm. A comparison of the original data and the reprocessed data shows that the reprocessed versions exhibit better vertical resolution, as well as improved lateral continuity of several important events.

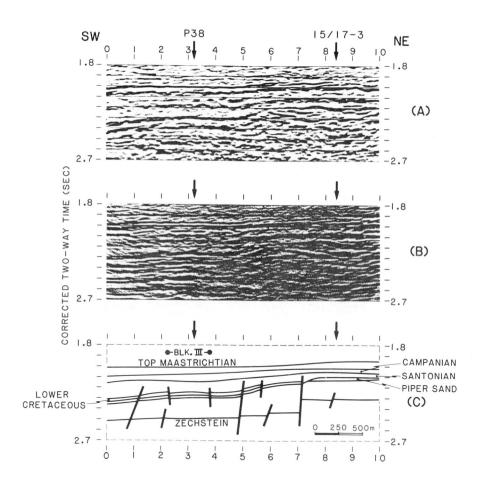

Figure 8-42   (A) Original version of surface seismic Line #2. (B) Surface seismic Line #2 after being reprocessed with deconvolution operators derived from the VSP data recorded in well P38. (C) Structural interpretation made from the reprocessed surface seismic data. Note the distance scale in the lower right corner. The fault picks made from these VSP-reprocessed data do not correspond to those mapped from surface data in Figure 8-37. (After Johnson, et al., 1982, Copyright SPE-AIME).

Fault locations, fault throws, and lateral correlation of several reflectors can now be estimated with greater confidence. These estimates are the critical information needed in order to decide how to properly develop the field.

The new structural interpretation made from each reprocessed line is shown in panel C. These structural cross-sections differ from the original interpretation of this portion of the field in that Blocks II and III are now separated by a zone having distinct dipping events and containing only a few small faults. One hypothesis, that another major fault existed between blocks II and III, was originally considered, but had to be discounted in lieu of these reprocessed data. Note that a fault with considerable throw at the Zechstein level at shotpoint 7 on each line only slightly displaces the Piper Sands. In fact, all faults shown in the reservoir unit now have rather small throws.

As a result of this VSP work, a sizeable hydrocarbon accumulation in Block III can possibly be added to the Piper Field reserves. The VSP results justify that an additional well be drilled to confirm this interpretation (Johnson, et al., private communication).

From the analyses in Figures 8-32 through 8-42, it is obvious that vertical seismic profiles executed in deviated wells are extremely valuable since the recorded data can provide such detailed horizontal and vertical imaging of the subsurface. In addition, the value of VSP data in a deviated well is enhanced by the fact that logging tools are rarely lowered down an uncased, highly deviated borehole because of the high probability that they will jam and result in expensive fishing efforts. Only limited log data, primarily gamma ray counts, can be recorded after a deviated well is cased, but the casing is not a detriment to recording VSP data. Deviated wells are more numerous in marine oil and gas production than are vertical wells, and vertical seismic profiles conducted in these deviated wells are highly recommended as a means for improved mapping of reservoir facies and reservoir seals. The recording of three-component VSP data in deviated wells should also be a valuable way by which the movement of secondary recovery processes can be seismically monitored.

CHAPTER 9

THE FUTURE OF VERTICAL SEISMIC PROFILING

Vertical seismic profiling is now an established exploration technique and reservoir evaluation procedure, but some aspects of the technology are being altered in order to meet stricter data requirements demanded by users and so that a wider range of VSP applications can be provided. For example, several VSP service contractors are now in the process of converting to borehole instrumentation designed specifically for vertical seismic profiling rather than velocity surveying. Some major differences in this second generation of VSP equipment are that the downhole geophone systems have improved locking arms that create better geophone-to-formation coupling, the surface recorders have more channels and larger dynamic range, and the energy sources create more uniform wavelets throughout the duration of a VSP survey. Additional improvements in VSP equipment are also being considered by several geophysical groups, and some of these equipment changes will be described in this chapter.

A significant trend in vertical seismic profiling is that a wider variety of exploration applications of VSP data is now being considered by explorationists. Most active VSP data users are aware that the full potential of vertical seismic profiling has not been realized, and good quality VSP data are now viewed as an avenue by which the basic physics of seismic wave propagation in the earth can be rigorously studied and applied in exploration seismology. As a result, vertical seismic profiling should play a valuable role in the development of surface shear wave measurements, as well as supporting surface 3-D seismic recording and interpretation. New VSP applications in these seismic technologies will no doubt be heavily emphasized in the next few years. Other VSP applications that are more in the realm of research possibilities, but which may lead to practical exploration and production uses, will also be described in this chapter.

Multiple-Depth-Level Geophone Systems

One feature of vertical seismic profiling that discourages some explorationists from conducting VSP surveys is the length of time required to record the data. Generally, economic considerations are the ultimate reason why only a limited amount of time can be reserved for performing a vertical seismic profile. Companies who record VSP data charge several tens of thousands of dollars for their service, and these service charges are roughly proportional to the amount of time needed to record the data. More importantly, if data are recorded while a drill rig is still on site, then drill rig standby charges must also be paid. These standby charges can be quite large, particularly for offshore drill rigs. Too often, the cost of acquiring VSP data cannot be fit into the monies budgeted for a well.

The economics of vertical seismic profiling would be more attractive if the amount of time needed to record data could be reduced. Presently, it can require 24 hours or more to record data in some VSP experiments. Since the cost of VSP data is controlled by the amount of time consumed in executing the measurement more than by any other factor, changes need to be made, either in equipment or in operating procedures, so that the data can be collected in a time frame of 5 or 6 hours, which is commensurate with the time required to record most well logs.

One effective way to reduce VSP field recording time is to record data at more than one depth for each seismic shot. Ideally, one would like to have geophones stationed vertically throughout an entire drilled stratigraphic section so that data from a single seismic shot could be recorded at all depth levels simultaneously. This objective would mean that perhaps as many as 100 geophones would have to be locked in a borehole at closely spaced depth intervals. It is unlikely that a VSP geophone system this large could be fabricated, and certainly it could not be easily installed in a study well even if it existed. However, recording VSP data at even two or three depth levels simultaneously would result in an appreciable reduction of recording time over that now required to collect data with current single-point recording systems.

Multi-level VSP geophone systems with non-retractable locking arms have been used in the Soviet Union for several years; none appear to be available elsewhere. A three-level geophone system, such as will be needed in future VSP work is shown in Figure 9-1. This system has retractable locking arms, not fixed arms. Russian researchers have established that the recording cable above the top geophone of a multi-level system should be slacked when recording data, but that it is not necessary to slack the cable connecting successive geophones (Gal'perin, private communication). This fact simplifies the operation of a multi-level system, when compared to the difficulties of having to slack cable above each individual geophone.

A major problem with a multi-level VSP geophone system is the difficulty that is encountered when one tries to lower a long flexible assembly of geophones to the bottom of a well without them becoming wedged. A jammed geophone tool not only terminates a VSP experiment, but it also jeopardizes a borehole that may represent an investment of millions of dollars. Several possible designs of multi-level geophone systems are being

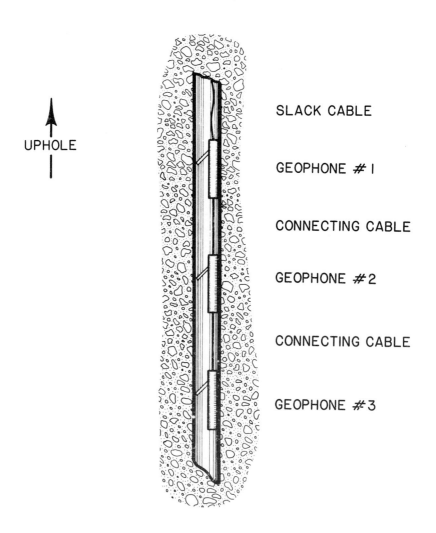

UPHOLE

SLACK CABLE

GEOPHONE #1

CONNECTING CABLE

GEOPHONE #2

CONNECTING CABLE

GEOPHONE #3

Figure 9-1   A multiple-depth-level VSP geophone system.

considered by American and European geophysical companies. The major concern is to make a tool that can be safely lowered into either cased or uncased boreholes. If the system is flexible, like the one shown in Figure 9-1, then accelerometers or some type of motion sensing devices need to be incorporated into each geophone package, and their responses monitored at the surface as the system is lowered down a borehole in order to detect when the relative rates of movement of the individual geophone packages differ, and wedging is imminent.

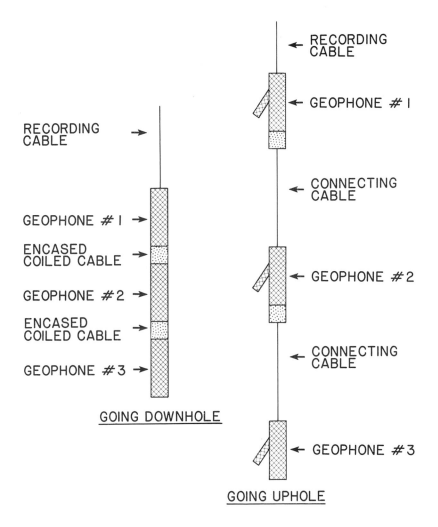

Figure 9-2   A possible design of a multi-level VSP geophone system that minimizes tool-wedging as the device is lowered downhole.

A promising design concept is to fabricate a system that is short and rigid as it goes downhole, but alters to a long, flexible system as it is brought uphole. This concept is shown in Figure 9-2. The latches holding individual geophones together into a single rigid assembly can be released by electrical commands from the surface once the tool is at the deepest recording depth.

One capability provided by a multiple-depth-level geophone system is the ability to record multifold common reflection point VSP data. This field recording option can be demonstrated by the recording geometry shown in Figure 9-3. The sequence of three vertically connected dots shown in ascending order on the right side of the illustration represents the upward movement of a three-depth-level geophone system. The geophone assembly is first locked in the well so that the deepest geophone is at depth Z_1, and the shallowest geophone at depth Z_3. The seismic response generated by one or more shots fired at source location 1 is recorded, and the assembly is then moved up so that the deepest geophone is at depth Z_2, where data are again recorded. This operation is repeated by successively raising and locking the geophone assembly at the indicated depths, Z_1, Z_2, ... Z_6, and recording data at each depth level. The overlapping geophone

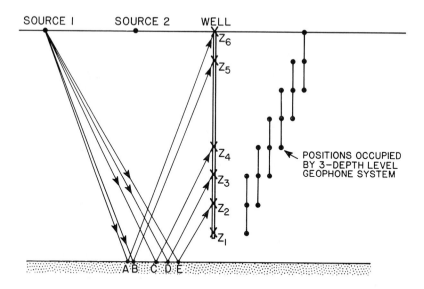

Figure 9-3   A multiple-depth-level geophone system provides a convenient and economical way by which common reflection point VSP data can be recorded. For example, the three-depth-level system shown here can record 3-fold CRP data.

positions drawn on the right show that single fold data are recorded at depth Z_1, two-fold data at Z_2, and three-fold data from Z_3 all the way up to Z_5, where the recording fold decreases to two-fold, and then to one-fold data. As a result, points B and E are imaged by two-fold data, and all points between B and E by three-fold data. The signal-to-noise properties and overall quality of VSP data should improve as a result of this CDP stacking capability, just as do these same aspects of surface reflection data when higher fold surface stacks are created.

A geophone system comprised of N geophone levels can provide N-fold common reflection point data when the surface source remains at location 1 during the data recording. If data from M different source locations are recorded at each of the indicated geophone positions, then the stacking fold can increase to the product of M times N. Some care must be taken when sorting the data traces into CRP gathers. For example, the three-fold image of point D recorded at depth Z_3, when the source is at location 1 must be combined with a second three-fold image of D, which is recorded at a shallower depth when the source moves to location 2.

Measuring VSP Geophone Orientation In Situ

The full potential of vertical seismic profiling can be realized only if three-component data are measured by a VSP geophone tool so that the complete physics of all propagating seismic P-wave and S-wave modes can be studied. Some present geophone systems record satisfactory three-component data, but many applications using these data are difficult to perform because the geophone orientation is unknown when the data are recorded. This equipment deficiency is well known to VSP data users, and corrective steps are being planned by some VSP service companies. Two types of orientation measurement packages that are being considered are shown in Figure 9-4. In uncased boreholes, a combination of north seeking magnetometers and gravity sensitive accelerometers can be used. Adequate systems which employ these components, and which are compatible with VSP geophone packaging requirements, are commercially available.

The more difficult borehole environment for in situ measurements of VSP geophone orientation is a cased hole, where magnetometers become ineffective. Gyroscopic systems are one way to obtain orientation measurements in cased holes, but mechanical gyroscopes appear to be too fragile and require too much stabilization time to be considered a satisfactory system for VSP surveys requiring several hours of continuous data collection. The newly developed laser gyroscope is a possible solution to this dilemma.

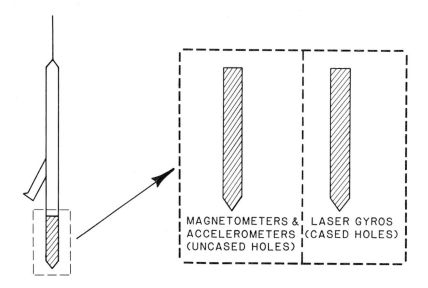

Figure 9-4   Orientation equipment needed for three-component VSP data recording.

Compared to mechanical gyroscopes, some advantages of laser gyros are that they have no mass imbalance and no structural instabilities, they are low cost, and they are fully operational within milliseconds after start-up. The basic elements of a laser gyro are shown in Figure 9-5. The instrument consists of two optical oscillators (i.e. lasers) of identical length. These oscillators can be a system of reflecting mirrors or optical waveguides. Energy travels clockwise in one oscillator and counterclockwise in the second in such a way that both oscillators enclose the same area, A. The shape of the area circumscribed by the laser radiation is arbitrary; it can be a triangle, a polygon, or a circle as shown in this sketch.

Although the clockwise and counterclockwise lasing paths have identical lengths when the gyro is stationary, the paths are unequal when the area A is rotated in inertial space. This path difference is an effect of general relativity and is not caused by Doppler shifting. Killpatrick (1967) shows that the change in path length, ΔL, is given by

$$\Delta L = \frac{4A\omega}{c} \tag{1}$$

where ω is the inertial rotation rate of A, and c is the speed of light. Because each oscillator length, L, must be an integer multiple of the wavelength, λ, of the laser

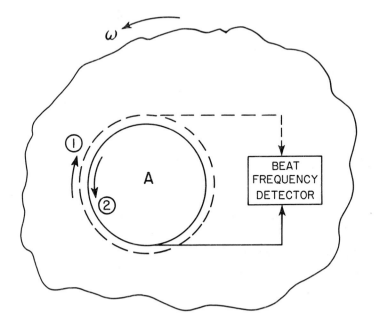

Figure 9-5   The basic elements of a laser gyroscope. Two laser beams having the same frequency, f, travel around an area, A, in opposite directions along paths 1 and 2, which have identical lengths. The enclosed area, A, is shown as a circle, but it can be any arbitrary shape. The angular rotation rate, ω, of the gyroscope is determined by measuring the beat frequency created by summing these two laser beams.

radiation, this equation can be written as

$$\Delta f = \frac{4A\omega}{\lambda L} \qquad\qquad (2)$$

where Δf is the frequency shift of the monochromatic laser radiation created by the rotation. The frequency shift is proportional to both the size of the area circumscribed by the oscillators and to its rotation rate in inertial space. The frequency change, Δf, can be detected by extracting small percentages of each laser beam via mirrored surfaces, and then combining these two radiation samples into a single beam. The beat frequency of this composite beam, which can be measured by photo-electronic circuitry, is Δf. The rotation ω can be continuously and accurately determined once Δf is known.

Specific designs of laser gyroscopes are described by Goldstein, et al. (1978), Ljung (1979), Gamertsfelder and Ljung (1980), Smith and Dorschner (1981), and Anderson, et al. (1981). Some of the basic theory of laser gyroscopes are provided by Killpatrick (1967) and Hecht (1982). Presently, no public announcement has been made of a commercial laser gyroscope system miniaturized to the scale needed for VSP geophone packaging. One major concern is that considerable lasing energy can be lost when optical waveguides are bent to fit inside small diameter VSP geophone tools.

Multi-Axis Geophone Tools

Only a modest amount of three-component VSP data has been illustrated in this text because non-Soviet geophysicists are just beginning to acquire and analyze data recorded with multi-axis borehole geophones. The bulk of VSP data analysis outside the Soviet Union has involved data recorded by geophones oriented only along the longitudinal axis of a geophone tool, or by single-axis gimbaled geophones. Only a few examples of multi-component VSP data have been publicly discussed, and even fewer have been published in English language geophysical literature (Lash, 1980, 1982; Omnes, 1980; DiSiena, et al., 1981; Parrott, 1980; Toksoz, et al., 1980; Turpening, et al., 1980; Wuenschel, 1976). Many three-component VSP recording experiments performed by American geophysicists have failed, or been marginally successful, because the first-generation borehole geophones used to record these data did not adequately couple non-vertical geophone elements to the borehole wall. This equipment deficiency is due in large part to the fact that these geophone packages were originally built only for the purpose of collecting velocity check-shot data via vertically oriented geophone elements, and in the rush to investigate the various possibilities and applications of vertical seismic profiling, the devices were not modified so that they properly coupled non-vertical geophones to the formation.

Several geophysical research laboratories and VSP service contractors are now building improved borehole geophones that should record three-component data of sufficient quality for exploration and production applications. These new tool designs have focused on constructing locking arm mechanisms that create large laterally directed locking forces, using two locking arms rather than one, developing shorter geophone tools, and moving the geophone elements closer to the locking arms. All of these modifications improve the coupling of non-vertical geophones to a formation. Because of the strong interest in developing shear wave seismology into a viable exploration technology, one area of vertical seismic profiling that should certainly be emphasized in the next few years is the continued improvement of multi-axis borehole geophone tools.

Some results of recent field tests of a newly developed multi-axis borehole geophone system are shown in Figures 9-6, 9-7, and 9-8 (B. Seeman, F. Mons, and V. de Montmollin, Schlumberger, private communication). The objectives of these tests were to determine the effects of tool mass and locking force on the quality of VSP data. The test data in these illustrations document some of the effects that these two tool parameters have on the vertical coupling of the tool to a formation. Tests are continuing to determine the effects on horizontal coupling, and hopefully these data will be published soon.

VSP data recorded at a depth of 1390 meters in an uncased borehole are shown in Figure 9-6. The lithology at this depth is shale, which will be called a soft formation. For plotting purposes, the traces are adjusted so that they have the same maximum amplitude; the relative trace-to-trace maximum amplitudes are noted along the right margin. Two sets of trace data are shown. The top group of traces represents the response of the basic geophone tool which has a mass of 60 kg. The bottom traces were recorded at the same depth with an additional 40 kg mass added to the tool. The surface energy source used in this experiment was an airgun operated in a water-filled pit. The source behavior was carefully monitored to insure that consistent shot wavelets were generated for the test data presented here.

The amount of force applied to press the tool against the borehole wall cannot be directly measured until further modifications are made. The force depends on the angle of contact between the pivoting locking arm and the formation and on the amount of current applied to the motor that extends the arm. The contact angle with the soft formation is unknown, but the motor current is measurable and is plotted along the vertical axis of each display. Eventually, a pressure transducer will be installed on the pad that contacts the formation in order to obtain direct measurements of the locking force at each recording level. The tool was unclamped and then reclamped to the formation each time the motor current was increased. The tool had to be brought to the surface in order to add additional mass to it, and then lowered back to the same recording depth. As a result, there may be slight variations in the depth position of the tool between the top and bottom set of recordings. Any wavelet effects caused by the tool not being at the same depth level are ignored in this analysis.

The amplitude of the P-wave first arrival appears to increase as the motor current (locking force) increases from 200 mA to 400 mA. Indeed, the amplitude does increase somewhat, but the increase is not as great as shown in this display because different gains are used to plot each wiggle trace. (The actual increase in P-wave amplitude is shown in

408

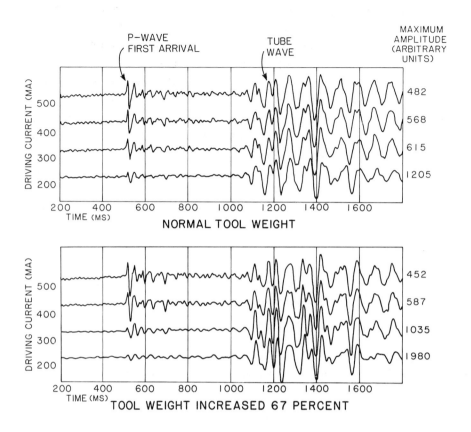

Figure 9-6   Waveforms recorded by a vertically oriented geophone in a soft formation at a depth of 1390 meters for various tool weights and locking arm forces. The locking force increases as the motor driving current increases. (Courtesy of Schlumberger).

Figure 9-8, Part C). However, this trace plotting technique does correctly show that the ratio of the amplitude of the P-wave to the amplitude of the tube wave becomes larger as the locking force increases, which is a desirable result. This data behavior is further evidence that a geophone that is firmly bonded to a formation tends to reject tube waves, and that a loosely bonded geophone emphasizes tube waves.

Several tests of this nature were made in both soft and hard formations for various combinations of tool mass and locking force. The effect of locking force on the frequency bandwidth of VSP data recorded by a vertically oriented geophone in a soft formation is shown by the spectral plots of the P-wave first arrivals in Figure 9-7, Part A. The tool mass was the same in all of the tests. The frequency content of the data between 50 and 100 Hertz significantly increases as the locking force is increased by incrementing the drive motor current from 200 mA to 600 mA. Thus, the wavelet shape recorded in a soft formation is strongly affected by the magnitude of the coupling force.

An important difference in VSP data behavior occurs if the formation is hard, rather than soft. As shown in Part B, the bandwidth of the P-wave direct arrival recorded in a hard carbonate formation, which is at a depth of 1340 meters in this well, changes very little as the locking force increases between the limits generated by motor currents of 200 mA and 600 mA. The spectra calculated for several values of driving current overlay almost exactly, so only one spectrum is plotted. The P-wave spectrum in the hard formation is also much whiter than the spectrum observed in a soft formation, regardless of the amount of locking force used to bond the tool to the soft formation. Obviously, the effect of locking force should be considered when investigating body wave attenuation behavior with VSP data, or when associating any frequency variations of VSP data with the mechanical properties of rocks.

The influences of tool mass and locking force on other dynamic properties of VSP data recorded in a soft formation are summarized in Figure 9-8. Each curve in these displays is an average of several tests. One can conclude from the data in Part A, that the spectral bandwidth of P-wave signals increases as tool mass decreases and as the locking force increases. Thus, an ideal VSP geophone tool should be lightweight and have a locking arm that generates a large laterally directed locking force.

Since tube waves are commonly occurring noise modes that camouflage body wave signals, these test data were examined to determine how tube wave and body wave amplitudes vary relative to each other as tool mass and locking force are altered. These measurements are shown in Part B. For a vertically oriented geophone, the tube wave amplitude decreases relative to the amplitude of a P-wave event as the tool mass decreases and as the locking force increases. This same behavior is shown by the wiggle trace data plotted in Figure 9-6. Russian investigators have also noted that tube wave amplitudes diminish as clamping force increases (Gal'perin, 1974, p. 21). These results again suggest that lightweight geophone tools which have robust locking mechanisms are desirable.

Even though the ratio of P-wave to tube wave amplitudes improves as the tool mass decreases, the absolute amplitude of compressional body waves actually increases as tool mass increases (Part C). However, for a given tool mass, the P-wave amplitudes are not significantly affected by variations in locking force for motor currents greater than 300 mA, which is a result not obvious from the data shown in Figure 9-6.

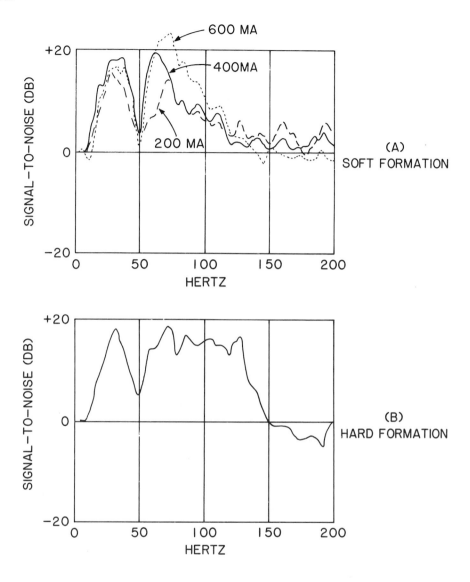

Figure 9-7 (A) Effect of locking force on the signal-to-noise spectrum of a P-wave first arrival wavelet as measured by a vertically oriented geophone in a soft formation. The indicated current is the number of milliamps supplied to the motor that extends the locking arm. Recording depth is 1390 meters.

(B) Effect of locking force on the signal-to-noise spectrum of a P-wave first arrival wavelet as measured by a vertically oriented geophone in a hard formation. Only one spectrum is shown since the spectrum changed very little as the motor current was varied. Recording depth is 1340 meters. (Courtesy of Schlumberger).

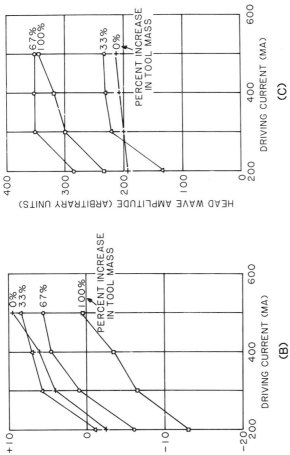

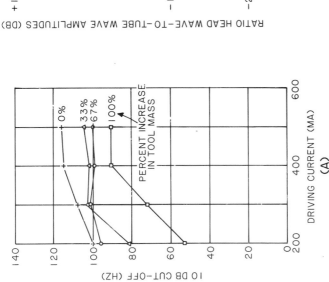

Figure 9-8   Effect of tool weight and locking force on some dynamic properties of VSP data recorded in a soft formation. (A) High frequency cutoff of the P-wave first arrivals increases as tool mass decreases and locking force increases. (B) Ratio of amplitude of P-wave first arrival to amplitude of tube wave becomes more favorable as tool mass decreases and locking force increases. (C) Amplitude of P-wave first arrival increases as tool mass increases but changes only slightly as locking force increases.   Data were measured by a vertically oriented geophone. (Courtesy of Schlumberger).

This type of documentation is vitally needed in order to fully understand the physics of the seismic wavefields that are recorded by downhole geophone systems, and particularly by multi-axis geophones. The effect of tool mass and locking force on the horizontal coupling of a VSP geophone tool now needs to be documented in this same detail. Eventually, tests need to be performed in various formations and in different borehole conditions to confirm whether the relationships shown in these initial tests are local behaviors or are valid generalizations that apply in a wide range of geological conditions.

Converted Shear Waves

The most common elastic waves used in exploration seismology are compressional waves. Onshore, seismic energy sources are often chosen on the basis of whether or not a large portion of their energy output propagates as compressional waves rather than as shear waves. In marine exploration, all energy initially created by marine seismic sources is, of course, compressional energy. Likewise, seismic receiving devices are often deliberately designed to preferentially react to compressional waves and reject shear waves. For example, vertically oriented, spike-planted surface geophones reject SH shear modes, and hydrophones reject all shear modes. In addition, seismic data processing usually accentuates compressional reflections and attenuates shear reflections by means of the compressional velocity functions used to stack and migrate recorded data.

Geophysicists have long recognized that a considerable amount of compressional wave energy is converted into shear wave modes whenever a propagating compressional wave encounters a reflecting interface at an oblique angle of incidence. These converted shear wave modes are potentially just as valuable for interpreting subsurface geological conditions as are shear modes deliberately created by shear wave energy sources. Converted shear modes are particularly valuable seismic measurements when used in concert with compressional data to interpret elastic constants of rocks or to predict the types of pore fluids in rock units. One special virtue of converted shear waves is that they can be created in a marine environment where shear wave energy sources cannot function.

Some facts that must be known in order to interpret converted shear wave seismic cross-sections are:

1. A knowledge of the depths of all interfaces where compressional-to-shear conversion occurs.

2.   Accurate estimates of compressional and shear wave velocities throughout a prospective stratigraphic section.

3.   Recognition of the arrival times of converted shear wave modes at the surface.

4.   Knowledge of the reflection polarity and strength associated with both compressional and converted shear wave events.

All of this information can be obtained via a vertical seismic profile that records three-component subsurface particle motions.   VSP wells will not always be available in exploration areas where shear wave data are needed; however, some reliance on three-component vertical seismic profiling seems unavoidable if converted shear wave seismology is to be developed as an exploration technique.

VSP data that demonstrate P-wave to S-wave conversions are illustrated in Figure 9-9 (Courtesy of Total and B. Seeman, F. Mons, V. de Montmollin, Schlumberger, private communication).   These data were recorded with a geophone tool similar to that which recorded the test data illustrated in Figures 9-6 through 9-8.   The internal geophone geometry was a triaxial, 54 degree arrangement with a fourth vertically oriented geophone element.   The surface energy source was a compressional vibrator offset 500 meters from the wellhead.   Thus, downgoing events intersect horizontal interfaces at rather large angles of incidence between the recording depths of 280 and 605 meters.

The data in the top panel of Figure 9-9 are the result of mathematically rotating the responses of the triaxial geophone system so that they represent the output of a single geophone oriented along the raypath of the P-wave first arrival at each recording level (see Equations 23 and 24, Chapter 7; DiSiena, et al., 1981).   The data in the bottom panel represent the response that a geophone would record if it were positioned in a vertical plane containing the P-wave first arrival raypath and then oriented in this plane so that it is normal to the P-wave raypath.   These data thus contain the full response of those downgoing SV modes which travel along the same raypath as does the P-wave direct arrival, partial responses of SV modes which arrive at the triaxial geophone arrangement along raypaths that differ from the P-wave raypath (which will be true for the majority of downgoing SV modes since they do not refract the same as P-wave modes do), and partial responses of later arriving downgoing and upgoing P-wave events whose raypaths intersect the geophone assembly at various angles of inclination.

A major impedance change occurs at depth Z_1 (415 meters) in this well.   The data show an obvious P-to-SV conversion at A, since the first downgoing SV mode can be traced upward to intersect the downgoing P-wave first arrival at this point.   The second SV mode can be traced upward to 280 meters, and one cannot be sure whether it is an SV mode

414

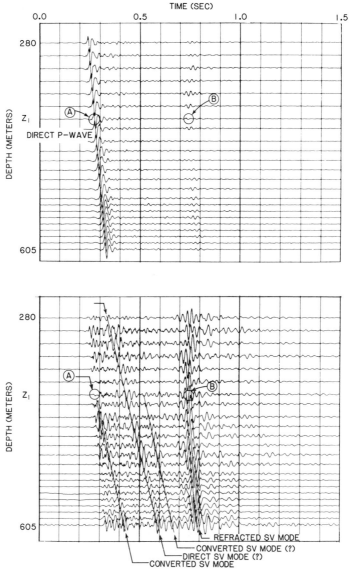

Figure 9-9   These data were recorded with a triaxial, 54 degree VSP geophone system and then mathematically rotated in space to emphasize selected wave modes. (Top) Particle velocity motion orientated along the raypath associated with the P-wave first arrival. This motion defines downgoing P-wave first arrivals and downgoing P-wave multiples generated in the shallow part of the stratigraphic section. (Bottom) Particle motion normal to the P-wave raypath and lying in the vertical plane containing the P-wave raypath. Downgoing SV events will be emphasized by this response. Some obliquely traveling P-wave events may occur at later recording times. (Courtesy of Total and Schlumberger).

created directly at the source, or a P-SV conversion occurring shallower in the section.

The event between 0.7 and 0.8 seconds also appears to have been created at the Z_1 boundary since it reaches the geophone at depth Z_1 before it reaches the geophone positions above or below Z_1. Also, the moveout behavior of the event is symmetrical about depth Z_1. There is some uncertainty as to exactly what type of wave mode this event is, since no downgoing disturbance can be seen in these displays which could have obviously generated it. Because the event appears to travel horizontally and have most of its particle displacement occurring in the SV plane, it will be assumed to be a refracted SV mode. If a refracted SV mode travels horizontally into the vertical geophone array, its apparent velocity will be much higher than the velocity of a downgoing SV mode. The shortest travel path from the refraction point to a recording geophone should occur along the horizontal refracting interface, which it does in this case since the apex, B, of the event is located at depth Z_1. Since a horizontally traveling SV refraction event is not perpendicular to the raypath of the downgoing P-wave first arrival, not all of its particle displacement is in the SV plane defined in the bottom panel. As a result, portions of this SV mode can be seen in the geophone responses plotted in the top panel. The geology immediately around this well is not provided, therefore, other interpretations of this event may be possible.

Measurements like these will be essential in order to properly interpret surface-recorded P-wave and P-SV data. The depths at which P-SV conversions occur can certainly be defined with good quality VSP data, and careful amplitude studies of the incident P-wave and the transmitted P and SV modes can determine P-SV reflection coefficients. Another VSP example of a P-to-SV conversion that demonstrates these possibilities has been published by Krug, et al. (1981). SV velocities can also be measured in these types of experiments so that proper moveout corrections can be made when stacking surface-recorded data for the purpose of revealing SV events. Obviously, good quality three-component VSP data will be essential if converted shear wave seismology is to be developed into a viable exploration tool. A study by Graves (1979) provides some insights into possible P-SV exploration techniques.

Development Drilling and Reservoir Monitoring

Since a vertical seismic profile cannot be performed unless a borehole exists, it should follow that the preponderance of VSP data would be recorded where wells are most heavily concentrated. Development wells must be carefully positioned in a reservoir unit in order to achieve optimum production, and particularly so in offshore drilling, where platforms have a restricted number of well slots available. Thus, vertical seismic profiling should assume an important role in development drilling. In many reservoirs, development wells often create considerable confusion about the stratigraphic relationships and facies changes within a reservoir interval rather than providing definitive answers or confirmations of initial reservoir models. High resolution VSP data should thus become a standard data measurement in development wells so that improved reservoir models can be constructed.

As worldwide hydrocarbon exploration areas diminish, more and more monies are being directed toward improving secondary and tertiary reservoir recovery processes in established oil fields. Any drive mechanism used to recover immobile oil must reach all sectors of a reservoir unit and push extracted hydrocarbons toward preselected producing wells. The success or failure of these recovery techniques depends on accurately monitoring the horizontal and vertical paths taken by the propagating recovery fluid, and taking corrective action when the drive does not extend into appropriate regions of a reservoir. Vertical seismic profiles, employing a source-receiver geometry that allows high resolution horizontal stacks to be constructed from the VSP data, can be an effective way to map the progress of any recovery process that alters the acoustical reflection properties of reservoir rocks. For optimum interpretation of the efficiency of a recovery process, both compressional and shear wave modes should be recorded. There are presently no case histories that establish the value of vertical seismic profiling in secondary recovery, but it seems that the modest expense of acquiring VSP data, compared with the overall cost of some secondary and tertiary recovery processes, warrents testing the concept.

Three-Dimensional Surface Transmission Seismology

Historically, the majority of surface-recorded seismic data used in exploration has been measurements of reflected seismic wavefronts. Consequently, surface seismic exploration techniques have largely been efforts to extract subsurface geological information from reflected seismic signals. The equally valuable question, "What information about subsurface geology is contained in transmitted seismic wavefields?", has been largely ignored.

An adaptation of vertical seismic profiling that may allow areal recording of seismic transmission data is shown in Figure 9-10. This geometry reverses the typical positions of the source and receiver used in vertical seismic profiling; i.e., the source is in the

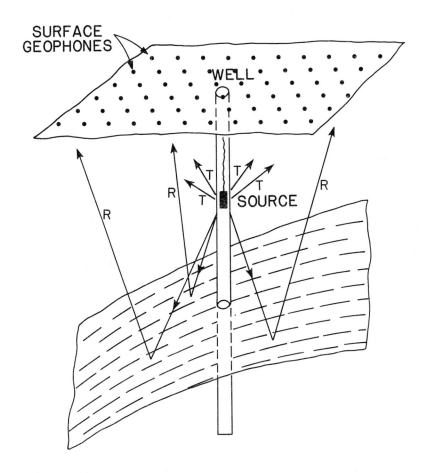

Figure 9-10 Extending VSP concepts to surface areal coverage of geophones for 3-D seismic transmission studies of the subsurface.

borehole and the geophone is on the earth's surface. An experiment of this nature has been reported by Balch and Lee (1982c). However, only 48 recording channels were used in their experiment, whereas, an extensive areal surface coverage of geophones employing more than 48 geophone groups is suggested in Figure 9-10. This recording geometry allows the transmitted compressional and shear wavefields traveling directly from the source to the geophone positions to be recorded. These transmitted raypaths, labeled "T", create a three-dimensional tomographic image of the stratigraphic section between the source and the surface. In addition, reflected rays from interfaces below the source position will be recorded, as shown by the raypaths labeled "R". Some researchers think that this type of seismic tomography can be developed into a practical reservoir imaging technique. VSP studies of this nature should be tried so that the limits and practicality of the technique can be established.

Interpretive Processing of Surface-Recorded Reflection Data

VSP data are widely used to interpret surface-recorded reflection data after the surface data have been processed into a final seismic display, but VSP data are seldom used to optimize the procedures for processing surface-recorded data. This situation is unfortunate, but it should occur less in future seismic exploration. VSP data provide many parameters which describe real earth seismic wave propagation and which can thus be used to improve the processing of surface-recorded reflection data near a VSP well. For instance, VSP data allow a seismic processor to work with just the downgoing wavefield in order to design deconvolution operators that attenuate upgoing multiples. VSP data also provide a means by which more precise relative reflection amplitudes can be extracted from surface records near a VSP well since incident, transmitted, and reflected wavelets are captured at each interface penetrated by the well. In addition, VSP measurements indicate what frequency filters should be applied in certain reflection time windows, since frequency attenuation can be determined from spectral studies of downgoing and upgoing VSP wavelets. These processing guidelines, plus others that can be provided by VSP data, are valuable when processing any surface-recorded data, but will be particularly important when developing shear wave processing procedures. Good quality shallow VSP data should also be sought in areas where near-surface static corrections or strong near-surface multiples are a problem, so that processing procedures can be developed to remove these detrimental effects from surface-recorded data.

On-Site VSP Data Processing

Whenever VSP data are recorded in a well for the explicit purpose of acquiring information that is to be used to decide how that well is to be completed (e.g., should the wellbore be deviated in order to intersect a target that has been missed), then it is essential that these data be converted into display formats suitable for making the required interpretation as soon as possible after they are recorded. Any drilling delay caused by having to wait for lengthy VSP data processing to be completed increases the cost of a well, and perhaps more importantly, increases the likelihood that a borehole may be lost or seriously damaged. The time required to transport VSP data by common land and air carriers from a well site to a data processing center usually represents a large portion of the total time delay that occurs before VSP data can be used to influence a drilling decision. Satellite transmission of data to a distant processing center, where powerful computers and a complete suite of software are available, is a more attractive way to achieve quick processing turnaround. Interfacing satellite links into VSP data acquisition and processing will be a necessity if the full potential of VSP technology is to be realized.

Instead of transporting VSP field data to a processing center, which may be hundreds of miles away, an alternative solution is to transport a data processing center to the well site. Typical VSP data sets are not large in terms of the number of traces and sample points involved. Likewise, many VSP processing requirements make modest computational demands that do not require large computers. Thus, most VSP processing procedures can be performed by mini-computers, which can be packaged in transportable containers that can be carried to either onshore or marine drill sites. However, VSP service companies are presently not so much concerned about the technical problems of assembling a computer system at a work site as they are that properly trained people cannot be provided to operate such systems.

A modest amount of on-site VSP data processing can be justified from a field data quality control point of view, even in cases where immediate processing turnaround is not needed in order to assist a drilling decision. Unfortunately, quality control people responsible for acquiring high quality VSP field data are usually handicapped while in the field by the limitation of typical field systems being able to display only the last recorded VSP trace. Better quality control decisions can be made if one can see the total accumulated data at any time during a VSP experiment, and if the data can be numerically manipulated, when necessary, in order to verify that an experiment's objectives are being accomplished. Both of these requirements can be met if some type of data processing capability exists at a well site. Such a processing system would have to perform only simple procedures such as editing, summing, time-shifting, and frequency filtering of data stored on disk or tape, plus make hard copy plots of the data in wiggle trace form. Landgren and Grubbs (1983) describe one system which performs these data manipulations, and also provides a velocity filtering capability.

Day-to-Day Exploration, Drilling, and Production Problems

Some of the preceding sections of this chapter describe rather new frontiers in exploration seismology where vertical seismic profiling can play a valuable supportive role. However, the dominant use of vertical seismic profiling should continue to be to provide solutions to the problems that explorationists and engineers encounter in their day-to-day work of searching for energy reserves and investigating the physics of the earth. These problems are usually one of the following:

- How can stratigraphic boundaries be accurately identified in surface-recorded seismic data?

- What is the correct subsurface structural picture around a borehole?

- Where are the boundaries of a productive reservoir unit located?

- What rock conditions exist below the bottom of a well?

- How can multiple reflections be better removed from surface-recorded seismic data?

- How can the subsurface conditions extending laterally from a well be imaged with better resolution?

The ability of vertical seismic profiling to address these issues, plus others of similar nature, are discussed in Chapters 7 and 8. The realization that VSP data can provide new insights that allow these commonly occurring questions to be answered with more rigor, better resolution, and increased confidence should insure a continued aggressive development of all aspects of vertical seismic profiling.

REFERENCES

This reference section is rather exhaustive, and some entries are listed that are not noted in the text. Technical articles in the English language are emphasized. There has been no attempt to document the large amount of vertical seismic profiling information that exists in Soviet geophysical literature.

A few of the references will be difficult to obtain, since they existed only for a limited time as preprints distributed at specific technical meetings and were never published in a geophysical journal. Even so, they are still listed so that their authors can be credited for contributing to some aspect of VSP technology, and in the hope that interested readers will know some of these authors and contact them for further information.

A review of United States patent literature pertaining to vertical seismic profiling equipment, or to any type of VSP field procedure, is also included. Some non-U.S. patents are listed, but no extensive effort was made to search out all non-U.S. patents on VSP technology. It should be emphasized that a patent's title often gives no clue about the technical claims contained in the patent. For example, the titles, SEISMIC PROSPECTING SYSTEM or SEISMIC EXPLORATION METHOD, occur often in patent literature, and such patents have to be carefully examined in order to determine if some aspect of vertical seismic profiling is involved in their claims. Hopefully, the patents listed here can save much future patent search time by those wanting to know the patent art in vertical seismic profiling.

Aki, K., Fehler, M., Aamodt, R. L., Albright, J. N., Potter, R. M., Pearson, C. M., and Tester, J. W., 1982, Interpretation of seismic data from hydraulic fracturing experiments at the Fenton Hill, New Mexico, hot dry rock geothermal site: J. Geophy. Res., v. 87 (B2), p. 936-944.

Albright, J. N. and Pearson, C. F., 1980, Location of hydraulic fractures using microseismic techniques: Paper SPE 9509, 55th Annual Fall Tech. Conf. and Exhib. of Soc. Petrol. Eng. of AIME, Dallas.

Alexander, W. A., 1965, Seismic method of earth exploration: U.S. Patent No. 3,208,549.

Allen, T. O. and Atterbury, J. H. Jr., 1954, Effectiveness of gun perforating: Trans. AIME, v. 201, p. 8-14.

Anderson, D. B., August, R. R., Thompson, D. E., and Yao, S., 1981, Interferometer gyroscope formed on a single plane optical waveguide: U. S. Patent No. 4,273,445.

Anderson, D. L. and Archambeau, C. B., 1964, The anelasticity of the earth: J. Geophy. Res., v. 69, p. 2071-2084.

Angeleri, G. P. and Loinger, E., 1982, Amplitude and phase distortions due to absorption in seismograms and VSP: Paper presented at 44th annual meeting of EAEG.

Annis, M. R. and Monaghan, P. H., 1962, Differential pressure sticking - laboratory studies of friction between steel and mud filter cake: J. Pet. Tech., v. 2, p. 537-543.

Anstey, N. A., 1974, The new seismic interpreter: International Human Resources Development Corporation, Boston, 614 pages.

Anstey, N. A., 1980, Seismic delineation of oil and gas reservoirs using borehole geophones: Gr. Brit. Patents 1,569,581 and 1,569,582. Canadian patents 287,178 and 375,890-7. (According to Seismograph Service Ltd., these patents were assigned to Seismograph Service Ltd., July, 1981.)

Anstey, N. A., 1982, Simple seismics: International Human Resources Development Corporation, Boston, 168 pages. (Chapters 6, 7, and 8 discuss VSP).

Athy, L. F. and Prescott, H. R., 1940, Seismic method of logging boreholes: U.S. Patent No. 2,207,281.

Audet, J. and Garotta, R., 1982, The detection of porosity changes using shear waves: Tech. Paper S34, p. 1799-1825, 51st Annual International Meeting of SEG.

Bailey, J. R., 1977, Continuous bit positioning system: U. S. Patent No. 4,003,017.

Baird, C. and Plum, W. B., 1976, Oil well survey tool: U. S. Patent No. 3,980,986.

Balch, A. H., Lee, M. W., Miller, J. J., and Ryder, R. T., 1980a, The use of vertical seismic profiles and surface seismic profiles to investigate the distribution of aquifers in the Madison Group and Red River Formation, Powder River Basin, Wyoming-Montana: Preprint SPE 9312, 55th Annual Fall Tech. Conf. and Exhib. of the Soc. of Petrol. Eng. of AIME, Dallas.

Balch, A. H., Lee, M. W., and Muller, D. C., 1980b, A vertical seismic profiling experiment to determine depth and dip of the Paleozoic surface at drill hole U10bd, Nevada test site, Nevada: USGS Open-File Report 80-847.

Balch, A. H., Lee, M. W., Miller, J. J., and Ryder, R. T., 1981a, Seismic amplitude anomalies associated with thick First Leo sandstone lenses, eastern Powder River basin, Wyoming: Geophysics, v. 46, p. 1519-1527.

Balch, A. H., Miller, J. J., Lee, M. W., and Ryder, R. T., 1981b, Processed and interpreted U. S. Geological Survey seismic reflection profile and vertical seismic profile, Powder River and Custer counties, Montana: USGS Chart OC-108.

Balch, A. H., Lee, M. W., Miller, J. J., and Ryder, R. T., 1982a, The use of vertical seismic profiles in seismic investigations of the earth: Geophysics, v. 47, p. 906-918.

Balch, A. H., 1982b, A seismic stratigraphic investigation of the Madison Group and Red River Formation and its implications for groundwater exploration in the Powder River Basin, Montana-Wyoming: Part I - Vertical seismic profiles and stratigraphic framework: U.S. Geological Survey Professional Paper. (Several chapters on methods, acquisition, processing, and interpretation of vertical seismic profiles authored by R. T. Ryder, M. W. Lee, and R. M. Turpening. A. H. Balch editor. In press.)

Balch, A. H. and Lee, M. W., 1982c, Some considerations on the use of downhole sources in vertical seismic profiles: Paper presented at 35th Annual SEG Midwestern Exploration Meeting.

Bardeen, T. and Williams, R. W., 1955, Pressure-sensitive deep well seismograph detector: U.S. Patent No. 2,717,369.

Barton, D. C., 1929, The seismic method of mapping geologic structure: Geophy. Prosp. (Amer. Inst. Min. and Mat. Eng.), v. 1, p. 572-624.

Bazhaw, W. O., 1955, Method of deep well surveying: U.S. Patent No. 2,718,930.

Becquey, M., Lavergne, M., and Willm, C., 1979, Acoustic impedance logs computed from seismic traces: Geophysics, v. 44, p. 1485-1501.

Bednar, J. B., 1982a, Applications of median filtering to deconvolution, pulse estimation, and statistical editing of seismic data: Paper presented at 35th Annual SEG Midwestern Exploration Meeting.

Bednar, J. B., 1982b, Three robust procedures and their application to deconvolution, pulse estimation, and statistical editing of seismic data: Paper S6.8, 52nd Annual International Meeting of SEG, Technical Program Abstracts, p. 83-84.

Beers, R. F., 1941a, Method of and means for analyzing and determining the geological strata below the surface of the earth: U.S. Patent No. 2,244,484.

Beers, R. F., 1941b, Means for analyzing and determining geological strata: U.S. Patent No. 2,249,108.

Bernhardt, T. and Peacock, J. H., 1978, Encoding techniques for the Vibroseis system: Geophys. Prosp., v. 26, p. 184-193.

Berryman, J. G., 1979, Long-wave elastic anisotropy in transversely isotropic media: Geophysics, v. 44, p. 896-917.

Beynet, P. A., Farr, J. B., and Pottorf, N., 1977, Downhole seismic source: U.S. Patent No. 4,040,003.

Biot, M. A., 1952, Propagation of elastic waves in a cylindrical bore containing a fluid: J. Appl. Physics, v. 23, p. 997-1005.

Blair, D. P., 1982, Dynamic modeling of in-hole mounts for seismic detectors: Geophys. Jour. Roy. Astron. Soc., v. 69, p. 803-817.

Blanchard, A., 1948, Apparatus for producing pressure pulses in bore holes: U.S. Patent No. 2,451,797.

Bois, P., LaPorte, M., LaVergne, M., and Thomas, G., 1971, Computerized determination of seismic velocities between well shafts: Geophys. Prosp., v. 19, p. 42-83. (In French).

Bois, P., LaPorte, M., LaVerne, M., and Thomas, G., 1972, Well-to-well seismic measurements: Geophysics, v. 37, p. 471-480.

Brekhovskikh, L. M., 1960, Waves in layered media: Academic Press, New York, p. 87-100.

Brewer, H. L. and Holtzscherer, J., 1958, Results of subsurface investigations using seismic detectors and deep bore holes: Geophys. Prosp., v. 6, p. 81-100.

Burkhard, N. R., 1980, Resolution and error of the back projection technique algorithm for geophysical tomography: Report UCRL-52984, Lawrence Livermore Laboratory, 60 pages.

Butler, D. K. and Curro, J. R., Jr., 1981, Crosshole seismic testing - procedures and pitfalls: Geophysics, v. 46, p. 23-29.

Byun, B. S., 1982, Seismic parameters for media with elliptical velocity dependencies: Geophysics, v. 47, p. 1621-1626.

Carlson, R. C., Stearns, R. T., Berens, H. B., and Hearst, J. R., 1968, High-resolution seismic uphole surveys at the Lawrence Radiation Laboratory: Geophysics, v. 33, p. 78-87.

Chelminski, S. V., 1974, Displaceable diaphram structures for use in seismic inpulse transmission: U.S. Patent No. 3,800,907.

Chelminski, S. V., 1976, Pressurized gas discharging apparatus for use as a down-bore seismic impulse source: U.S. Patent No. 3,997,021.

Chelminski, S. V., 1978, Seismic land source: U.S. Patent No. 4,108,271.

Cheng, C. H. and Toksoz, M. N., 1981a, Elastic wave propagation in a fluid-filled borehole and synthetic acoustic logs: Geophysics, v. 46, p. 1042-1053.

Cheng, C. H. and Toksoz, M. N., 1981b, Tube wave propagation and attenuation in a borehole: Paper presented at Massachusetts Institute of Technology Industrial Liaison Program Symposium, Houston.

Cheng, C. H., Toksoz, M. N., and Willis, M. E., 1981c, Velocity and attenuation from full waveform acoustic logs: Paper O, Trans. SPWLA 22nd Annual Logging Symposium, Vol. I.

Cheng, C. H., Keho, T., and Toksoz, M. N., 1982a, Analysis of tube wave data in shear wave VSP: Paper S12.6, 52nd Annual International Meeting of SEG, Technical Program Abstracts, p. 161-162.

Cheng, C. H., Toksoz, M. N., and Willis, M. E., 1982b, Determination of in situ attenuation from full waveform acoustic logs: Jour. Geophy. Res., v. 87, p. 5477-5484.

Cheng, C. H. and Toksoz, M. N., 1982c, Generation, propagation and analysis of tube waves in a borehole: Paper P, Trans. SPWLA 23rd Annual Logging Symposium, Vol. I.

Cherry, J. T. and Waters, K. H., 1968, Shear-wave recording using continuous signal methods, part I, early development: Geophysics, v. 33, p. 229-239.

Cholet, J. and Pauc, A., 1980, Device for generating seismic waves by striking a mass against a target member: U.S. Patent No. 4,205,731.

Chun, J., Stone, D. G., and Jacewitz, C. A., 1982, Extrapolation and interpolation of VSP data: Seismograph Service Companies Report, Tulsa, Oklahoma, 26 pages.

Clifford, E. L., Redding, V. L., and Ording, J. R., 1958, Method of surveying a borehole: U. S. Patent No. 2,842,220.

Cowles, C. S., 1981, Method and means for measuring and identifying up-and-down travelling waves in underground formations: Gr. Brit. Patent 1,584,503.

Crampin, S., McGonigle, R., and Bamford, D., 1980, Estimating crack parameters from observations of P-wave velocity anisotropy: Geophysics, v. 45, p. 345-360.

Crawford, J. M., Doty, W. E. N., and Lee, M. R., 1960, Continuous signal seismograph: Geophysics, v. 25, p. 95-105.

Cunningham, A. B., 1979, Some alternate vibrator signals: Geophysics, v. 44, p. 1901-1921.

Deeming, T., 1979, Synthetic seismograms and vertical seismic profiles: Paper presented at 49th Annual International Meeting of SEG.

Demidenko, Yu. B., 1969, Vertical seismic profiling: International Geology Review, v. 11, p. 803-824.

Dennison, A. T., 1960, The response of velocity-sensitive well geophones: Geophys. Prosp., v. 8, p. 68-84.

Der, Z. A., 1970, Some data processing results for a vertical array of triaxial seismometers: Geophysics, v. 35, p. 337-343.

Diedrich, R. P., 1981, The effect of fractures on the compressional wave velocity of Paleozoic carbonate rock: M.S. Thesis, Bowling Green State University.

Dines, K. A. and Lytle, R. J., 1979, Computerized geophysical tomography: Proc. IEEE, v. 67, p. 471-480.

DiSiena, J. P., Byun, B. S., and Fix, J. E., 1980, Vertical seismic profiling - a processing analysis case study: Paper R-19, 50th Annual International Meeting of SEG.

DiSiena, J. P., Gaiser, J. E., and Corrigan, D., 1981, Three-component vertical seismic profiles - orientation of horizontal components for shear wave analysis: Tech. Paper S5.4, p. 1990-2011, 51st Annual International Meeting of SEG.

DiSiena, J. P. and Gaiser, J. E., 1983, Marine vertical seismic profiling: Paper OTC 4541, Offshore Technology Conference, Houston, Texas, p. 245-252.

Dix, C. H., 1939, The interpretation of well-shot data (Part I): Geophysics, v. 4, p. 24-32.

Dix, C. H., 1945, The interpretation of well-shot data (Part II): Geophysics, v. 10, p. 160-170.

Dix, C. H., 1946, The interpretation of well-shot data (Part III), Geophysics, v. 11, p. 457-461.

Dobrin, M. B., Ingalls, A. L., and Long, J. A., 1965, Velocity and frequency filtering of seismic data using laser light: Geophysics, v. 30, p. 1144-1178.

Dobrin, M. B., 1976, Introduction to geophysical prospecting: McGraw-Hill Book Company, New York, 630 pages.

Douze, E. J., 1964, Signal and noise in deep wells: Geophysics, v. 29, p. 721-732.

Dunoyer de Segonzac, P. and Leherrere, J., 1959, Application of the continuous velocity log to anisotropy measurements in Northern Sahara - results and consequences: Geophys. Prosp., v. 7, p. 202-217.

Durschner, Von H. and Jentsch, M., 1982, Vertical seismic profiling-example of the Betzendorf Z-1 deep well: Erdoel-Erdgas-Zeitschrift, v. 98, No. 4, p. 135-139 (in German).

Ebbersten, E. P., 1966, Determining the position and quality of bedrock: U.S. Patent No. 3,260,992.

Edelmann, H. A. K. and Werner, H., 1982, The encoded sweep technique for Vibroseis: Geophysics, v. 47, p. 809-818.

Elkington, W. B., 1978, Determining the locus of a processing zone in an in situ oil shale retort by sound monitoring: U.S. Patent No. 4,082,145.

Ellis, L. G., 1959, Seismographic exploration: U.S. Patent No. 2,900,037.

Embree, P., Burg, J. P., and Backus, M. M., 1963, Wide-band velocity filtering - the pie slice process: Geophysics, v. 28, p. 948-974.

Erickson, E. L., Miller, D. E., and Waters, K. H., 1968, Shear-wave recording using continuous signal methods, part II - later experimentation: Geophysics, v. 33, p. 240-254.

Evans, J. F., 1962, Subsurface seismic surveying: U.S. Patent No. 3,061,037.

Evans, J. R., 1981, Fortran computer programs for running median filters and a general despiker: U.S. Geol. Survey Open File Report 81-1091, 19 pages.

Evans, J. R., 1982, Running median filters and a general despiker: Bull. Seis. Soc. Am., v. 72, p. 331-338.

Fail, J. P. and Grau, G., 1963, Les filtres en eventail: Geophy. Prosp., v. 11, p. 131-163.

Fair, D. W., 1964, Shear wave transducer: U.S. Patent No. 3,159,232.

Fair, D. W. and Brown, G. L., 1973, Bore hole seismic transducer: U.S. Patent No. 3,718,205.

Farr, J. B., 1977, Downhole seismic source: U.S. Patent No. 4,033,429.

Fehler, M., Turpening, R., Blackway, C., and Mellen, M., 1982, Detection of a hydrofrac with shear wave vertical seismic profiles: Paper S12.5, 52nd Annual International Meeting of SEG, Technical Program Abstracts, p. 159-161.

Fessenden, R. A., 1917, Method and apparatus for locating ore bodies: U. S. Patent No. 1,240,328.

Fitch, A. A., 1981, Vertical seismic profiling: Paper presented at the VSP short course sponsored by the Southeastern Geophysical Society in New Orleans.

Fitch, A. A. and Dillon, P. B., 1983a, Upward extension of the vertical seismic profile, the log of reflection coefficients, and the recovery of the earliest portion of the surface seismogram: Seismograph Service Ltd. has indicated that a patent application is pending in Great Britian and that any patent resulting from such application will be assigned to Seismograph Service Ltd.

Fitch, A. A. and Dillon, P. B., 1983b, The track filter: Seismograph Service Ltd. has indicated that a patent application is pending in Great Britian and that any patent resulting from such application will be assigned to Seismograph Service Ltd.

Fitch, A. A. and Dillon, P. B., 1983c, Removal of the reverberant tails from the reflections recorded in the vertical seismic profile: Seismograph Service Ltd. has indicated that a patent application is pending in Great Britian and that any patent resulting from such application will be assigned to Seismograph Service Ltd.

Frieden, B. R., 1976, A new restoring algorithm for the preferential enhancement of edge gradients: J. Opt. Soc. Amer., v. 66, p. 280-283.

Gaiser, J. E., and DiSiena, J. P., 1982a, VSP fundamentals that improve CDP data interpretation: Paper S12.2, 52nd Annual International Meeting of SEG, Technical Program Abstracts, p. 154-156.

Gaiser, J. E., Ward, R. W., and DiSiena, J. P., 1982b, Three component vertical seismic profiles - polarization measurements of P-wave particle motion for velocity analysis: Paper S12.7, 52nd Annual International Meeting of SEG, Technical Program Abstracts, p. 162-165.

Gallagher, J. N. and Lash, C. C., 1978, Seismic attenuation studies at Mounds, Oklahoma: Paper presented at 48th Annual International Meeting of SEG.

Gal'perin, E. I. and Frolova, A. V., 1961, Three-component seismic observations in boreholes, Parts I and II: Bull. Acad. Sci. USSR, Geophys. Ser., English Trans., p. 519-528 and 644-653.

Gal'perin, E. I., 1974, Vertical seismic profiling: Society of Exploration Geophysicists Special Publication No. 12, Tulsa, 270 pages.

Gal'perin, E. I., 1977, Polyarizatsionnyi metod seismicheskikh issledovanii (Polarization method of seismic prospecting): Nedra Press, Moscow, 280 pages.

Gal'perin, E. I., 1979, The polarizational method in vertical seismic profiling: Paper presented at the 49th Annual International Meeting of SEG.

Gamertsfelder, G. R. and Ljung, B. H. G., 1980, Ring laser gyroscope: U. S. Patent No. 4,190,364.

428

Ganley, D. C. and Kanasewich, E. R., 1980, Measurement of absorption and dispersion from check shot surveys: J. Geophys. Res., v. 85, p. 5219-5226.

Gardner, L. W., 1949, Seismograph determination of salt-dome boundary using well detector deep on dome flank: Geophysics, v. 14, p. 29-38.

Garotta, R., 1978, Seismic exploration using shear waves: Proc. Indonesian Pet. Assoc., p. 253-267.

Garotta, R., 1980, Land seismic shear waves: Technical Series Publication No. 507.78.05, Compagnie Generale de Geophysique, 16 pages.

Garotta, R. and Michon, D., 1982, Comparisons between P-S_h-S_v and converted waves: Paper S5.3, 52nd Annual International Meeting of SEG, Technical Program Abstracts, p. 61-63.

Garriott, J. C., 1981, Vertical seismic profile data acquisition: Paper presented at the VSP short course sponsored by the Southeastern Geophysical Society in New Orleans. (Also available as a 19 page report from the Seismograph Service Companies).

Geyer, R. L., 1969, The Vibroseis system of seismic mapping: Jour. Canadian Soc. Explor. Geophy., v. 6, p. 39-57.

Geyer, R. L., 1971, Vibroseis parameter optimization: Oil and Gas Jour., v. 68, No. 15, p. 116-123, and v. 68, No. 17, p. 114-116.

Giles, B. F., 1968, Pneumatic acoustic energy source: Geophys. Prosp., v. 16, p. 21-53.

Ginsburgh, I. and Papadopoulos, C. G., 1980, Method for determining the position and inclination of a flame front during in situ combustion of an oil shale retort: U. S. Patent No. 4,184,548.

Goetz, J. F., Dupal, L., and Bowler, J., 1979, An investigation into discrepancies between sonic log and seismic check-shot velocities: Aust. Pet. Explor. Assoc. Jour., v. 19, part 1, p. 131-141.

Goff, D. D. and O'Brien, J. T., 1981, Three component detector and housing for same: U.S. Patent No. 4,300,220.

Goldstein, R., Krogstod, R. S., Shorthill, R. W., and Vali, V., 1978, Double optical fiber waveguide ring laser gyroscope: U. S. Patent No. 4,120,587.

Goupillaud, P. L., 1976, Signal design in the Vibroseis technique: Geophysics, v. 41, p. 1291-1304.

Graves, J. E., 1979, A P-SV converted wave reflection seismic prospecting system: M. S. Thesis No. T-2224, Colorado School of Mines.

Gregory, A. R., 1976, Fluid saturation effects on dynamic elastic properties of sedimentary rocks: Geophysics, v. 41, p. 895-921.

Gustavson, C. A., Shutes, E. B. and Wuenschel, P. C., 1973, Clamped detector; U.S. Patent No. 3,777,814.

Gustavson, C. A., Shutes, E. B. and Wuenschel, P. C., 1974, Device for gripping and imparting slack in a cable: U.S. Patent No. 3,791,612.

Gustavsson, M., Israelson, H., Ivansson, S., Moren, P., and Pihl, J., 1982, The seismic crosshole method in crystalline rock: Paper EG.5, 52nd Annual International Meeting of SEG, Technical Program Abstracts, p. 471-472.

Halliday, D. and Resnick, R., 1960, Physics - Parts I and II: John Wiley and Sons, Inc., New York, 1324 pages.

Hampson, D. and Mewhort, L., 1983, Using a vertical seismic profile to investigate a multiple problem in western Canada: Paper presented at 36th Annual Meeting of the Midwest Society of Exploration Geophysicists, Denver, Colorado.

Hardage, B. A., 1981a, The user/interpreter awareness of VSP data: Paper presented at the VSP short course sponsored by the Southeastern Geophysical Society in New Orleans.

Hardage, B. A., 1981b, An examination of tube wave noise in vertical seismic profiling data: Geophysics, v. 46, p. 892-903.

Hardage, B. A., 1983, A new direction in exploration seismology is down: The Leading Edge, v. 2, No. 6, p. 49-52.

Hauge, P. S., 1981, Measurements of attenuation from vertical seismic profiles: Geophysics, v. 46, p. 1548-1558.

Hawes, W. S. and Gerdes, L., 1974, Some effects of spatial filters on signal: Geophysics, v. 39, p. 464-498.

Hawkins, L. V. and Whiteley, R. J., 1980, Seismic method of borehole-logging: Australian Patent No. 510,573.

Hawkins, L. V., 1981, High resolution downhole-crosshole seismic reflection profiling to resolve detailed coal seam structure: U. S. Patent No. 4,298,967.

Hawkins, L. V., Whiteley, R. J., Holmes, W. H., and Dowle, R., 1982, Downhole-crosshole high resolution seismic reflection profiling to resolve detailed coalseam structure: Paper C1.3, 52nd Annual International Meeting of SEG, Technical Program Abstracts, p. 423-426.

Hecht, J., 1982, Laser gyros - the guiding light: High Tech., v. 2, No. 3, p. 24-28.

Heelan, P. A., 1953, Radiation from a cylindrical source of finite length: Geophysics, v. 18, p. 685-696.

Helbig, K. and Mesdag, C. S., 1982, The potential of shear-wave observations: Geophys. Prosp., v. 30, p. 413-431.

Helmick, W. E. and Longley, A. J., 1957, Pressure-differential sticking of drill pipe: Oil and Gas Jour., v. 55, p. 132-136.

Henderson, J. B. H. and Brewer, R., 1953, Core hole velocity surveys: Geophysics, v. 18, p. 324-337.

Hicks, W. G., 1959, Lateral velocity variations near boreholes: Geophysics, v. 24, p. 451-464.

Hildebrand, S. T., 1982, Two representations of the fan filter: Geophysics, v. 47, p. 957-959.

Hildebrandt, A. B., 1949, Apparatus for seismic prospecting: U.S. Patent No. 2,483,770.

Hildebrandt, A. B., 1959, Geophone assembly: U.S. Patent No. 2,898,575.

Holste, W., 1959, Problems and results with refraction seismics in boreholes (determination of salt-flanks and other interfaces): Geophys. Prosp., v. 7, p. 231-240.

Hoover, G. M. and O'Brien, J. T., 1980, The influence of the planted geophone on seismic land data: Geophysics, v. 45, p. 1239-1253.

Horton, C. W., 1943, Secondary arrivals in a well velocity survey: Geophysics, v. 8, p. 290-296.

Howes, E. T., 1956a, Seismic prospecting system: U.S. Patent No. 2,740,945.

Howes, E. T. and Hoy, W. F., 1956b, Seismic prospecting apparatus: U.S. Patent No. 2,757,355.

Howes, E. T., 1957, Seismic prospecting apparatus: U.S. Patent No. 2,788,510.

Huang, C. F. and Hunter, J. A., 1981, The correlation of "tube wave" events with open fractures in fluid-filled boreholes: Current Research, Part A, Geological Survey of Canada, Paper 81-1A, p. 361-376.

Huang, T. S., Yang, G. J., and Tang, G. Y., 1979, A fast two-dimensional median filtering algorithm: IEEE Trans. Acoust., Speech, Signal Processing, v. ASSP-27, p. 13-18.

Hubbard, T. P., 1979, Deconvolution of surface recorded data using vertical seismic profiles: Preprint S-46, 49th Annual International Meeting of SEG. (Also available as a 22 page report from the Seismograph Service Companies).

Hubral, P. and Krey, T., 1980, Interval velocities from seismic reflection time measurements: Society of Exploration Geophysicists, Tulsa, 203 pages.

Ingard, U. and Kraushaar, W. L., 1960, Introduction to mechanics, matter, and waves: Addison-Wesley Publishing Company Inc., Reading, Massachusetts, 672 pages.

Itria, O. A., 1958, Determination of propagation characteristics of earth formations: U.S. Patent No. 2,865,463.

Jayant, N. S., 1976, Average and median-based smoothing techniques for improving digital speech quality in the presence of transmission errors: IEEE Trans. Commun., v. COM-24, p. 1043-1045.

Jenkins, F. A. and White, H. E., 1957, Fundamentals of optics: McGraw Hill Book Company, Inc., New York, 637 pages.

John, P. W., 1979, Process for determining the location and/or extent of rock cavities: U. S. Patent No. 4,158,963.

Johnson, R., Riches, H. and Ahmed, H., 1982, Application of the vertical seismic profile to the Piper Field: Paper EUR 274, European Petroleum Conference, p. 39-47.

Jolly, R. N., 1953, Deep-hole geophone study in Garvin County, Oklahoma: Geophysics, v. 18, p. 662-670.

Jolly, R. N., 1956, Investigation of shear waves: Geophysics, v. 21, p. 905-938.

Jolly, R. N., 1957, Lock-in geophone for boreholes: U.S. Patent No. 2,786,987.

Kallweit, R. S. and Wood, L. C., 1982, The limits of resolution of zero-phase wavelets: Geophysics, v. 47, p. 1035-1046.

Kan, T. K., Corrigan, D., and Huddleston, P. D., 1981, Attenuation measurement from vertical seismic profiles: Tech. Paper S5.3, p. 1950-1989, 51st Annual International Meeting of SEG.

Karus, E. V., Ryabinkin, L. A., Gal'perin, E. I., Teplitskiu, V. A., Demidenko, Yu, B., Mustafayet, K. A., and Rapaport, M. B., 1975, Detailed investigation of geological structures by seismic well surveys: Paper presented at the 9th World Petroleum Congress PD 9(4), v. 26, p. 247-257.

Keith, C. M. and Crampin, S., 1977a, Seismic body waves in anisotropic media-reflection and refraction at a plane interface: Geophys. J. Roy. Astr. Soc., v. 49, p. 181-208.

Keith, C. M. and Crampin, S., 1977b, Seismic body waves in anisotropic media-propagation through a layer: Geophys. J. Roy. Astr. Soc., v. 49, p. 209-223.

Keith, C. M. and Crampin, S., 1977c, Seismic body waves in anisotropic media-synthetic seismograms: Geophys. J. Roy. Astr. Soc., v. 49, p. 225-243.

Keller, G. R., Sturdivant, J. D., Kukas, E. E., Fowler, F. B., and Fallis, J. N., 1983, Synthetic vertical seismic profiles as a practical interpretation aid: Paper presented at 36th Annual Meeting of the Midwest Society of Exploration Geophysicists, Denver, Colorado.

Kelly, K. R., Alford, R. M. and Whitmore, N. D., 1982, Modeling - the forward method: Chapter 4 of Concepts and Techniques in Oil and Gas Exploration edited by K. C. Jain and R.J.P. deFigueiredo, Society of Exploration Geophysicists, Tulsa, 289 pages.

Kennett, P. and Ireson, R. L., 1971, Recent developments in well velocity surveys and the use of calibrated acoustic logs: Geophys. Prosp., v. 19, p. 395-411.

Kennett, P. and Ireson, R. L., 1973, Some techniques for the analysis of well geophone signals as an aid to the identification of hydrocarbon indicators in seismic processing: Paper presented at the 43rd Annual International Meeting of SEG.

Kennett, P. and Ireson, R. L., 1977, Vertical seismic profiling - recent advances in techniques for data acquisition, processing and interpretation: Paper presented at the 47th Annual International Meeting of SEG.

Kennett, P., 1979, Well geophone surveys and the calibration of acoustic velocity logs: Chapter 3 in Developments in Geophysical Exploration Methods, Applied Science Publishers, Ltd., London, 311 pages. Edited by A. A. Fitch.

Kennett, P., Ireson, R. L., and Conn, P. J., 1980, Vertical seismic profiles - their applications in exploration geophysics: Geophys. Prosp., v. 28, p. 676-699.

Kennett, P. and Ireson, R. L., 1981, The V. S. P. as an interpretation tool for structural and stratigraphic analysis: Paper presented at the 43rd Meeting of EAEG, Venice, Italy.

Kennett, P. and Ireson, R. L., 1982, Vertical seismic profiles: EAEG Continuing Education Short Course, 184 pages.

Killpatrick, J., 1967, The laser gyro: IEEE Spectrum, v. 4, p. 44-55.

Kinkade, R. R., 1980, Two-dimensional frequency domain filtering: U. S. Patent No. 4,218,765.

Kinsler, L. E. and Frey, A. R., 1950, Fundamentals of acoustics: John Wiley and Sons, Inc., New York, 516 pages.

Kokesh, F. P., 1952, The development of a new method of seismic velocity determination: Geophysics, v. 17, p. 560-574.

Kramer, F. S., Peterson, R. A., and Walter, W. C., 1968, Seismic energy sources 1968 handbook: Bendix United Geophysical Corporation, 57 pages.

Kraut, E. A., 1963, Advances in the theory of anisotropic elastic wave propagation: Rev. Geophys., v. 1, p. 401-448.

Kretzschmar, J. L., Kibbe, K. L., and Witterholt, E. J., 1982, Tomographic reconstruction techniques for reservoir monitoring: Preprint SPE-10990, 57th Annual SPE of AIME Fall Technical Conference, New Orleans.

Krey, T., 1969, Remarks on the signal to noise ratio in the Vibroseis system: Geophy. Prosp., v. 17, p. 206-218.

Krug, V., Baum, G., and Windgassen, W., 1981, Special borehole-seismic measurements used to study petrophysical parameters in the environment of deep borings: Z. Geol. Wiss., v. 9, No. 8, p. 851-862. (In German)

Lamer, A., 1970, Geophone-ground coupling: Geophys. Prosp., v. 18, p. 300-319. (In French)

Landgren, K. M. and Grubb, K., 1983, Acquisition and processing of Gulf Coast VSP data: Paper OTC 4539, Offshore Technology Conference, Houston, Texas, p. 229-236.

Lang, D. G., 1979a, Downhole seismic technique expands borehole data: Oil and Gas Jour., v. 77, No. 28, p. 139-142.

Lang, D. G., 1979b, Downhole seismic combination of techniques sees nearby features: Oil and Gas Jour., v. 77, No. 29, p. 63-66.

Lang, H. M., 1968, Gas igniting source for well bores: U.S. Patent No. 3,380,551.

Larner, K., Hale, D., Zinkham, S. M., and Hewlitt, C., 1982, Desired seismic characteristics of an air gun source: Geophysics, v. 47, p. 1273-1284.

Lash, C. C., 1980, Shear waves, multiple reflections, and converted waves found by a deep vertical wave test (vertical seismic profile): Geophysics, v. 45, p. 1373-1411.

Lash, C. C., 1982, Investigation of multiple reflections and wave conversions by means of vertical wave test (vertical seismic profiling) in southern Mississippi: Geophysics, v. 47, p. 977-1000.

Lee, M. W., Balch, A. H., Ryder, R. T., and Head, W. J., 1976, Progress report of a vertical seismic profile (VSP) recorded in the Bechtel ETSI 0-1 water well, southeast Powder River Basin, Wyoming: U.S. Geological Administration Report for WRD Division, 10 pages.

Lee, M. W., Miller, J. J., Ryder, R. T., and Balch, A. H., 1981, Processed and interpreted U. S. Geological Survey seismic reflection profile and vertical seismic profiles, Niobrara County, Wyoming: U. S. Geological Survey Oil and Gas Investigations Chart OC-114.

Lee, M. W. and Balch, A. H., 1982, Theoretical seismic wave radiation from a fluid-filled borehole: Geophysics, v. 47, p. 1308-1314.

Lee, M. W. and Balch, A. H., 1983, Computer processing of vertical seismic profile data: Geophysics, v. 48, p. 272-287.

Leslie, H. D. and Mons, F., 1982, Sonic waveform analysis and applications: SPWLA 23rd Annual Logging Symposium, Paper GG, Trans V2, 25 pages.

Levin, F. K. and Lynn, R. D., 1958, Deep hole geophone studies: Geophysics, v. 23, p. 639-664.

Levin, F. K., 1978, The reflection, refraction, and diffraction of waves in media with an elliptical velocity dependence: Geophysics, v. 43, p. 528-537.

Levin, F. K., 1979, Seismic velocities in transversely isotropic media: Geophysics, v. 44, p. 918-936.

Levin, F. K., 1980, Seismic velocities in transversely isotropic media, II: Geophysics, v. 45, p. 3-17.

Lindseth, R. O., 1979, Synthetic sonic logs - a process for stratigraphic interpretation: Geophysics, v. 44, p. 3-26.

Ljung, B. H. G., 1979, Ring laser gyroscope having wedge desensitizing optical means: U. S. Patent No. 4,167,336.

Lynn, R. D., 1963, A low-frequency geophone for borehole use: Geophysics, v. 28, p. 14-19.

Lytle, R. J. and Dines, K. A., 1980, Iterative ray tracing between boreholes for underground image reconstruction: IEEE Trans. Geosci. and Remote Sensing: v. GE-18, p. 234-240.

Mack, H., 1966, Attenuation of controlled wave seismograph signals observed in cased boreholes: Geophysics, v. 31, p. 243-252.

Malmberg, J. H., 1965, Geophysical borehole apparatus: U.S. Patent No. 3,221,833.

Mason, I. M., Buchanan, D. J., and Booer, A. K., 1980, Fault location by underground seismic survey: IEE Proceedings, v. 127, Part F, No. 4, p. 322-336.

McCollum, B. and Larue, W. W., 1931, Utilization of existing wells in seismograph work: Early Geophysical Papers, v. 1, p. 119-127. (Also Bull. Amer. Ass. Pet. Geol., v. 15, p. 1409-1417.)

McCollum, B., 1933a, Seismic method of profiling geologic formations: U.S. Patent No. 1,909,205.

McCollum, B., 1933b, Seismic method of profiling geologic formations: U.S. Patent No. 1,923,107.

McCollum, B., 1935, Seismic method for profiling geologic formations: U.S. Patent No. 2,021,943.

McCormack, M. D., Quay, R. G. and Verm, R. W., 1979, A study of the variant character of source signatures in boreholes: Paper presented at the 49th Annual International Meeting of SEG.

McCormack, M. D., Dunbar, J. A., and Sharp, W. W., 1982, A case study of stratigraphic interpretation using shear and compressional seismic data: Paper S2.5, 52nd Annual International Meeting of SEG, Technical Program Abstracts, p. 21-22.

Michon, D., 1976, Vertical seismic profiling: Paper presented at the 46th Annual International Meeting of SEG.

Michon, D. and Omnes, G., 1978, The applications of the vertical seismic profile: Paper presented at the 31st Annual Midwestern SEG Meeting.

Miller, G. F. and Pursey, H., 1954, The field and radiation impedance of mechanical radiators on the free surface of a semi-infinite isotropic solid: Proc. of Royal Soc. of London, Series A, v. 223, p. 521-530.

Miller, J. J., Ryder, R. T., Balch, A. H., and Lee, M. W., 1981, Processed and interpreted U. S. Geological Survey seismic reflection profile and vertical seismic profile, Powder River and Carter Counties, Montana: U. S. Geological Survey Oil and Gas Investigations Chart OC-110.

Mons, F., 1980a, Vertical seismic exploration and profiling technique: Gr. Brit. Patent No. 2,029,016.

Mons, F., 1980b, Method and equipment for vertical seismic exploration: France Patent No. 2,432,177.

de Montmollin, V., 1983, Three component vertical seismic profile-geometrical processing and wave identification: Paper to be presented at the 53rd Annual International Meeting of SEG, Las Vegas.

Moore, P. L., 1974, Drilling practices manual: Penn Well Books, Tulsa, 448 pages.

Moos, D. and Zoback, M. D., 1983, In situ studies of velocity in fractured crystalline rocks: Jour. Geophy. Res., v. 88, No. B3, p. 2345-2358.

Morris, J. L., Jr. 1979, Attenuation estimates from seismograms recorded in a deep well: Master of Science Thesis, Texas A&M University.

Morris, R. L., Grine, D. R., and Arkfeld, T. E., 1964, Using compressional and shear acoustic amplitudes for the location of fractures: J. Pet. Tech., v. 16., p. 623-632.

Mott-Smith, L. M., 1975, Multiple air gun array of varied sizes with individual secondary oscillation suppression: U.S. Patent No. 3,893,539.

Narasimhan, K. Y., Nathan, R., and Parthasarathy, S. P., 1980, System for plotting subsoil structure and method therefor: U.S. Patent No. 4,214,226.

Narvarte, P. E., 1946, On well velocity data and their application to reflection shooting: Geophysics, v. 11, p. 66-81.

Newman, P., 1973, Divergence effects in a layered earth: Geophysics, v. 38, p. 481-488.

Newman, P. J. and Worthington, M. H., 1982, In-situ investigation of seismic body wave attenuation in hetrogeneous media: Geophys. Prosp., v. 30, p. 377-400.

Nolting, R. P., 1961, Accurate depth determination of the velocity survey well phone: Geophysics, v. 26, p. 100.

O'Brien, P.N.S., 1957, The relationship between seismic amplitude and weight of charge: Geophys. Prosp., v. 5, p. 349-352.

O'Brien, P.N.S., 1960, Seismic energy from explosions: Geophy. Jour., v. 3, p. 29-44.

O'Brien, P.N.S. and Lucas, A. L., 1971, Velocity dispersion of seismic waves: Geophy. Prosp., v. 19, p. 1-26.

Omnes, G., 1978a, Vertical seismic profiles - a bridge between velocity logs and surface seismograms: Paper No. 53, 53rd Annual Conf. and Exhib. of Soc. of Petrol. Eng. of AIME, Houston.

Omnes, G., 1978b, Exploring with SH-waves: Canadian Society of Exploration Geophysicists Journal, v. 14, No. 1, p. 40-49.

Omnes, G., 1980, Logs from P and S vertical seismic profiles: Petrol. Tech., v. 32, p. 1843-1849.

Omnes, G., 1981, Vertical seismic profile-shear wave technique and methods: Paper presented at the VSP short course sponsored by the Southeastern Geophysical Society in New Orleans.

Ording, J. R. and Redding, V. L., 1953, Sound waves observed in mud-filled well after dynamite charges: J. Acoust. Soc. Am., v. 25, p. 719-726.

Outmans, H. D., 1958, Mechanics of differential pressure sticking of drill collars: Pet. Trans. AIME, v. 213, p. 265-274.

Paillet, F. L., 1981, Predicting the frequency content of acoustic waveforms obtained in boreholes: Paper SS, Trans. SPWLA 22nd Annual Logging Symposium, Vol. II.

Paillet, F. L. and White, J. E., 1982, Acoustic modes of propagation in the borehole and their relationship to rock properties: Geophysics, 47, p. 1215-1228.

Parrott, K. R., 1980, An investigation of the interior of a salt structure using the vertical seismic profiling technique: Master of Science thesis, Colorado School of Mines.

Peterson, E. W., 1974, Acoustic wave propagation along fluid-filled cyclinder: J. Appl. Phys., v. 45, p. 3340-3350.

Peterson, R. A., 1942, Method of making weathering corrections: U.S. Patent No. 2,276,335.

Peterson, R. A., 1957, Geophysical prospecting system: U.S. Patent No. 2,792,067.

Peterson, R. A., 1960, Well-shooting system: U.S. Patent No. 2,947,377.

Pickett, G. R., 1963, Acoustic character logs and their applications in formation evaluation: J. Pet. Tech., v. 15, p. 659-667.

Poster, C. K., 1982, Comparison of seismic sources for VSP's in a cased well: Paper presented at C.S.E.G. National Convention, Calgary.

Poster, C. K., 1983, Interpretation aspects of offshore VSP data: Paper OTC 4540, Offshore Technology Conference, Houston, Texas, p. 237-244.

Postma, G. W., 1955, Wave propagation in a stratified medium: Geophysics, v. 20, p. 780-806.

Quarles, M., 1978, Vertical seismic profiling - a new seismic exploration technique: Paper presented at the 48th Annual International Meeting of SEG.

Rabiner, L. E., Sambur, M. R., and Schmidt, C. E., 1975, Applications of a nonlinear smoothing algorithm to speech processing: IEEE Trans. Acoust., Speech, Signal Proc., v. ASSP-23, No. 6, p. 552-557.

Rademacher, F. J. C., 1965, Spudding-in-seismometers: U.S. Patent No. 3,186,502.

Rice, R. B., et. al., 1981, Developments in exploration geophysics, 1975-1980: Geophysics, v. 46, p. 1088-1099.

Riggs, E. D., 1955, Seismic wave types in a borehole: Geophysics, v. 20, p. 53-67.

Robertson, J. D., Pritchett, W. C., and Dees, J. L., 1979, Radiation patterns of a shear wave vibroseis source in a transversely isotropic medium: Paper presented at the 49th Annual International Meeting of SEG.

Roden, R. B., 1965, Horizontal and vertical arrays for teleseismic signal enhancement: Geophysics, v. 30, p. 597-608.

Roden, R. B., 1968, Seismic experiments with vertical arrays: Geophysics, v. 33, p. 270-284.

Roever, W. L., Rosenbaum, J. H., and Vining, T. F., 1974, Acoustic waves from an impulsive source in a fluid-filled borehole: J. Acoust. Soc. Am., v. 55, p. 1144-1157.

Ryder, R. T., Balch, A. H., Lee, M. W., and Miller, J. J., 1981, Processed and interpreted U. S. Geological Survey seismic reflection profile and vertical seismic profile, Carter County, Montana, and Crook County, Wyoming: U. S. Geological Survey Oil and Gas Investigations Chart OC-115.

Safer, M. H., 1976, The radiation of acoustic waves from an air-gun: Geophys. Prosp., v. 24, p. 756-772.

Salvatori, H., 1938, Seismic well logging: U.S. Patent No. 2,137,985.

Sax, R. L. and Hartenberger, R. A., 1964, Theoretical prediction of seismic noise in a deep borehole: Geophysics, v. 29, p. 714-720.

Scarascia, S., Colombi, B., and Cassinis, R., 1976, Some experiments on transverse waves: Geophys. Prosp., v. 24, p. 549-568.

Schaewtzer, T., 1967, Offset-shooting - a study of faulting and structures by well surveys and velocity logs: Paper presented at the 29th annual EAEG meeting.

Schoenberger, M. and Levin, F. K., 1974, Apparent attenuation due to intrabed multiples: Geophysics, v. 39, p. 278-291.

Schulze-Gattermann, R., 1972, Physical aspects of the airpulser as a seismic energy source: Geophys. Prosp., v. 20, p. 155-192.

Seavey, R. W., 1982, Borehole seismic unit: Sandia Report SAND82-0373, Sandia National Laboratories.

Seeman, B. and Horowicz, L., 1981, Vertical seismic profiling-separation of upgoing and downgoing acoustic waves in a stratified medium: Tech. Paper S5.1, p. 1882-1914, 51st Annual International Meeting of SEG.

Seeman, B., 1982, Seismic exploration process by the vertical seismic profile technique and apparatus for its implementation: France Patent 2,494,450 (In French).

Seismograph Service Ltd, 1980a, The V.S.P. modelling atlas: Holwood, Keston, Kent, England, 48 pages.

Seismograph Service Ltd, 1980b, Borehole geophysics worldwide service 1980/81: Holwood, Keston, Kent, England, 44 pages.

Sengbush, R. L. and Foster, M. R., 1968, Optimum multichannel velocity filters: Geophysics, v. 33, p. 11-35.

Seriff, A. J. and Kim, W. H., 1970, The effect of harmonic distortion in the use of vibratory surface sources: Geophysics, v. 35, p. 234-246.

Sharpe, J. A., 1942a, The production of elastic waves by explosion pressures - (I) theory and emperical field observations: Geophysics, v. 7, p. 144-154.

Sharpe, J. A., 1942b, The production of elastic waves by explosion pressures - (II) results of observations near an exploding charge: Geophysics, v. 7, p. 311-321.

Sheriff, R. E., 1973, Encyclopedic dictionary of exploration geophysics: Society of Exploration Geophysicists, Tulsa, 266 pages.

Shorthill, R. W. and Vali, V., 1979, Stimulated Brillouin scattering ring laser gyroscope: U. S. Patent No. 4,159,178.

Silverman, D. and Bailey, J. R., 1976a, Seismic method for determining the position of the bottom of a long pipe in a deep borehole: U. S. Patent No. 3,979,724.

Silverman, D. and Bailey, J. R., 1976b, Seismic method for logging position of a deep borehole in the earth: U. S. Patent No. 3,979,140.

Silverman, D., 1979, Bit positioning while drilling system: U. S. Patent No. 4,144,949.

Simaan, M. and Jovanovich, D., 1980, Optimum filters for vertical seismic array data processing: Paper R-20, 50th Annual International Meeting of SEG.

Slotnick, M. M., 1941, Method of geophysical investigation: U.S. Patent No. 2,268,130.

Slotnick, M. M., Brooks, J. A., and Redding, V. L., 1950, A note on well-shooting data and linear increase of velocity: Geophysics, v. 15, p. 663-666.

Smith, I. W. and Dorschner, T. A., 1981, Laser gyroscope system: U. S. Patent No. 4,284,329.

Smith, S. G., 1975, Measurement of airgun waveforms: Geophy. J. Roy. Astr. Soc., v. 42, p. 273-280.

Sparks, N. R., 1958, Receiving seismic waves directionally: U.S. Patent No. 2,846,662.

Spencer, T. W. and Morris, J. L., 1979, Seismic attenuation estimated from a vertical seismic profile: Paper S-8, 49th Annual International Meeting of SEG.

Spencer, T. W., Sonnad, J. R., and Butler, T. M., 1982, Seismic Q - stratigraphy or dissipation: Geophysics, v. 47, p. 16-24.

Stewart, R. R., Turpening, R. M., and Toksoz, M. N., 1981a, Study of a subsurface fracture zone by vertical seismic profiling: Geophy. Res. Lett., v. 8, p. 1132-1135.

Stewart, R. R. and Toksoz, M. N., 1981b, Statistical velocity analysis of vertical seismic profiles: Tech. Paper S5.5, 51st Annual International Meeting of SEG.

Stewart, R. R., Huddleston, P., and Kan, T. K., 1982, Traveltime analysis and vertical seismic profiles: Paper S12.4, 52nd Annual International Meeting of SEG, Technical Program Abstracts, p. 158-159.

Stinson, L. B., 1975, Method and system for determining the position of an acoustic generator in a borehole: U. S. Patent No. 3,876,016.

Stone, D. G., 1980, Statistical relations between borehole and surface data: Paper R-18, 50th Annual International Meeting of SEG.

Stone, D. G., 1981, VSP - the missing link: Paper presented at the VSP short course sponsored by the Southeastern Geophysical Society in New Orleans.

Stone, D. G., 1982, Prediction of depth and velocity on VSP data: Seismograph Service Companies Report, Tulsa, Oklahoma, 22 pages.

Stoneley, R., 1949, Seismological implications of aeolotropy in continental structure: Monthly Notices of Royal Astronomical Society, v. 5, p. 343-353.

Swan, M., 1957, Wiring system: U.S. Patent No. 2,810,118.

Tango, G. J., 1981, Vertical seismic profiling - an overview: Paper included in the VSP short course notes prepared by the Southeastern Geophysical Society in New Orleans.

Tariel, P. and Michon, D., 1982, Comments on VSP Processing: Paper presented at 44th Annual Meeting of EAEG.

Tariel, P., Michon, D., Naville, C., and Omnes, G., 1982, Processing of zero offset vertical seismic profile data: Paper S12.1, 52nd Annual International Meeting of SEG, Technical Program Abstracts, p. 152-154.

Tatham, R. H. and Stoffa, P. L., 1976, Vp/Vs - A potential hydrocarbon indicator: Geophysics, v. 41, p. 837-849.

Tatham, R. H., 1982, Vp/Vs and lithology: Geophysics, v. 47, p. 336-344.

Telford, W. M., Geldart, L. P., Sheriff, R. E., and Keys, D. A., 1976, Applied geophysics: Cambridge University Press, Cambridge, 860 pages.

Temme, P. and Muller, G., 1982, Numerical simulation of vertical seismic profiling: J. Geophys, 50, p. 177-188.

Toksoz, M. N., Turpening, R. M., and Stewart, R. R., 1980, Assessment of an Antrim oil shale fracture zone by vertical seismic profiling: Fossil Energy Topical Report FE-2346-91, U.S. Department of Energy.

Toksoz, M. N. and Johnson, D. H., 1981, Seismic wave attenuation - Geophysics reprint series No. 2: Society of Exploration Geophysicists, Tulsa, 459 pages.

Tongtaow, C., 1980, Transient response of an acoustic logging tool in transversely isotropic media: Master of Science Thesis, Colorado School of Mines.

Tongtaow, C., 1982, Wave propagation along a cylindrical borehole in a transversely isotropic medium: PhD thesis, Colorado School of Mines.

Tooley, R. D., Spencer, T. W., and Sagoci, H. F., 1965, Reflection and transmission of plane compressional waves: Geophysics, v. 30, p. 552-570.

Treitel, S., Shanks, J. L., and Frasier, C. W., 1967, Some aspects of fan filtering: Geophysics, v. 32, p. 789-800.

Tsang, L. and Rader, D., 1979, Numerical evaluation of the transient acoustic waveform due to a point source in a fluid-filled borehole: Geophysics, v. 44, p. 1706-1720.

Tullos, F. N. and Reid, A. C., 1969, Seismic attenuation of Gulf Coast sediments: Geophysics, v. 34, p. 516-528.

Turpening, R. M., Stewart, R. and Liskow, A., 1979, A shallow shearwave vertical seismic profiling experiment: Paper R-29, 49th Annual International Meeting of SEG.

Turpening, R. M., Liskow, A., and Thomson, F. S., 1980, Seismic investigations of Antrim shale fracturing: Dow Chemical Co., Midland, Michigan, U.S. Dept. of Energy Publication FE-2346-90, 35 pages.

Uhrig, L. F. and Van Melle, F. A., 1955, Velocity anisotropy in stratified media: Geophysics, v. 20, p. 774-779.

Vaage, S., Ursin, B., and Haugland, K., 1982, Interaction between air guns: Paper S14.3, 52nd Annual International Meeting of SEG, Technical Program Abstracts, p. 185-187.

Vail, P. R., Todd, R. G., and Sangree, J. B., 1977, Seismic stratigraphy and global changes of sea level, part 5 - chronostratigraphic significance of seismic reflections: p. 99-116 of AAPG Memoir 26, Seismic stratigraphy - applications to hydrocarbon exploration, Tulsa, 516 pages.

van Sandt, D. R. and Levin, F. K., 1963, A study of cased and open holes for deep hole seismic detection: Geophysics, v. 28, p. 8-13.

vander Stoep, D. M., 1966, Velocity anisotropy measurements in wells: Geophysics, v. 31, p. 900-916.

Walling, D. and Savit, C. H., 1957, Interpretation method for well velocity surveys: Geophys. Prosp., v. 5, p. 69-78.

Ward, R. W. and Hewitt, M. R., 1977, Monofrequency borehole traveltime survey: Geophysics, v. 42, p. 1137-1145.

Washburn, H. and Wiley, H., 1941, The effect of the placement of a seismometer on its response characteristics: Geophysics, v. 6, p. 116-131.

Waters, K. H., 1941, A numerical method of computing dip data using well-velocity information: Geophysics, v. 6, p. 64-73.

Waters, K. H., 1978, Reflection seismology - a tool for energy resource exploration: John Wiley & Sons, New York, 377 pages.

Weatherby, B. B., 1936, Method of making subsurface determinations: U.S. Patent No. 2,062,151.

Weiss, O., 1955, Seismic method of geological exploration: U.S. Patent No. 2,718,929.

White, J. E. and Sengbush, R. L., 1953, Velocity measurements in near-surface formations: Geophysics, v. 18, p. 54-69.

White, J. E. and Sengbush, R. L., 1963, Shear waves from explosive sources: Geophysics, v. 28, p. 1101-1019.

White, J. E., 1964, Motion product seismograms: Geophysics, v. 29, p. 288-298.

White, J. E., 1965, Seismic waves - radiation, transmission, and attenuation: McGraw-Hill, New York, 302 pages.

White, J. E. and Zechman, R. E., 1968, Computed response of an acoustic logging tool: Geophysics, v. 33, p. 302-310.

White, J. E. and Tongtaow, C., 1981, Cylindrical waves in transversely isotropic media: J. Acoust. Soc. Am., v. 70, p. 1147-1155.

White, J. E., 1982, Computed waveforms in transversely isotropic media: Geophysics, v. 47, p. 771-783.

Wolf, A., 1944, The equation of motion of a geophone on the surface of an elastic earth: Geophysics, v. 9, p. 29-35.

Woodward, J. H., 1965, Well geophone: U.S. Patent No. 3,188,607.

Wuenschel, P. C., 1972, Reproducible shot hole: U.S. Patent No. 3,693,717.

Wuenschel, P. C., 1974, Reproducible shot hole apparatus: U.S. Patent No. 3,812,912.

Wuenschel, P. C., 1975, Precision seismology: U.S. Patent No. 3,876,971.

Wuenschel, P. C., 1976, The vertical array in reflection seismology-some experimental studies: Geophysics, v. 41, p. 219-232.

Wyatt, K. D., 1981a, Synthetic vertical seismic profile: Geophysics, v. 46, p. 880-891.

Wyatt, K. D. and Wyatt, S. B., 1981b, The determination of subsurface structural information using the vertical seismic profile: Tech. Paper S5.2, p. 1915-1949, 51st Annual International Meeting of SEG.

Wyatt, K. D. and Wyatt, S. B., 1982a, Determination of subsurface structural information using the vertical seismic profile: Paper presented at 35th Annual SEG Midwestern Exploration Meeting.

Wyatt, K. D. and Wyatt, S. B., 1982b, Downhole vertical seismic profile survey reveals structure near borehole: Oil and Gas Jour., v. 80, No. 42, p. 77-82.

Wyatt, S. B., 1979, The propagation of elastic waves along a fluid-filled annular region: Master of Science Thesis, University of Tulsa.

Yamakawa, N., 1962, Scattering and attenuation of elastic waves, Part I - Scattering of elastic waves by various kinds of obstacles: Geophys. Mag. (Tokyo), v. 31, p. 63-103.

Yoshimura, M., Fujii, S., Tanaka, K., and Morita, K., 1982, On the relationship between P and S-wave velocities in soft rock: Paper S11.3, 52nd Annual International Meeting of SEG, Technical Program Abstracts, p. 143-145.

442

Zavalishin, B. R., 1975, The size of the region that forms a reflected wave at a boundary: Prikladnaya Geofizika, v. 77, p. 67-74 (in Russian). (Translated into English in Stanford Exploration Project Report 28, October, 1981, p. 345-353. This translation is available only to those people who financially support the Stanford Exploration Project.)

Zavalishin, B. R., 1982, Improvements in constructing seismic images using the method of controlled directional reception: Paper S9.6, 52nd Annual International Meeting of SEG, Technical Program Abstracts, p. 121-122.

Ziolkowski, A., Parkes, G., Hatton, L., and Haugland, T., 1982a, The signature of an air gun array: Geophysics, v. 47. p. 1413-1421.

Ziolkowski, A., 1982b, An air gun model which includes heat transfer and bubble interactions: Paper S14.4, 52nd Annual International Meeting of SEG, Technical Program Abstracts, p. 187-189.

Ziolkowski, A., 1982c, The concept of high primary-to-bubble ratio in the deconvolution of air gun data: Paper RW-3, 52nd Annual International Meeting of SEG, Technical Program Abstracts, p. 514.

Ziolkowski, A., 1982d, The concept of high primary-to-bubble ratio in the deconvolution of air gun data: Paper RW-3, 52nd Annual International Meeting of SEG, Technical Program Abstracts, p. 514.

Zwart, W. J. and Baer, K. E., 1982, Index of wells shot for velocity: Society of Exploration Geophysicists, Tulsa, 111 pages.

AUTHOR AND NAME INDEX

446

SUBJECT INDEX